Lecture Notes in Computer Sc

Edited by G. Goos, J. Hartmanis and J. va

D0867796

Springer
Berlin
Heidelberg
New York
Barcelona
Hong Kong
London
Milan
Paris
Singapore
Tokyo

Sathya Rao   Kaare Ingar Sletta (Eds.)

Next Generation Networks

Networks and Services
for the Information Society

5th IFIP TC6 International Symposium,
INTERWORKING 2000
Bergen, Norway, October 3-6, 2000
Proceedings

 Springer

Series Editors

Gerhard Goos, Karlsruhe University, Germany
Juris Hartmanis, Cornell University, NY, USA
Jan van Leeuwen, Utrecht University, The Netherlands

Volume Editors

Sathya Rao
Telscom AG
Sandrainstr. 17, 3007 Bern, Switzerland
E-mail: rao@telscom.ch

Kaare Ingar Sletta
Telenor Research and Development
P.O. Box 83, 2027 Kjeller, Norway
E-mail: kaare-ingar.sletta@telenor.com

Cataloging-in-Publication Data applied for

Die Deutsche Bibliothek - CIP-Einheitsaufnahme

Next generation networks : network services for the information
society ; 5th IFIP TC6 international symposium, Bergen, Norway,
October 3 - 6, 2000 ; proceedings / Sathya Rao ; Kaare Ingar Sletta
(ed.). - Berlin ; Heidelberg ; New York ; Barcelona ; Hong Kong ;
London ; Milan ; Paris ; Singapore ; Tokyo : Springer, 2000
  (Lecture notes in computer science ; Vol. 1938)
  ISBN 3-540-41140-2

CR Subject Classification (1998): C.2, H.3, H.4, D.2, D.4.4, K.4-6

ISSN 0302-9743
ISBN 3-540-41140-2 Springer-Verlag Berlin Heidelberg New York

Springer-Verlag Berlin Heidelberg New York
a member of BertelsmannSpringer Science+Business Media GmbH
© 2000 IFIP International Federation of Information Processing, Hofstrasse 3, 2361 Laxenburg, Austria
Printed in Germany

Typesetting: Camera-ready by author
Printed on acid-free paper   SPIN 10781268   06/3142   5 4 3 2 1 0

Networks and Services for the Information Society

The International Symposia on Interworking received their initial impetus from work done in several projects of the ACTS (Advanced Communication Technologies and Services) Program, which was part of the 4th Framework Program of the European Union. This year's Symposium was the fifth of a series of biannual conferences, and was now back in Europe again, following Bern (in 1992), Nice (1994), Nara, Japan (1996), and Ottawa (1998).

"Networks and Services for the Information Society" was the overall theme of this year's Interworking 2000 symposium hosted by Telenor, Norway, in Bergen - a beautiful city on the west coast of Norway. A major reason for this choice: Bergen is the European City of Culture for the year 2000.

Interworking 2000 aimed to provide a platform for exchanging views on heterogeneous communication network issues including their concepts, evolution, services, equipment and user requirements. It has highlighted the importance of interoperability between equipment, protocols, signalling and network management for the provision of end-to-end services at an international level.

Contributors to the Symposium include technology and industry leaders gathered from all aspects of telecommunications networking, as well as experts from universities and research institutes.

The research and development work around the globe and the deployment of new technologies in the field has been heading towards the realisation of a virtual society based on information and communication technologies. With the evolution towards this information society, high-speed networks and infrastructures, based on formal international or proprietary standards, will have to interwork with each other in order to provide end-to-end interconnection and operation, independent of the underlying network infrastructure.

Recent years have seen the innovation in both the hardware and software of high-speed communications, using advanced new technologies including ATM, IP networking (IPv6), mobile communication (UMTS) and intelligence in networks, services and operational systems. At the same time, the applications for collaborative working, virtual presence and secured electronic commerce – all steps towards the information society of the 21st century – are becoming a reality.

The success of any service will depend upon the ability of networks, application and service platforms to inter-operate with each other in supporting the end-to-end applications with guaranteed QoS. This publication aims to give the reader an idea of how this can be achieved.

The Interworking 2000 conference has integrated 'Next Generation Networks (NGN)' projects of the Information Society Technologies (IST) Programme launched by the European Commission. This has facilitated greater dissemination of results from European NGN initiatives to the wider audience at large.

We sincerely hope that this volume will grant its readers an overview of the technologies involved in Networks and Services as foreseen for the information society of the coming century.

August 2000   The Editors

Acknowledgements

This Volume could not exist without the contributors of its papers. We would like to thank them on behalf of the Symposium organisers, for their support in making this a very successful conference. The editors would also like to thank all reviewers for their help in selecting quality papers.

Organising such international events is not easy without the support of sponsors. We would like to thank TELENOR, which was very generous in accepting to host this conference under its Patronage. Our sincere thanks also go to all industrial sponsors and to the members and staff of the European Commission, who provided support of various kinds. In particular we would like to thank Dr. Paulo de Sousa of the European Commission, who helped us integrating the NGN concertation activity into the conference, and Ms. May Krosby of Telenor, who took care of the Secretariat.

Last but not least, our sincere thanks to committee members who provided timely help in realising this conference and to our publishers Springer-Verlag for bringing out an excellent volume in time for the conference.

Interworking 2000 Organisation

Chairman of the conference:
Mr. R. Haugen
Telenor, Norway
rolf-bjorn.haugen@telenor.com

Organisation Committee Chairman:
Mr. K. I. Sletta
Telenor, Norway
kaare-ingar.sletta@telenor.com

Co-Chairman:
Prof. A. Casaca
INESC, Portugal
augusto.casaca@inesc.pt

Symposium Management:
Dr. P. de Sousa
European Commission
paulo.desousa@cec.eu.int

Technical Committee Chairman:
Dr. S. Rao
Telscom, Switzerland
rao@telscom.ch

Co-Chairman:
Mr. Terje Ormhaug
Telenor, Norway
terje.ormhaug@telenor.com

Technical Committee Members:
Mr. E. Demierre, Swisscom, Switzerland
Dr. A. Profumo, Italtel, Italy
Dr. H. Uose, NTT, Japan
Mr. M. Potts, Martel, Switzerland
Dr. I. S. Venieris, NTUA, Greece
Mr. B. F. Koch, Siemens, Germany
Dr. L. Rodrigues, ITU, Geneva
Prof. P. Van Binst, ULB, Belgium
Mr. S. A. Wright, BellSouth, USA
Mr. G. A. Hendrikse, KPN, The Netherlands
Mr. H. K. Pathak, Lucent Tech., USA
Prof. J. Quemada, DIT-UPM, Spain
Mr. T. Rybczynski, Nortel, Canada
Mr. J. Clarke, Lake Comm., Ireland
Dr. K.-O. Detken, Optinet, Germany
Prof. L. G. Mason, INRS, Canada
Dr. J. Ashworth, Salford University, UK
Mr. J. Ruutu, Nokia, Finland
Prof. C.-H. Youn, ICU, Korea
Mr. O. Baireuther, DT, Germany
Mr. D. Nyong, Cable & Wireless, UK
Mr. P. Vincent, ENIC, France
Dr. J. Pitts, QMW College, UK
Prof. A. Casaca, INESC, Portugal
Mr. V. Lagarto, Portugal Telecom, Portugal
Mr. J. Pritchard, ETSI, France
Mr. P. Stollenmayer, Eurescom, Germany
Mr. C. E. Joys, Alcatel, Norway

Table of Contents

The Eurescom Project HINE (Heterogeneous In-House Networking Environment): A Cooperation between Telecom Operators for Advanced Home Networking

Paolo Pastorino[1], Steve Brown[2], Pierre-Yves Danet[3], Gabriele Goldacker[4], José Gonzales Torres[5], Torvald Konstali[6], Frederic Phytoud[7]

[1]CSELT, Via Reiss Romoli 274, 10148 Torino - ITALY,
[1]paolo.pastorino@cselt.it
[2]British Telecom, UK
[3]France Télécom R&D, France
[4]GMD-Fokus, Germany
[5]Telefonica I/D, Spain
[6]Telenor, Norway
[7]Swisscom, Switzerland

Abstract. An overview of an integrated of home networking environment from a Telecom Operator and Service Provider's point of view is presented. This is a result of a cooperative European project (Eurescom P915 HINE) integrating heterogeneous home networks. It gives an insight of the basic requirement of an integrated home network that can enable the delivery of advanced telecommunication services, furthermore the results of a European distributed demonstrator are reported.

1   Introduction and basis

The scenario of people seamlessly being able to control all elements of their heterogeneous home environment remotely via the Internet - like their video, heating, delivery of shopping and so on, is fast becoming reality. Over the past year, seven European Telecom Operators have been actively collaborating with each other and component vendors in the Eurescom project HINE (Heterogeneous In-house Networking Environment) [1] to fully understand the potential for the home networking environment and whether there is a market out there.

The current target environments are mainly the in-house, business and home areas with wired and wireless systems. In addition to classical Customer Premises Equipment such as telephone, fax, PC and so on, there are many largely standalone microelectronics-controlled equipment (stereo / video products, home theatre, white goods, light control, burglar alarm, person identification / location systems etc.) that are applied in households.

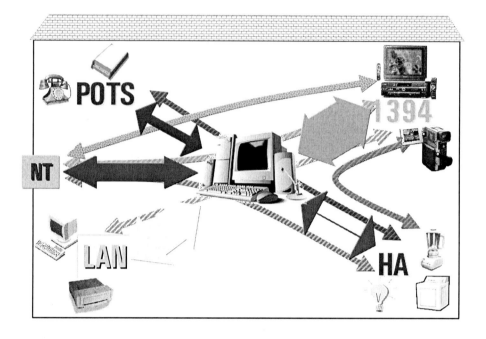

Fig. 1. The Heterogeneous home network

Integrating these devices in an homogeneous and interworking communication environment using mainly existing or easy-to-install networks (e.g. power-line, wireless systems), will allow major enhancements to existing and the creation of completely new applications and services.

The project has focused on analyzing the key technologies for the realization of heterogeneous in-house networking systems for the home and office environment, providing a platform for communication, automation, control and facility management applications and services and operation of an Internet integrated HINE test-bed for commercial and pre-commercial components, applications and services.

2 The theoretical work

Beginning with existing or "easy to install" wired or wireless physical media and transmission systems, there is already a huge variety of possible means to access those devices. Some standards even allow the simultaneous use of the physical media for different transmission technologies, while other standards require special topologies, which might no be well suited for some countries. On top of the transmission technologies, it will be necessary to provide interoperability between devices in the same system and even across system boarders. Due to the focus on private customers and consumers, plug and play capabilities are a must as well. In addition, the Telecom Operators will have to provide complete solutions as a service to their customers.

In-house technologies have been compared and conclusions so far suggest that the commercial reality of the HINE environment is likely to comprise a mix of both wired and wireless networking. The constraints, approaches and impacts for the gateway and application programming interfaces to hook the in-house network into the access network have been evaluated, along with the influences of home networks on access networks. Furthermore scenarios have to be developed to allow customers to buy complete solutions in a one-stop-shopping manner from a Telecom Operator.

Therefore one goal of the EURESCOM Project HINE has been to analyse the currently available solutions and future trends in the area of home communication and automation with an explicit focus on the possibility to attach theses systems to the access networks provided by the Telecom Operators [2-4]. The theoretical work, that has been carried out by the Project members provided an overview over possible solutions, which have taken into account the above mentioned criteria. In addition this work has been the basis for the development and implementation of the HINE demonstrator.

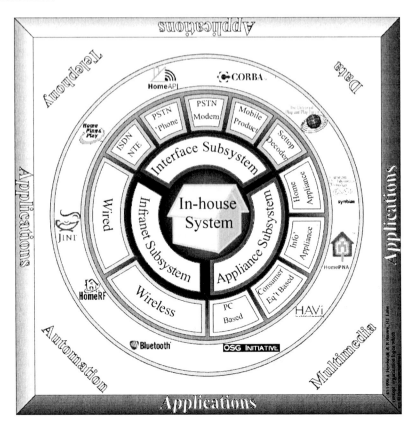

Fig. 2. The In-house system structure

2   Residential Gateway & APIs

The Residential Gateway (RG) is the key component in the home for connecting the various internal networks to different external access networks. However, it also serves as a platform for deploying different kinds of services to the end user.

The requirements necessary to put on the Gateway are very hard to fulfill – the Gateway should be cheap, have a nice looking, make no noise, be easy to use, easy to manage, reliable, secure, and should be able to perform several challenging tasks simultaneously.

To achieve this, it is essential that the several actors to some extent follow the same approaches. This is not the case today, as there exists several initiatives that work in parallel on different solutions on the several aspects relating to this.

To make a RG within acceptable cost limits, the number of simultaneously available network interfaces has to be reduced to the minimum that is required to provide an overall set of services. To provide flexibility it is essential to use a standard internal interface for network interface cards, and hot plugging is a desire. The RG might be single physical unit, but might also be a distributed system containing several more or less interconnected units.

Alarm services are potential services for this system, but introducing such services puts constraints on the availability and security features of the system. The RG has to be placed in a restricted area, and has to have a back up power supply.

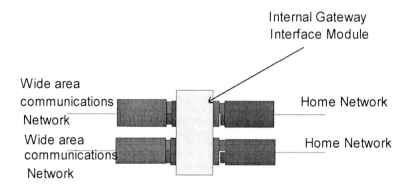

Fig. 3. The centralized Residential Gateway

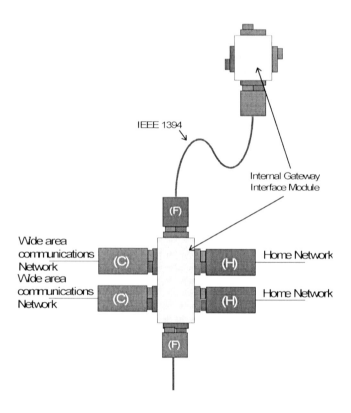

Fig. 4. The distributed Residential Gateway

Today existing home applications are based on a number of different APIs that are often proprietary, incompatible and normally just address subsets of the devices that exists in a home. This leads to a demand for a general API giving standardized access to all home devices independent of the network type used. Such an API should have functionality for invoking methods, provide status, do management and acquire specific attributes for devices. In addition persistence, security, automation and resource handling has to be taken into account. There are several initiatives to define the necessary API. Among them we have the Home API (now merged with Universal Plug and Play) and the OSGi (Open Service Gateway Initiative) initiatives. Both of them are in an early stage of development and it is too soon to judge the importance of these efforts. Nonetheless OSGi has succeeded in releasing its first specification and many manufacturers and providers are starting the first implementation of OSGi compliant platforms.

Another very important aspect of the home system is the "the plug and play" capabilities. Also here there are several initiatives; Microsoft leads the work on "Universal Plug and Play" and Sun promotes Jini. "Universal Plug and Play" is still in an early stage of development and not much information is available (a Device Architecture specification has just been released in June 2000). Jini is a Java based resource discovers system that may be a component in other systems as HAVi (an initiative coming from the consumer electronics market) and OSGi. It is not clear today if any solution will win over the others, and in that case, which it will be. Because of this several equipment manufacturers works with several solutions in parallel. For example, companies like Echelon and Axis seems to promote both Jini and Universal Plug and Play.

The overall situation in this area is rather chaotic, and there is really a need to establish some overall guidelines to get a more smooth evolution. Telecom Operators and Service Providers can be puzzled when it comes the time for choosing the right platform.

3   The practical work

Besides the important theoretical work, the main goal of the Project has been to realise a distributed HINE demonstrator, interconnected via the Internet, broadband access networks and ISDN or POTS. All HINE Project partners have set up local scenarios to demonstrate the functionality and usability of global accesses to a variety of home communication systems. Due to the large number of standards, transmission technologies and physical media, the resulting demonstrator is thus completely heterogeneous and it gives useful and necessary experiences to all partners involved for future product and service developments. The distributed HINE demonstrator comprises a mix of online applications and hardware demonstrators. Due to the distributed physical approach of the demonstrator, the Internet has been used to interconnect the different demonstrator sites. This allowed creating a central demonstrator site for the access through the WWW.

4   The distributed demonstrator

The development and operation of a distributed demonstrator was a central goal of HINE. The expectation that realistic experiences with home networking technologies can best be made in this way became absolutely true.

Scope of demonstrator evaluation

After having separately tested and evaluated several technologies relevant or promising for the home networking area, the scope of the demonstrator evaluation was focused on the capabilities of the overall systems developed.

The major issues identified are: man-machine interface, reliability, security, API implementations and access network basic requirements. Besides these areas, several other conclusions of broad relevance have been drawn. The major observations are presented in the following sections.

The distributed demonstrator concept

Each HINE partner has set up a self-contained demonstrator part realizing several scenarios on a common technical basis. All demonstrator parts are accessible via a common World Wide Web access point (a web portal) that is partly limited by a password access control (some of the HINE Project results a confidential and reserved to the shareholders only). Additionally, dedicated access points for voice telephony, SMS and WAP are provided.

In a real life environment, a portal (a web portal is the more straightforward way of doing it) is a comfortable way to realize secure remote access without explicit authentication per location. It can be for example implemented at the home site of a given user or be provided by a (security) service provider (the basic assumption for a web portal is that the user needs to have for most of the scenarios an always-on connection).

The distributed demonstrator made it possible to realize similar scenarios using different technologies and thus to directly compare their suitability for a certain purpose.

Demonstrator scenarios

A broad selection of scenarios covering all areas of home networking - home automation, multimedia /entertainment and communication - has been realized. Here in the following a short description is included. We have not included on purpose broadband video services (e.g. digital video broadcasting, Video on Demand) because they are not aligned with the main focus of the project (heterogeneous networking).

Home automation
It includes in-home control, external control by the occupants, automatic provision of information to in-home and external places (e. g. police department) and external access and/or control by third parties like energy suppliers or emergency staff.

Table 1. Home Automation scenarios

Telemetric energy control
Telemetric security
Telemetric utility meters
Baby-care
Help to aged persons
Automatic disaster reaction
External heating/air conditioning control
External access to real-time video data
Self surveillance
Utility contact
Security
Web cam images for security via E-mail (WISE)
Remote access to kitchen appliances and storage

| facilities |
| Collection and use of environmental data |
| Mood setting |
| Services for disabled people |
| Home networking in a rented home |
| Multimedia control network with TV interface for domotic services |

Multimedia / entertainment

It includes control and exchange of medium to high volume data like video, animation or high-quality audio, in-home and/or with external locations.

Table 2. Multimedia/Entertainment scenarios

| The wireless WebPAD |
| VCR programming using TV guide accessed via Web |
| External recording control of video camera data |
| Self surveillance |
| Internet games |
| CD sharing |
| E-mail video orders (EVO) |
| Multimedia control network with TV interface for domotic services |

Communication

It especially covers speech communication (including man-machine interaction) and low volume data exchange.

Table 3. Communication scenarios

| The wireless WebPAD |
| Baby-care |
| Help to aged people |
| Utility contact |
| Smart answering machine (SAM) |
| Home networking in a rented home |

Man-machine interface experiences

The man-machine interfaces have been identified as the major factor for end user acceptance. They have to be easy to use, shall quickly be operational after start-up and have to provide convincing confirmations (live pictures instead of test messages).

Java applets can provide excellent functionality, but this has to be carefully balanced against the corresponding downloading time. Java applications may be the better solution for frequent access to invariable functions from a given location.

Voice interfaces are very promising, but commercially available speech recognition software is still far from being accurate and easy to use for everybody, especially when e. g. GSM voice quality is not good enough.

SMS, remote controls and TV screens have reasonable areas of application, but they are less suited for complex interaction.

WAP is expected to provide a useful addition to a full Web interaction. In the demonstrator, it has successfully been used for environmental status data presentation and remote home appliances control.

Specifically designed hardware and/or software solutions for disabled people can enable these persons to highly benefit from home networking. These solutions have to go far beyond the adaptation of typical I/O devices as demonstrated by mouse operation controlled by a head movement detection system.

Reliability

Hardware and software reliability is an important issue especially in the typically non-expert home environment. Most of the commercial equipment used to implement the demonstrator sufficiently fulfilled this requirement.

Microsoft Windows 95/98 during our tests showed not to be stable enough for a home server, Linux or Windows NT are better suited.

Concurrent applications may be a problem for the computers typically used in home environments, distribution of these applications onto several hosts may be necessary to reach the required stability (at a higher cost, though).

Powerline transmission may be a good solution for home networking as it does not require additional cabling, but less expensive variants like X10 do not fulfil the reliability requirements of critical applications, e. g. alarm handling. Powerline transmission for data (e.g. PC to PC) is currently not commercially available and should take into account EMC compatibility and regulatory issues.

Security

Security against unauthorized access to their home is a point where people are generally fairly sensitive about. Nevertheless, typical firewall concepts currently hinder intended access too much to be applicable.

The security service provider concept introduced for the distributed demonstrator seems much more appropriate and could be enhanced by e. g. virus protection mechanisms. More complex access control mechanisms based on origin of access or time of day also easily be integrated.

API implementations

Existing home applications are based on a number of different APIs that are often proprietary, incompatible and normally just address subsets of the devices that exist in a home.

Due to the non-availability of API approaches like HomeAPI or OSGi at demonstrator implementation time, CORBA and Java have been used for individual solutions by several HINE partners.

In general, CORBA is suitable for the demonstrator goals, but some mechanisms do not comply with firewalls. Java applets have to be carefully designed to meet man-machine interface requirements (see above).

All partners have used IP as a low level basis. This should be taken into account by unification approaches for the vast number of different low level home networking protocols used today.

Access network basic requirements

Access network requirements highly depend on the applications used in a given home. Assuming that all scenario categories and information media (text, pictures, video sequences, sound...) will more and more be used in home networking environments, xDSL availability and always-on support will be a growing requirement of residential customers.

xDSL is the only access networking technology provided today for residential customers which can sufficiently adapt to the highly varying bandwidth requirements of home networking applications.

Always-on support is necessary for sufficiently fast and easy external access to home locations.

Overall conclusions

Besides the specific technical problems described in the sections above, it has generally been concluded that comfortable home networking which covers all relevant areas is possible with today's technologies, but can be still too complex to install and too expensive with reference to certain advanced scenarios. The integration into a single hardware platform of advanced functionalities should be weighted with respected to the scenario to be implemented. It is probably unrealistic the possibility of having flexibility, computing power, low cost, future proofness in a single box, but it is likely to be able in the short term to have a home networking equipment capable of integrating multimedia and data streams (e.g. with IEEE1394 and Ethernet interfaces with the use of IP for everything) and of hosting new bundles of services.

References

1. http://www.eurescom.de/public/projects/P900-series/P915/P915.htm
2. EURESCOM Project P915-PF Deliverable 1, "Analysis of In-house System Technologies", http://www.eurescom.de/public/projectresults/p900-series/915d1.htm

3. EURESCOM Project P915-PF Deliverable 2, "Getting the in-house network hooked up to the outside world - constraints, approaches, impacts", http://www.eurescom.de/public/projectresults/p900-series/915d2.htm
4. EURESCOM Project P915-PF Deliverable 4, "Description and Evaluation of the HINE Demonstrator", http://www.eurescom.de/public/projectresults/p900-series/915d4.htm

The Generic Network Model - an ITU Approach for Interoperability

Terje Henriksen, Telenor R&D,

PO box 83, Kjeller, Norway
tel: +47 63 84 86 82, fax: +47 63 81 00 76,
email: terje-fredrik.henriksen@telenor.com

Abstract. Network interoperability may be achieved by a number of different means. This paper addresses the usage of management mechanisms to realize this goal. Where interoperability by means of management systems is concerned, standardization bodies like ITU are frequently associated with standardized network element (NE) models. This paper presents another option, interoperability on the network level. The network level view is more abstract and comprises fewer details, simplifying many management tasks considerably. A natural consequence of the proposal is the relaxation of the requirements for standardized NE models. The generic network model defined in the G.85x series is the core element of the proposal. The functionality of the current version as well as the planned extensions (access networks, connection-less networks including IP) are described and so are the underlying mechanisms, the generic network architecture defined in G.805, the Logical Layered Architecture in M.3010 and the enhanced Reference Model for Open Distributed Processing (RM-ODP) in G.851.1.

Keywords: Generic network architecture, generic network model, interoperability, interworking, logical layered architecture, management architecture, network level management, RM-ODP.

1 Introduction

The proposal made in this paper is for telecom network interoperability by means of management systems to be addressed on the network level rather than the network element level, for three reasons:

1. A widely accepted generic network architecture, the ITU-T rec. G.805 "Generic Functional Architecture of Transport Networks", exists already.
2. Interoperability at the network level is easier to achieve because the network view is more abstract and hence comprises fewer details.

3. Focus[1] for the management of network technologies should be be shifted from the network element level to the network level anyway, to reduce complexity and thus, operational expenses.

A generic network level model is a prerequisite for the application of this approach. Such a model is being developed in Question 18 of Study Group 4 within ITU-T, in brief, Q.18/4. After a review of the generic network architecture and the Logical Layered Architecture this paper explains the applicability of the enhanced RM-ODP (Reference Model for Open Distributed Processing) framework to the development of the generic network model. The creation of a networkTTP (Trail Termination Point) is used as a modeling example throughout the RM-ODP viewpoints.

Multiple network technologies from multiple vendors spanning multiple network provider domains may interoperate following the approach outlined in this paper.

The existing model is applicable to connection-oriented core network technologies like SDH, WDM and ATM. Work has started in Q.18/4 to expand the scope by including connectionless technologies such as IP as well as access network technologies.

2   The G.805 Generic Network Architecture

In the early nineties, influenced by the emerging object oriented methodology, SG XVIII (later SG13) in ITU developed the generic network architecture [1]. It provides a high level view of the generic network function based on a small set of architectural entities (functional blocks) interconnected via reference points, see Fig.1. Two main network representations exist:
- Topology[2] in terms of links, subnetworks and access groups.
- Connectivity in terms of trails, link connections, subnetwork connections, ports and reference points.
  In an implementation, a visible reference point becomes an interface.

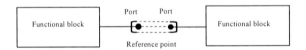

Fig. 1. Functional architecture, basic elements

The topological view describes the geographical distribution of the resources of a layer network. The access group is a container for a number of co-located access points. The subnetwork represents the routing capabilities within a site or grouping of sites and the link represents the transport capacity between subnetworks. Layering is a methodology for splitting the overall transport function into a hierarchy of layer networks on top of each other, each of which utilizing the service from the server layer to provide its own service. A topological view of a network consisting of three sites is shown in Fig.2.

---

[1] From the point of view of the network operator
[2] of a layer network

The topological view is well suited when searching for candidate routes for connections. By introducing selection criteria such as the number of hops, the average spare capacity of the links, etc. in addition to the topological information, the search process may be automated.

All the architectural entities are available (but not necessarily visible) in the connectivity view, i.e., the ports, the reference points, the link connections, the subnetwork connections and trails as well as the topological entities.

The trail preserves the integrity of the information transported between access points at the boundary of the layer network by adding overhead at the transmit end and extracting and analyzing it at the receive end.

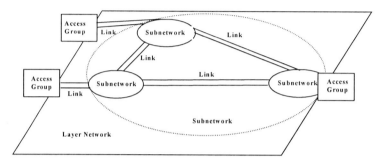

Fig. 2. Network architecture – the topological view

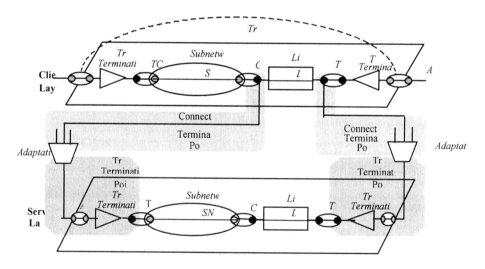

Fig. 3. The connectivity view – client and server layer networks

The subnetwork and the associated subnetwork connections provide the capability of flexibly routing the incoming traffic of a site to the appropriate outgoing links. The routing control may be realized by management or signaling.

The interlayer relationship is provided by the adaptation function as shown in Fig.3. The capability of individual management for each layer is a general requirement. To make this possible, the link connections are there to provide a virtual[3] representation in the client layer of the server layer support.

Partitioning of subnetworks implies that a composite subnetwork has an internal structure so that it may be recursively decomposed into component subnetworks interconnected by links. An example of subnetwork partitioning is illustrated in Fig.2. The outer subnetwork is a composite subnetwork comprising three component subnetworks.

In a similar fashion, subnetwork connections may be partitioned into component subnetwork connections interconnected by link connections. Partitioning of subnetworks and subnetwork connections may be utilized to represent parts of a layer network at different levels of abstraction.

The approach taken when developing G.805 is generic in the sense that it is not referencing any particular technology. Most of the architectural entities defined are applicable to a broad range of technologies, eventually with the behavior slightly modified. At the moment, the following network architectures exist:

- G.803 – SDH [2]
- G.872 – WDM [3]
- I.326 - ATM [4]
- G.902 – Access networks [5]
- g.cls – Connectionless communication [6]

The last one, g.cls, is currently under development. It will cover connectionless technologies such as IP.

3 The Logical Layered Architecture – Relaxed Requirements for Standardised Ne-Models

The Logical Layered Architecture in Fig.4 is part of the TMN but has been applied in a number of non-TMN applications as well.

---

[3] The actual server layer support is the sequence of the adaptation function , the server layer trail and the other adaptation function in Fig.3.

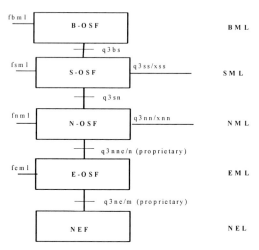

Fig. 4. The Logical Layered Architecture

As with the network architecture, the LLA consists of functional blocks bound together via standardized (q, x, f) or proprietary[4] (m, n) reference points. On top of the basic information transport function, the Network Element Function (NEF), there is a hierarchy of management layers[5] providing management functions (OSFs) in a layered fashion, i.e., the functions at one layer rely on the service of its server layer to perform its own service. The layers are frequently referred to as different levels of management abstraction, the higher the layer, the higher the level of abstraction.

The Element Management Layer (EML) is performing management operations pertaining to each individual NE. No knowledge of the relationship with other NEs is presumed.

The execution of an operation on the EML may result in interactions between the EML and the NEF across the q3ne/m reference point. The protocol and model governing these interactions are often proprietary as they depend on equipment manufacturer properties to a large extent.

Network level operations like setting up connections are carried out at the Network Management Layer (NML). As a result, connection commands are sent to the EML across the appropriate q3nne reference points based on the topological knowledge of the network structure held by the NML. These operations are, however, normally not visible across the q3nn or q3sn reference points.

The Service Management Layer (SML) is managing services running on top of the NML.

The Business Management Layer is managing business processes within the organization of the network provider and, in some cases, his customers.

Management systems may be recursively nested within the NML and the SML. They may also interoperate with other management systems within the same administrative domain (same network provider) across q3nn or q3ss reference points,

---

[4] Strictly speaking, proprietary reference points are not part of the LLA.
[5] One may also speak about a certain level, such as the network management level.

or in other administrative domains across xnn or xss reference points. The f-reference point for operator access may be provided in all layers.

Network technologies may be combined to provide more comprehensive functions. Of particular interest is the case where IP is running on top of different access and core network technologies to provide a unified end-to-end network representation to all services in the SML. This is the so-called "Full Service Network" (FSN) concept. The access portion of the FSN is often denoted the "Full Service Access Network" (FSAN).

In principle, every management task should be handled at the highest possible level of abstraction because, potentially, that makes the task simpler. For example, instead of setting up a connection at the NML, the connections in each node may be set up individually at the EML but that takes more time.

When an alarm occurs in an NEF it is detected by the EML and mapped to the NML so that the faulty network connections are identified. Depending on the importance of the traffic and the protection and restoration mechanisms available the traffic may or may not be rerouted by the NML. The task of repairing the fault is left to the EML - when time and priority permits.

The other usage of the EML is for the configuration of the NEF during installation or upgrade. There seems to be a trend towards limiting operator intervention to the setting of configuration options that may neither be done in the factory nor automated.

A consequence of dealing with interoperability at the NML is that the requirements for standardized NE models at the EML may be relaxed, i.e., the q3nne and q3ne reference points may be replaced by proprietary n and m reference points. To allow the parties involved (the management system manufacturer, the network provider, third party implementers) to perform the implementation in a controlled and orderly manner, however, n and m must be well specified. Taken together with increased automation of EML functions wherever possible, reduction in complexity and thus in operational expenses should result.

If interoperability is more attractive on the NML what about even higher layers, i.e., the SML? Up to now, standardizing SML models have been regarded as interfering with proprietary service implementations. Certain initiatives have been provided such as the standardization of the leased line and the BVPN services in the ITU and the definition of the service management architecture in the TMF [8]. This may change with the forecasted extensive deployment of new services such as e-commerce on top of the IP network.

4   Modelling Methodology – the Enhanced RM-ODP Framework

Due to a number of limitations, in particular the absence of mappings to management requirements and the lack of support for distribution, Q.18/4 concluded that OSI Management did not have sufficient modeling capabilities to become the modeling methodology of choice for the development of the network model. Instead, it was decided that an enhanced version of the Reference Model for Open Distributed Processing (RM-ODP) should be developed. The resulting methodology [9], [10] and the model for an example, subnetwork connection management [11], [12], [13] were developed. They became approved in 1996.

RM-ODP is defined in [14], [15], [16] and [17]. It consists of five viewpoints:
- The Enterprise Viewpoint
- The Information Viewpoint
- The Computational Viewpoint
- The Engineering Viewpoint
- The Technology Viewpoint

Each viewpoint is a self-contained specification of the system from a particular perspective. In addition, certain mappings between the viewpoints need to be maintained for the integrity of the overall system to be preserved.

Because the target is a model for management, only the system aspects subject to management, the management requirements, need to be modeled. They are expressed as actions with associated policies in the Enterprise Viewpoint. This implies that the requirements become an integrated part of the model.

4.1 The Enterprise Viewpoint

When compared with the original version of RM-ODP, most of the changes pertain to the Enterprise Viewpoint. A finer modeling granularity is allowed to correspond with the granularity of the network resource types in G.805. All the constructs (in all viewpoints) are provided with unique labels for backward traceability from the model elements in the Information-, Computational-, and Engineering Viewpoints to the functional requirements in the Enterprise Viewpoint. This is a fundamental mechanism for the support of conformance testing and also when estimating the cost of implementation for particular requirements.

Functional requirements are modeled as actions invoked by a caller role[6] and carried out by a provider role. Actions are enforced or restricted by action policies, Permissions, Prohibitions or Obligations respectively, to provide additional level of detail to the action definition.

Actions are grouped into functional units called Communities. An ordered series of actions combined to provide more comprehensive functions is called an Activity. By taking advantage of existing communities, substantial reuse of specification and eventually implementation may be achieved.

The Service Contract is essentially a listing of the functional contents of a community. Due to the labels it provides unique references to the functional elements already defined, simplifying reuse. The service contract may be used to define the technical part of the Service Level Agreement (SLA), the Service Level Specification (SLS).

In addition to capturing the functional requirements, the Enterprise Viewpoint also serves as the roadmap towards the other viewpoints. Actions in the Enterprise Viewpoint map to interface operations in the Computational Viewpoint. The client and provider roles map to computational objects. Enterprise actions are normally concerned with the manipulation (create, delete, associate, etc.) of G.805 network

---

[6] in RM-ODP, a role is a fraction of the behavior of an object, in this case, an Enterprise object

resources such as trails, access groups etc. Network resources map to objects, attributes or relationships in the Information Viewpoint.

When passing from the less formal architectural description of network resources in G.805 to a formal network model, additional behavior needs to be settled. Rec. G.852.2 "Transport Network Enterprise Model [18] is doing that for the elements already defined in G.805 and is also providing definitions for some important elements currently missing.

An extract from the Enterprise Viewpoint for trail management showing the role, action and service contract definitions for trail termination point creation is provided below.

Community trail management

Role

tm_caller

This role reflects the client of the actions defined within this community. One and only one caller role occurrence must exist in the community.

tm_provider

This role reflects the server of the actions defined within this community. One and only one provider role occurrence must exist in the community.

layer network domain

This role represents the layer network domain resource defined in Recommendation G.852.2. One and only one layer network domain role occurrence may exist in the community.

trail termination point

This role reflects the trail termination point resource as defined in Recommendation G.852.2. Zero or more trail termination point role occurrences may exist within this community.

ACTION

Create trail termination point

This action is used for the creation of a trail termination point. The caller has the ability to provide a unique user identifier to identify the trail termination point that has been created.

ACTION_POLICY

OBLIGATION inputDirectionality

The caller shall specify the directionality of the trail termination point to be created.

PERMISSION inputUserId

The caller may provide a user identifier for the requested trail termination point.

OBLIGATION rejectUserIdNotUnique

If PERMISSION inputUserId is part of the contracted service and if the user identifier is not unique in the provider context, then the provider shall reject the action.

OBLIGATION provideUserId

If PERMISSION inputUserId is part of the contracted service, then the provider shall use the user identifier as the unique identifier when communicating with the caller.

OBLIGATION successReturnId

If PERMISSION inputUserId is not part of the contracted service, the provider shall, upon success of this action, return the unique identifier for the created trail termination point.

PERMISSION inputUserLabel

The caller may provide a user label for the requested trail termination point.

SERVICE CONTRACT tm_src
ROLE
tm_caller, tm_provider, layer network domain, trail termination point
ACTION
Create trail termination point {OBLIGATION inputDirectionality, rejectUserIdNotUnique, provideUserId, successReturnId; PERMISSION inputUserId};

4.2   The Information Viewpoint

The Information Viewpoint describes the static behavior of the system in terms of information elements, i.e., objects, attributes and relationships. The behavior of the information objects is made up of invariants, attributes and mandatory relationships. It may either be provided as structured English text or be described using the formal language Z [19].

Rec. G.853.1 "Common Information Viewpoint"[20] is a library constituting the information elements that may be directly defined on the basis of G.805 and G.852.2, that is, without taking functional requirements pertaining to any specific management area into account. When producing an Information Viewpoint specification for a specific functional area the elements from the library are being used as superclasses and specializations are provided taking the additional functional requirements into account. New elements may be added as well.

Information elements are mapped to the corresponding Enterprise Viewpoint elements, eventually to elements in G.852.2.

The Information Viewpoint is the ultimate source for the definition of the information elements within the system. This is reflected by the Parameter Matching clause in the Computational Viewpoint which maps the input and output parameters to the corresponding information elements.

An extract from the Information Viewpoint for trail management showing the UML class diagrams, the information objects and the relationships for networkTTP creation is shown below.

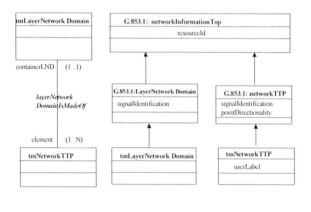

Relationship diagram   **Inheritance diagram**

Fig. 5. UML class diagrams for trail termination point creation

tmNetworkTTP
<COMMUNITY: trail management, ROLE: trail termination point>
DEFINITION
"This object class is derived from <networkTTP>."
ATTRIBUTE
<userLabel>
"<COMMUNITY: trail management, ACTION: create trail termination point,
ACTION_POLICY: inputUserLabel>
This attribute is imported from G.853.1 and is used as a user friendly label for the
networkTTP."
RELATIONSHIP
<accessGroupIsMadeOfNetworkTTPs>
<trailIsTerminatedByPointToPoint>
<subnetworkTPIsRelatedToExtremity>
<layerNetworkDomainIsMadeOf>

tmLayerNetworkDomain
<COMMUNITY: trail management, COMMUNITY_POLICY: signalId>
DEFINITION
"This object class is derived from <layerNetworkDomain>."
ATTRIBUTE
-- none additional
RELATIONSHIP
<layerNetworkDomainIsMadeOf>

layerNetworkDomainIsMadeOf (imported from G.853.1)
DEFINITION
"The layerNetworkDomainIsMadeOf relationship class describes the relationship that exists between a layerNetworkDomain and the objects that compose it."
ROLE
containerLND
"Played by an instance of the <layerNetworkDomain> information object type or subtype."
element
"Played by an instance of the subtype of the <networkInformationTop> information object type."
INVARIANT
inv_containerLNDRoleCardinality
"One and only one instance of the role *containerLND* must participate in the relationship."
inv_elementLNDRoleCardinality
"One or more instances of the role *element* must participate in the relationship."
inv_signalIdentification
"The *containerLND* and the *element* must contain the same signalIdentification information."

4.3   The Computational Viewpoint

The dynamic system behavior is described as interactions between computational objects in the Computational Viewpoint. Operational as well as notification interfaces are provided by these objects. The computational objects represent the finest granularity of objects in the mapping[7] to the Engineering Viewpoint.

For each computational object, the mapping is provided to the appropriate caller and provider roles in the Enterprise Viewpoint.

The major part of the Computational Viewpoint deals with the specification of the operations belonging to each interface. The methodology used is commonly known as "Design by contract" [21]. An operation is invoked providing a number of input parameters and upon the successful execution, a number of output parameters are returned. Each parameter has a name, a type specifier and a value assigned. Every parameter is mapped to the corresponding element in the Information Viewpoint in the Parameter Matching clause.

The invariant state of the system before and after the execution of an operation is defined by a set of pre- and post-conditions. They are defined by the state of the relevant relationships and attributes. Whenever an invariant is violated, a specific exception is raised. For each exception, an explanatory text as well as a type specifier is provided.

A report notification is defined for every operation. The report may contain additional information for the recipients to take advantage of the outcome of the successful event.

---

[7] This is a formal mapping defined in RM-ODP

The operations are described in a communication protocol neutral fashion. Protocol specific constructs for the communication protocol chosen are added in the Engineering Viewpoint. The parameters are defined in the ASN.1 language.

Operations are mapped to actions in the Enterprise Viewpoint.

An extract from the Computational Viewpoint for trail management describing the create networkTTP operation and the associated ASN.1 types, is shown below.

Create networkTTP
<COMMUNITY: trail management, ACTION: create trail termination point>
OPERATION createNetworkTTP {
INPUT_PARAMETERS
layerND: LayerNetworkDomainChoice;
pointDir: PointDirectionality;
suppliedUserIdentifier: UserIdentifier;
-- *zero length string or 0 implies none supplied.*
suppliedUserLabel: GraphicString;
-- *zero length implies none supplied.*

OUTPUT_PARAMETERS
networkTTP: NetworkTTPChoice;

RAISED_EXCEPTIONS
userIdentifierNotUnique: UserIdentifier;
failureToCreateNetworkTTP: NULL;
failureToSetUserIdentifier:NULL;

BEHAVIOUR
SEMI_FORMAL
PARAMETER_MATCHING
layerND: INFORMATION OBJECT: tmLayerNetworkDomain>;
suppliedUserIdentifier: <INFORMATION ATTRIBUTE: resourceId>;
pointDir: <INFORMATION ATTRIBUTE: pointDirectionality>;
networkTTP: <INFORMATION OBJECT: tmNetworkTTP>;
suppliedUserLabel: <INFORMATION ATTRIBUTE: userLabel>;

PRE_CONDITIONS
inv_uniqueUserIdentifier
"**suppliedUserIdentifier** shall not be equal to *resourceId* of any *element* in a <*layerNetworkDomainIsMadeOf*>   relationship where **layerND** refers to *containerLND*."
- POST_CONDITIONS
- inv_existingNetworkTTP
- "networkTTP and layerND must respectively refer to *element* and *containerLND* in a <*layerNetworkDomainIsMadeOf*> relationship."
- inv_agreedUserIdentifier
"*resourceId* of *tmNetworkTTP* referenced by **networkTTP** is equal to **suppliedUserIdentifier**, if it is supplied."

EXCEPTIONS
IF PRE_CONDITION inv_uniqueUserIdentifier NOT_VERIFIED
RAISE_EXCEPTION
userIdentifierNotUnique;
IF POST_CONDITION inv_existingNetworkTTP NOT_VERIFIED
RAISE_EXCEPTION
failureToCreateNetworkTTP;
IF POST_CONDITION inv_agreedUserIdentifier NOT_VERIFIED
RAISE_EXCEPTION
failureToSetUserIdentifier;
}

Supporting ASN.1 productions
LayerNetworkDomainChoice ::= CHOICE {
tmLayerNetworkDomainQueryIfce TmLayerNetworkDomainQueryIfce,
userIdentifier   UserIdentifier };

NetworkTTPChoice::= CHOICE {
tmNetworkTTPQueryIfce TmNetworkTTPQueryIfce,
userIdentifier   UserIdentifier};

4.4 The Engineering Viewpoint

The Engineering Viewpoint describes the operations for specific interfaces based on a given communication protocol. Specifications exists already for CMIP [22] [23] and Corba communication [24], [25], [26] and others will be provided. When utilizing CMIP, Managed Object (MO) classes representing network resources are mapped to the Enterprise Viewpoint resources in G.852.2. Actions are mapped to operations in the Computational Viewpoint and name bindings and attributes are mapped to the corresponding elements in the Information Viewpoint. The mapping scheme for Corba communication is not fully decided yet.

Distribution is another important feature for future applications. Within RM-ODP, there are mechanisms included to implement a number of distribution types by supporting the corresponding transparencies.

The functionality describing the network TTP creation for a CMIP interface is spread across a number of constructs, the MO class definition, the conditional packages, the name binding and the error parameters. It is not readily separable from a number of other functional elements either and, therefore, will not be presented here.

4.5 The Technology Viewpoint

The Technology Viewpoint is concerned with implementation issues only and will not be discussed here.

5 Existing Functionality

Following the approval of the enhanced RM-ODP framework in 1996, a range of recommendations applicable to technologies such as SDH, WDM and, to a certain extent ATM, were approved in 1999. The following functional areas were covered:
- Topology management [27], [28], [29].
- Pre-provisioned adaptation management [30], [31], [32].
- Pre-provisioned link connection management [33], [34], [35].
- Pre-provisioned link management [36], [37], [38].
- Trail management [39], [40], and [41].

Work is in progress to include partitioning, protection, routing, and failure propagation and also completing the model for ATM.

When assessing the applicability of a new technology to the model, the specific requirements of the technology are analyzed. In some cases the non-matching requirements may be modeled on the NE level, while others may lead to technology specific extensions on the network level. For example, the assessment of ATM resulted in two major changes, the redefinition of link capacity to be either bandwidth or the number of link connections, and the notion of dynamic creation of termination points during connection setup.

The applicability of the model to WDM technology was proven by the ACTS project Mephisto and demonstrated at the public presentation in October 1999 [42]. The setup of a protected trail in the OCH layer on a prototype system consisting of three OADMs in a ring structure was shown. The network model used in Mephisto is based on the generic network model. The configuration manager developed provides the operator with one operation for each action in the communities supported, that is, topology management, pre-provisioned link connection management, subnetwork connection management and trail management.

It is argued in the Mephisto project [43] that the generic model may be utilized for the management of SDH over WDM, provided that an enhanced version of the route discovery function is used in the OCH layer. An alternative solution to limit noise accumulation is the addition of a digital frame like the digital wrapper on top of the OCH layer combined with Forward Error Correction (FEC) inside each network provider domain involved.

So far, only technologies for the core network have been modeled. There are, however, increasing concerns to provide an end-to-end network view including access network technologies as well. A combined network and network element model for ATM PONs will be approved this year in ITU-T [44].

6 Modelling Connectionless Communication

The current versions of G.805 and the technology specific extensions all presume connection – oriented communication, that is, prior to traffic flowing, a connection has to exist. With connectionless communication there is no connection setup in advance. To cope with that, a novel network architecture with the working title "g.cls" is currently being defined in SG.13. The scope is not limited to connectionless technologies, so there is a potential for g.cls to become the replacement of G.805 rather than a complement. Developing a network model for connectionless technologies is part of the work plan for Q.18/4.

If IP is to become the major network technology of the future, enhancements in the routers as well as in the applications running in the hosts are necessary. A comprehensive network model is a powerful tool for analyzing as well as managing such an enhanced IP based network.

One of the major enhancements relates to the communication paradigm. It is widely agreed that the current "Best Effort" approach should not be the only level supported. Traffic Engineering is a tool that are being proposed to provide differentiation to the communication paradigm and thus to the QoS[8], utilizing mechanisms such as RSVP [45], Intserv [46], [47], Diffserv [48], [49], MPLS [50], [51] and COPS [52], [53], [54]. In addition, a control framework is needed to verify compliance between planned and actual traffic and also for specifying how to deal with excess traffic. At the same time, one is trying to preserve the structural simplicity of the existing Internet.

Differentiated service quality should lead to differentiated charging. Consequently, new charging models must be developed to complement the existing flat rate charging.

The second argument for an extension of the model is the interworking between IP and other technologies. Primarily, client-server schemes with IP in the client role are foreseen, but peer configurations are expected too. A variety of interworking schemes need to be taken into account, ranging from gateways supporting one specific management function to the full interworking between IP and the server technologies.

Another concern is that of handling the vast forecasted increase in traffic [55]. This traffic will be aggregated and handled by large capacity routers. Unfortunately, data traffic on the one hand and voice and video traffic on the other poses quite different requirements to the network in terms of packet loss, delay and delay variation. This makes the traffic grooming function more complex. There are issues relating to the work split between routing on the client level (IP) and routing on the server level (OTN/WDM) too.

6.   A Multi-Technology/MULTI-Domain Example

Two networks based on compatible technologies (for the traffic) are exchanging traffic across a domain boundary, confer Fig.6.

---

[8] QoS should be interpreted broadly including every kind of parameter affecting the perception of the end user of the service provided.

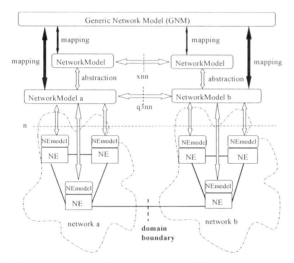

Fig. 6. Scenario for multi-technology, multi-domain interoperability

The network models[9] in both domains are based on the generic model but differences may exist due to manufacturer, equipment type/-version, technology specific properties, etc., catered for by the term "mapping" in Fig.6. The two domains may be operated by the same network provider, in which case there is a standardized q3nn reference point available for network level management across the domain boundary. With two different providers, there is a xnn reference point instead based on a more abstract network model.

The interactions between the network element models and the network models take place via proprietary n reference points. As long as the network models in the two domains are being consistently updated whenever changes occur on the NE level and vice versa, the capability of providing well defined end to end functions is maintained.

The generic network model (GNM) is providing the operators with an overall view of the networks in the two domains. On this basis the operators may find a route for an end-to-end trail and input a trail request to the GNM. The trail request is converted into SNC requests to the network models in the two domains (x or q) and further to cross connection requests to the appropriate NE models. Finally, crossconnections are established in the NEs to complete the setup of the end-to-end trail.

7   Conclusion

Interoperability is an attractive feature in a heterogeneous telecom network environment. A number of different approaches are being proposed for achieving that. This paper suggests that interoperability at the network level is a natural choice and

---

[9] All the   models in Fig.6 including the generic network model are physical models implemented in databases.

proceeds by describing the underlying tools and the existing version of the generic network level model as defined in the G.85x series of recommendations.

The consequence of addressing interoperability at the network level is that the requirements for a standardized NE model become relaxed. Taken together with increased automation of NE functionality, substantial savings in operational expenses are expected.

Technologies like SDH, WDM and, partly, ATM are supported by the current version of the model. Work has started to expand the scope by including technologies for the access network and for connectionless communication such as IP. This will make the generic network model well suited for modeling end-to-end trails across multiple network provider domains and multiple technologies spanning the access network as well as the core network.

References

1. ITU-T rec. G.805 "Generic functional architecture of transport networks", Geneva 11/95
2. ITU-T rec. G.803 "Architecture of transport networks based on the synchronous digital hierarchy (SDH)", Geneva 06/97
3. ITU-T draft rec. G.872 "Architecture of optical transport networks", Geneva 02/99
4. ITU-T rec. I.326 "Functional architecture of transport networks based on ATM", Geneva 11/95
5. ITU-T rec. G.902 "Framework recommendation on functional access networks (AN). Architecture and functions, access types, management and service node aspects", Geneva 11/95
6. ITU-T draft. rec. g.cls "G.cls functional model", Kyoto 03/00
7. ITU-T draft.rev.rec. M.3010 "Principles for a telecommunications management network architecture", Geneva, January 2000
8. TMF, Telecom Operations Map, GB910, version 1.1, April 1999
9. ITU-T rec. G.851.1 "Management of the transport network – application of the RM-ODP framework", Geneva, 03/99
10. Varma, E.L., et al "Achieving global information networking", Artec House, Boston/London, 1999
11. ITU-T rec. G.852.1 "Management of the transport network – Enterprise viewpoint for simple subnetwork connection management", Geneva, 11/96
12. ITU-T rec. G.853.2 "Subnetwork connection management information viewpoint", Geneva, 11/96
13. ITU-T rec. G.854.1 "Management of the transport network – Computational interfaces for basic transport network model", Geneva, 11/96
14. ITU-T rec. X.901 "Basic reference model for Open Distributed Processing- Part 1: Overview"
15. ITU-T rec. X.902 "Basic reference model for Open Distributed Processing- Part2: Foundations"
16. ITU-T rec. X.903 "Basic reference model for Open Distributed Processing- Part 3: Architecture"
17. ITU-T rec. X.904 "Basic reference model for Open Distributed Processing- Part 4: Architectural Semantics"
18. ITU-T rec. G.852.2 "Management of transport network- Enterprise viewpoint description of transport network resource model", Geneva, 03/99
19. Potter, B. et al "An introduction to formal specification and Z", Prentice Hall, 1992

20. ITU-T rec. G.853.1 "Management of transport network- Common elements of the information viewpoint for the management of a transport network", Geneva, 03/99

21. Meyer, B. "Object oriented software construction", Prentice Hall 1997

22. ITU-T rec. M.3100 amd.1 "Generic network information model Amendment 1", Geneva, 03/99

23. ITU-T rec. G.855.1 "Management of transport network- GDMO engineering viewpoint for the generic network level model", Geneva, 03/99

24. ITU-T draft. rec. X.780 "TMN guidelines for defining CORBA Managed Objects", London, 05/00

25. ITU-T draft. rec. Q.816 "CORBA based TMN services", London, 05/00

26. ITU-T draft. rec. M.3120 "CORBA generic Network and NE level information model", London, 05/00

27. ITU-T rec. G.852.3 "Management of transport network- Enterprise viewpoint for topology management", Geneva, 03/99

28. ITU-T rec. G.853.3 "Management of transport network- Information viewpoint for topology management", Geneva, 03/99

29. ITU-T rec. G.854.3 "Management of transport network- Computational viewpoint for topology management", Geneva, 03/99

30. ITU-T rec. G.852.8 "Management of transport network- Enterprise viewpoint for pre-provisioned adaptation management", Geneva, 03/99

31. ITU-T rec. G.853.8 "Management of transport network- Information viewpoint for pre-provisioned adaptation management", Geneva, 03/99

32. ITU-T rec. G.854.8 "Management of transport network- Computational viewpoint for pre-provisioned adaptation management", Geneva, 03/99

33. ITU-T rec. G.852.10 "Management of transport network- Enterprise viewpoint for pre-provisioned link connection management", Geneva, 03/99

34. ITU-T rec. G.853.10 "Management of transport network- Information viewpoint for pre-provisioned link connection management", Geneva, 03/99

35. ITU-T rec. G.854.10 "Management of transport network- Computational viewpoint for pre-provisioned link connection management", Geneva, 03/99

36. ITU-T rec. G.852.12 "Management of transport network- Enterprise viewpoint for pre-provisioned link management", Geneva, 03/99

37. ITU-T rec. G.853.12 "Management of transport network- Information viewpoint for pre-provisioned link management", Geneva, 03/99

38. ITU-T rec. G.854.12 "Management of transport network- Computational viewpoint for pre-provisioned link management", Geneva, 03/99

39. ITU-T rec. G.852.6 "Management of transport network- Enterprise viewpoint for trail management", Geneva, 03/99

40. ITU-T rec. G.853.6"Management of transport network- Information viewpoint for trail management", Geneva, 03/99

41. ITU-T rec. G.854.6 "Management of transport network- Computational viewpoint for trail management", Geneva, 03/99

42. http://www.infowin.org/ACTS/NEWS/CONTEXT_UK/990899fr.htm "Invitation to the MEPHISTO public demonstration Marcoussis, France, 20-24 September 1999"

43. Bertelon, L. et al "OTN management interworking with SDH- Specifications, ACTS project no. AC209 Mephisto, deliverable D20 (restricted), the Mephisto Consortium, October 1999.

44. ITU-T draft rec. m.xxxc "Management services, object models and implementation ensembles for ATM-PON system", Geneva, January 2000.

45. Braden, R., et al "Resource ReSerVation Protocol (RSVP) – version 1 Functional Specification", RFC 2205, September 1997

46. http://www.ietf.org/html.charters/intserv-charter.html

47. http://www.ietf.org/ids.by.wg/intserv.html
48. http://www.ietf.org/html.charters/diffserv-charter.html
49. http://www.ietf.org/ids.by.wg/diffserv.html
50. http://www.ietf.org/html.charters/mpls-charter.html
51. http://www.ietf.org/ids.by.wg/mpls.html
52. http://www.ietf.org/html.charters/rap-charter.html
53. http://www.ietf.org/html.charters/policy-charter.html
54 http://www.stardust.com/policy/whitepapers/qospol.htm1994.
55. Anderson, J., et al "Protocols and architectures for IP optical networking", Bell Labs Technical Journal, January-March 1999

Abbreviations

ACTS:	Advanced Telecommunication Technologies and Services
ASN.1:	Abstract Syntax Notation no.1
ATM:	Asynchronous Transfer Mode
AP:	Access Point
BML:	Business Management Layer
BVPN:	Broadband Virtual Private Network
COPS:	Common Open Policy Service
CP:	Connection Point
DiffServ:	Differentiated Services
CTP:	Connection Termination Point
CORBA:	Common Object Request Broker Architecture
CMIP:	Common Management Information Protocol
cls:	connectionless service
EML:	Element Management Layer
ETSI:	European Telecommunications Standards Institute
FEC:	Forward Error Correction
FSAN:	Full Service Access Network
FSN:	Full Service Network
GNM:	Generic Network Model
IDL:	Interface Description Language
IETF:	Internet Engineering Task Force
IntServ:	Integrated Services
IP:	Internet Protocol
ITU-T:	International Telecommunication Union- Telecommunications sector
LC:	Link Connection
LLA:	Logical Layered Architecture
Mephisto:	ManagemEnt of PHotonIc SysTems and netwOrks
MPLS:	MultiProtocol Label Switching
NE:	Network Element
NEF:	Network Element Function
NML:	Network Management Layer
OADM:	Optical Add and Drop Multiplexer
OCH:	Optical CHannel

OSF:	Operation System Function
OTN:	Optical Transport Network
PON:	Passive Optical Network
QoS:	Quality of Service
RM–ODP:	Reference Model –for Open Distributed Processing
RSVP:	resource ReSerVation Protocol
SLA:	Service Level Agreement
SLS:	Service Level Specification
SML:	Service Management Layer
SNC:	SubNetwork Connection
TCP:	Transport Connection Protocol
TMF:	Tele Management Forum
TMN:	Telecommunication Management Network
TTP:	Trail Termination Point
UML:	Unified Modeling Language
VCI:	Virtual Container Identifier
VPI:	Virtual Path Identifier
WDM:	Wavelength Division Multiplex

SIP for Call Control in the 3G IP-based UMTS Core Network

Delphine Plasse

BT Adastral Park, UK
delphine.plasse@bt.com

Abstract. This paper present an example of call control using the Session Initiation Protocol (SIP) for multimedia call management in the "all IP" architecture for the core network of UMTS. The first part of this publication gives an overview of the 3GPP IP-based UMTS core network. The second part introduces an interworking scenario between the UMTS core network and the SIP-enabled public Internet. The current main issues are highlighted.

1   Introduction

The 3rd Generation Partnership Program group (3GPP) [1] is in the process of specifying the core network architecture of UMTS phase 2 also known as Release 2000 (R00). A number of options have been proposed for the R00 architecture, one of which known as the UMTS "all-IP" architecture:

The "all IP" architecture is evolving from the GSM/GPRS architecture-based UMTS phase 1 (R99). It is intended to use IP components within the core network, with evolved GPRS elements supporting data, multimedia and voice over IP (VoIP) mechanisms for the delivery of, for example, speech and traditional circuit-switched type of features.

The "all IP" architecture is required to provide interworking solutions between UMTS and 2nd Generation mobile networks, the legacy PSTN/ISDN networks and external IP-based SIP/ H323-enabled networks. It is also required to support standardised services, Camel Service Switching Functions, call control and Media Gateway interrogation. The investigation of solutions for multimedia call management is an important issue in the 3GPP group. A possible solution is the application of IP-based signalling for call control in the "all-IP" architecture.

The first part of this paper introduces the UMTS R00 architecture. The second part will present a solution employing the SIP protocol [2], for multimedia call set-up and termination between UMTS/GPRS network and external multimedia IP-based networks.

2   The UMTS IP-based Architecture [3]

The UMTS "all-IP" architecture is based on the evolution of the GPRS core network standard, which will fully support advanced Voice over IP, data and multimedia applications.

Figure 1 shows a simplified overview of the Reference Architecture for Release 2000. The final approval of this architecture will depend on the resolution of a certain number of issues, which include the relationship between different call control models. The following section is describing the various functional elements:

CSCF: the Call State Control Function. It performs call control, service-switching functions, address translation functions and vocoder negotiation functions.

HSS: the Home Subscriber Server. It is the master database for a given user. It is responsible for keeping a master list of features and services (either directly or via servers) associated with a user, and for tracking of location and means of access for its users. It provides user profile information, either directly or via servers. It is a superset of the HLR functionality, since it also communicates via new IP-based interfaces.

MGCF: the Media Gateway Control Function. It is the point of interface for signalling between IP-based packet switched networks i.e. GPRS, and circuit switched networks. The MGCF communicates with the CSCF. It selects the CSCF depending on the routing number for incoming calls from legacy networks and

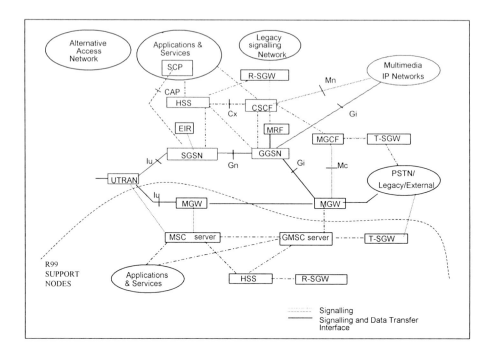

Fig. 1. Reference Architecture for Release 2000 [1]

performs protocol conversion between the Legacy (e.g. ISUP, R1/R2 etc.) and the R00 network call control protocols. Out of band information is assumed to be received in MGCF and may be forwarded to CSCF/MGW.

MGW: the Media Gateway function. This component is the PSTN/PLMN transport termination point for a defined network, and interfaces UTRAN with the core network over the Iu interface. A MGW may terminate bearer channels from a circuit switched network and may terminate media streams from a packet network (e.g., RTP streams in an IP network). It may support media conversion, bearer control and payload processing (e.g. codecs, echo canceller, conference bridge) for support of different Iu options for circuit switched services: AAL2/ATM based as well as RTP/UDP/IP based. The MGW interacts with the MGCF, the MSC server and GMSC server for resource control.

T-SGW: the Transport Signalling Gateway Function. This R00 network component is the PSTN/PLMN termination point for a defined network. It maps call related signalling from/to PSTN/PLMN on an IP bearer and sends it to/from the MGCF. It also needs to provide PSTN/PLMN to IP address mapping.

R-SGW: The Roaming Signalling Gateway function. It supports roaming to/from 2nd Generation/R99 circuit Switched to/from R00 UMTS GPRS domain. It provides signalling conversion between legacy SS7 based signalling and the IP based signalling to support roaming.

MSC server: The MSC server caters for R99 roaming users on the R00 network. It comprises the call control and mobility control parts of a GSM/UMTS R99 MSC, in an IP-based network. The MSC server would cater for VLR and control parts of MSC & GMSC.

GMSC server: the Gateway MSC Server. The GMSC comprises mainly the call control and mobility control parts of a GSM/UMTS GMSC.

3 Session / Call Control

In the UMTS IP-based core network, it is the role of the CSCF to perform session / call control. Although the CSCF functions are still being studied, the main CSCF functionality have been highlighted:

Call registration, routing of incoming calls, call set-up / termination, call states and event management, call event reports for billing & auditing, support of multi-party services, as well as address handling & cache management, communication with the HSS, MRF and MGW, and provision of service trigger mechanisms.

The SIP protocol is currently being studied as a possible protocol for multimedia call management in the "all – IP " UMTS architecture. SIP has been designed for the initiation and termination of any type of sessions across the Internet [4]:

Determination of the communicating end system(s), determination of the media parameters to be used, determination of the willingness of the called party to engage in communication, call transfer and termination of calls, and set-up of the call parameters at both called and calling party.

SIP signalling has many similarities with HTTP in terms of header field structure, encoding rules and authentication mechanisms, and SMTP in terms of addressing scheme, address resolution or server location mechanisms.

As shown in the following section, the SIP architecture enables a large number of call management functions [5]:

Application registration / authorisation with SIP REGISTER messages and events management messages such as the 100 message series. SIP supports multi-party services and address handling with its name to multiple address resolution schemes. Cache management and communication with location databases and gateways cater for the invocation of location-based services, while INVITE messages trigger all type of call services [6].

4   Using SIP to Support Call Control

Figures 2 and 3 shows an example of the usage of the SIP protocol for call control. The example is a successful UMTS originated call to a public Internet SIP-enabled terminal B. This call control scenario shows the UMTS terminal A registration with the CSCF / SIP server, then the scenario shows the initiation / termination of the multimedia session between a UMTS terminal and an Internet SIP-based terminal. A network resource reservation phase needs to take place in order to cater for real time traffic requirements (in our example, a VoIP session).

Certain assumptions have been made: the mobile user A is in his/her home PLMN. Terminal A has already discovered which CSCF to register with. User B is already registered with the network. SIP servers may be acting in proxy mode should they not be able to handle requests themselves.

The Registration phase either validates or invalidates a SIP client for user services provided by the SIP server:

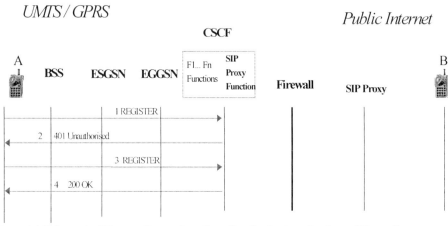

the establishment of the media session, then finally the termination of the call:

Fig. 2. Non-roaming user A registers with its home CSCF.

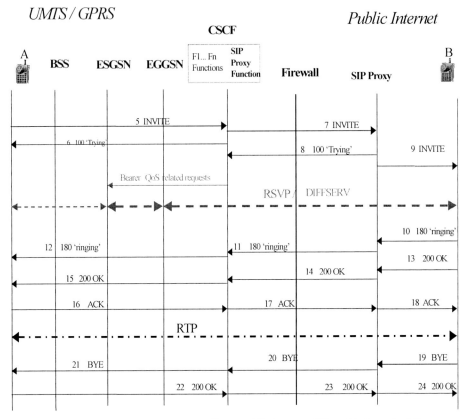

Fig. 3. UMTS originated VoIP call to a public Internet SIP enabled terminal B

1. User A sends a SIP REGISTER message to its home CSCF / SIP (proxy) server.
2. The SIP proxy server provides an authorization challenge to user A (the SIP digest format is described in RFC 2543).
3. User A enters his/her valid user Identifier and password, encrypted according to the challenge issued by the SIP server and sends the response to the SIP server.
4. The SIP proxy validates the user's credentials. It registers the user in its location database (which could be the user's UMTS HSS) and returns a 200 OK response to user A.

In the next phase, user A completes a call to user B. The successful call shows the initial signalling and the exchange of media information in the form of SDP payloads,

5. User A's client sends an INVITE message to its UMTS CSCF.
6. A's CSCF accepts the credentials and forwards the INVITE message to B's SIP proxy server. A 's CSCF is assumed to have been authenticated by B 's proxy. Both CSCF and proxy have inserted their Record-Route header into the INVITE message to ensure that it is present in all subsequent message exchanges.

7-12. "Trying" and "Ringing" event messages are sent to communicating parties, i.e. user A 's user agent client, and both CSCF and SIP proxies. Since SIP supports multi party call, the initial INVITE message may contains more than one "via" message header and may have been forked. Therefore user B client will insert a tag in message 10.

13-18. User B, its home proxy and the caller's CSCF accept the session invitation and receive communicating party acknowledgment messages. RTP streams are established between user A and user B and data transfer takes place.

19-24. User B hangs up with user A and the VoIP session terminates with a BYE message initiated by user B to its proxy server, and cascaded to user A via its CSCF. In return, user A confirms the session termination.

5 Open Issues

We assumed that users were not roaming. But what happens when a user registers from a visited network? Should the registration /session control logic be executed in the visited network or in the home network?

Once the CSCF decides to admit the call, network resources need to be reserved in order to support the call traffic requirements (Voice traffic). A priori, the network resources ought to be established before end users answer the call. A dialog between the CSCF and the calling terminal, or perhaps, with a GPRS resource manager node (as shown in Figure 3) needs to take place in order to trigger the GPRS bearer resource allocation. It is also necessary for resource managers within the external IP-based network and within the UMTS network to interwork in order to perform end to end resource management. Similarly, once the session is finished (BYE messages are sent), they need to be freed up. How these mechanisms could be handled are open issues. SIP session signalling does not provide resource reservation, but it can convey the appropriate information to the appropriate network resource managers.

6 Conclusion

This paper presented an overview of the UMTS R00 architecture. It also showed that SIP signalling, message interactions and architecture offer sets of functionality for call control services that can be used in the "all-IP" UMTS architecture. As a simple example, a successful case of a UMTS originated VoIP session to a public Internet SIP- enabled terminal B was presented. A number of open issues have also been introduced. As work progresses in this area other issues will emerge. Future work will try and answer them. It will tackle mobile originated, mobile terminated calls for roaming users, in both successful and failure cases.

References

1. www.3gpp.org
2. "SIP: Session Initiation Protocol", M. Handley, H. Schulzrinne, E. Schooler, and J. Rosenberg, RFC 2543, March 1999.
3. "Architecture Principle for Release 2000", TR 23.821 v 0.1.0, 3GPP SA.
4. "www.ietf.org
5. H. Schulzrinne and al, "SIP Telephony Call Flow Examples", Internet draft, March 00, work in progress.
6. "H. Schulzinne *et al.*, "Implementing Intelligent Network Services with the Session Initiation Protocol", Tech-Report Number CUCS-002-99, Columbia University.

Short Glossary

E-GPRS: Enhanced General Packet Radio Service.
HTTP: Hyper Text Transfer Protocol.
IP: Internet Protocol.
PLMN: Public Lan Mobile Network
SIP: Session Initiation Protocol.
SMTP: Simple Message Transfer Protocol.
RTP: Real Time Protocol.
UMTS: Universal Mobile Telecommunication Services.

Integration of IN and Internet: The Balance between Idealism and Current Products Realism

Geir Gylterud[1], Michael Barry[2], Valerie Blavette[3], Uwe Herzog[4], Telma Mota[5]

[1]Telenor R&D, Otto Nielsens v.12,
7004 Trondheim, Norway
[2]Broadcom Eireann Research Ltd., Kestrel House, Clanwilliam Place,
Dublin 2, Ireland
[3]CNET-DAC/ARP, Technopole Anticipa, 2 avenue Pierre Marzin,
22307 Lannion Cedex, France
[4]T-Nova, Technologiezentrum,
D-64307 Darmstadt, Germany
[5] Portugal Telecom Inovação, Rua Eng. José F. Pinto Basto,
3810 AVEIRO, Portugal
geir.gylterud@telenor.com
mgb@broadcom.ie
valerie.blavette@cnet.francetelecom.fr
uwe.herzog@telekom.de
telma@ptinovacao.pt

Abstract. Since the beginning of the 90's Intelligent Networks (IN) have been deployed in telecommunication networks worldwide for the realisation of value-added services in circuit switched networks. In today's marketplace, evolutionary trends in IN and the growth of the Internet, have the potential to make a wide range of advanced services available to consumers. To facilitate this, a new service architecture providing an environment for the development of integrated IN-Internet services is proposed here. This new service architecture combines the best parts of IN and the Internet. It will evolve from current industry initiatives and developments and from existing products focussed on the integration of IN and Internet such, that today's gateways and switches will provide access to tomorrows services and platforms.

1   Introduction

The business environment for classical telecom / IN services is undergoing rapid change. New trends of both technical nature and social character have emerged in the area of telecommunications. Firstly, there is a clear trend for distribution of telecom hardware and software over the network. Secondly, deregulation in the telecom market is driving the opening network interfaces to other service providers and competitors. Lastly, the success of the Internet means that it is becoming a big competitor for the classical telephone network and IN.

More and more customers have access to the Internet and are being provided with new Internet based services. Some of these services are developed by Internet service providers and are multimedia and interactive in nature, others are typical telephony services for which the Internet is used as transport media, e.g. IP telephony. Huge growth in the IP telephony market is forecast in the near future. According to a Frost & Sullivan report the IP telephony revenues are predicted to rise from 260 million US$ in 1998 to 13,8 billion US$ in 2005 [1]. This and similar examples [2] underline the shift from circuit switched networks towards packet switched networks. In some countries two years ago the amount of packet traffic already exceeded the circuit switched traffic.

Despite the growth of IP based telephony services, IN services will continue to exist for several years. IN has already solved problems that Internet is still fighting with, including quality, availability, reliability and the creation of revenue. A practical solution is to converge IN and the Internet in order to exploit the benefits they offer and compensate their weaknesses. Such an integrated IN-Internet architecture enables new classes of service that are characterized by a mix of Internet and Telecom functionality, using the synergy of both. These integrated services will make use of the ubiquitous telephony / IN service and of the user friendliness and openness of the Internet, providing wide ranges of choice and value to the consumer. The introduction and provision of these new services over heterogeneous networks (PSTN and Internet) requires generic but practical solutions for the evolution of Network Intelligence. In particular, solutions must be provided for the following questions:

1. How should the PSTN/IN and Internet be connected?
2. How can a joint service platform spanning IN and Internet services be realised?
3. How can 3rd party service providers be given access to the network preserving network integrity?

EURESCOM project P909, running from January 1999, is investigating answers and solutions for the above issues of IN-Internet convergence. These investigations focus on a common Reference Architecture, defined by the project, and implemented on middleware-based on off-the-shelf products and/or prototypes. In particular the project addresses:

- new IN services and integrated IN-Internet services
- evolution of IN architecture (Business Model and architecture elements)
- use of existing concepts to evolve IN towards an open architecture (e.g. TINA) and to define open APIs (e.g. á la Parlay)
- control of special resources (e.g. VoIP gateways, PABXs, e-mail servers, etc.)
- investigation and integration of existing products/prototypes from the telecom and IT industry

This paper presents solutions to the questions posed above. It starts by introducing a Reference Architecture for IN-Internet integration. Following this, some of the main features and components to be developed for the architecture are described. An overview of existing technologies and products in the area of IN-Internet integration that are being integrated into the reference architecture is presented. Finally some conclusions on state-of-the art for IN/Internet integration are drawn.

2 Putting IN and Internet Together

In general interworking between the PSTN and Internet can be broken into two main areas:
1. Interworking at bearer control level,
2. Interworking at service control level.

In the case of *bearer control* there has been a lot of work in developing protocols and standards within the ITU and the IETF. Voice over the Internet (VoIP) is now widely commercially available using TCP/IP and new protocols such as H.323 [3]. Such internetworking is facilitated by a gateway function which translates between the internet and PSTN bearer protocols. Another protocol which is currently being standardised by IETF, for the control of media gateways, is the Media Gateway Control protocol (MeGaCo)[4].

With respect to *Service and service control* there are quite a number of reasons for communicating at this level:
- Providing end-to-end service over a heterogeneous PSTN/VoIP call
- Providing IN services to internet users,
- Facilitating an internet entity (e.g. web server) to access data in the PSTN,
- Facilitating an internet entity to start a call in the PSTN.
- To allow web-based customer-control access to an IN service.
- Facilitating the IN to access internet data (e.g. directory information)

The interface to the existing IN systems may be at the SMF level for non-call-related purposes such as user service profile modifications, at the SCF level for setting up PSTN calls or at the SSF level by using basic IN switching capabilities and triggering mechanisms.

This paper proposes the integration at the service control level through the definition and implementation of a unique service platform, able to support joint services spanning both IN and the Internet. A Reference Architecture has been developed, utilising features of existing initiatives and products in its specification and implementation. The following sections describe in more detail the realisation of the platform.

3 The Reference Architecture

The architectural framework is presented in Fig.1. It identifies three distinct layers:
- the **service control layer** (Service Platform) providing telecommunication-related services and service features, e.g., IN-like services; such services can be either offered/sold to end users or third party service providers that belong to a different domain;
- the **resource layer**, which contains all the network (public and private) and special resources controlled by the control and service components;
- the **adaptation layer** contains adapters that interface the service control platform functionality with the underlying resource layer.

The different layers inter-operate through well defined and public API's and/or protocols) and a Distributed Processing Environment (DPE), which provides the communication capabilities among the functionality that are supported within the service platform.

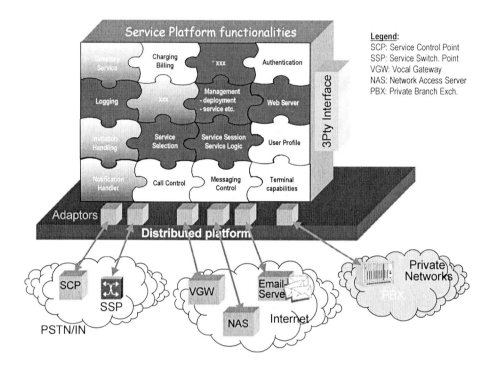

Fig. 1. Architectural framework

Fig. 1 depicts also a *3Pty interface* offered by the platform to 3rd parties in an external domain; e.g. consumers or third party providers. In some cases, the external domain will support a subset of the functionality supported by the service platform, such as subscription or accounting functionality. In other cases, it will support only service specific functionality and rely on the service platform for all the service generic functionality. The platform is flexible enough to offer to third parties different kinds of services and APIs at different levels. The functionality of platform should be used "on-demand".

The service control layer is a collection of functionality that interacts over the DPE. Composition and interaction of functionality allows service designers to build added value telecommunication services, while being abstracted away from the underlying network and special resources. In addition, this provides "service features" to services and applications in the external domain. These service features are offered by means of secured/controlled interfaces in order to ensure secure, robust and accountable interactions.

Particular interest is given to the *Call control component* in the service platform. This component offers a common call control interface to all services and provides the mechanisms to control calls across heterogeneous networks. Similarly, the *Message Control component* provides an abstract interface to services in the Service Control Layer. It maps towards messaging equipment and special peripherals providing support for e.g. email, fax, text-to-speech, speech recognition and other features.

The network and special resources can activate services implemented in the services layer by means of events that are sent to the service platform. Resource adapters, or wrappers, are used to abstract the upper layer from the details of specific protocols (e.g. INAP, Megaco/H.248).

To ensure that the service and network architecture is versatile enough to fulfil the needs and requirements identified for integrated IN-Internet services; it has been validated through the specification of use cases and services scenarios comprising of Click-to-Dial, Internet Call Waiting, Distributed & Enhanced Call-Centre, Virtual Presence, Universal Messaging and Meeting Scheduler services.

4   Major Components of the Reference Architecture

To build the proposed architecture and APIs existing concepts, specifications and products from existing standards bodies and initiatives are reused, particularly Parlay[5] and TINA[6].

The Parlay Industry Working Group has specified a new, open, network Application Programming Interface that allows new applications and services to be developed and deployed on today's and tomorrows networks[7]. The API is technology and network independent. It helps the network operator to maximise the value of their network technology by directly passing on that functionality to third parties in a safe and controlled manner, whilst hiding much of the complexity of network signalling from the developer.

The work of the TINA Consortium has also been reused in the platform TINA has defined and validated a consistent and open architecture for distributed telecommunications software applications [8]. Useful TINA concepts include the independence of service and transport networks, integration of control and management, adoption of object orientation and distributed processing techniques and reuse of existing standards wherever possible. Using TINA principles services can be developed and deployed in a network independent manner, making it a desirable candidate architecture for the integration of PSTN and Internet networks.

The service platform is composed of typical TINA components and other components above a Parlay-like interface. These present a common view of the communication capabilities typically offered by IN and other networks, allowing the service platform to be deployed over heterogeneous network environments. The main components that abstract the service platform from the underlying networks are the "Call Control", "User Interaction Control" and "Message Control" components. These components , depicted in Fig.2, offer a set of abstract operations to be used by advanced telecommunications services. They can be used to access all networks and special resources in the underlying network.

Authentication, Service Selection, Invitation Handling and the Service Session/Logic are also interesting functionality from the architecture point of view. These "service features" may be offered to third parties as part of the service platform.

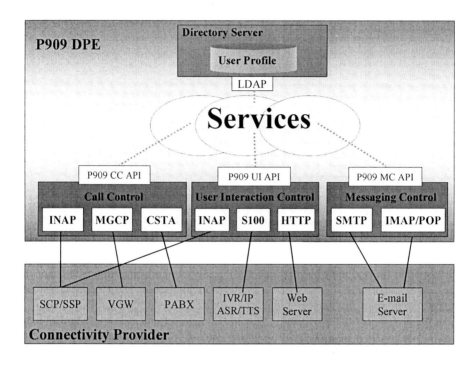

Fig. 2. Major components of the service platform

4.1   Call Control, Messaging Control and User Interaction

The *Call Control component* offers a common call control interface to all service applications. The definition of this Call Control interface is strongly based on Parlay. The Call Control component is responsible for the mapping between the set of operations offered to the services and the protocol specific operations supported by the underlying networks. It forwards notifications of new incoming calls to the service if a request has been made for such Notification. Call Control acts as a mediator for hybrid call legs connected to actors from different networks domains, such as the PSTN and Internet. This applies to all calls that require some intelligent routing to be executed and where call legs of different types are involved.

The *Call Control* functionality has two parts. A generic Call Control functionality providing services and a generic call control API with adapters towards different networks/protocols. The generic part includes a call and session model, simplified so that it can cover the features of the call/session model common to all underlying networks. This call/session model is, if needed, refined in the adapter for each underlying network. These adapters also translate internal operations to the protocol dependent operations needed to control network resources. The resources covered include: SCP, SSP, VGW and PBX. The supported protocols consist of, amongst others, INAP, CSTA, MeGaCo, SIP and proprietary SCP interfaces.

The *Message Control component* provides an abstract interface to services in the Service Control Layer. It maps towards messaging equipment and special peripherals providing support for e.g. voicemail, email, fax, SMS and other features. It forwards notifications of new incoming messages to the service if a request has been made for such notification. The *Messaging Control* like the Call Control is divided into two parts: generic and adaptation. The generic part is responsible of providing services with a generic messaging API and adapters are also needed here towards different messaging systems involved comprising of servers for e-mail, voice-mail, fax and SMS.

The *User Interaction Control* component provides an abstract User Interaction interface to the service components. It is responsible for mapping between the set of operations offered to the services and the protocol specific operations supported by the underlying network resources. It supports functionality for playing messages, collecting information, speech recognition and text-to-speech translation. As the other components this component is divided in two parts: A generic User Interaction functionality and adapters towards different resources. The resources may be IVR, ASR and TTS equipment and Web-servers supporting protocols like INAP, S100 and HTTP.

4.2 User Profile

The *User Profile component* stores an individual user profile for each user of a service. It contains service independent information like name and address, service specific information in the form of a user service profile for each subscribed service and registration information e.g. for UPT like services. Each user has his individual service profile for each service they are a user of. Service profiles contain information about constraints imposed by the provider or retailer such as restrictions according to service contract and the subscriber e.g. credit limits, restrictions in service options as well as user preferences and other selectable options e.g. language, user level.

The User profile is stored in a Directory Server and the services use LDAP to access the information stored in a users profile.

5 Supporting IN through Interworking

Interworking the service platform with IN poses unique problems. In particular, IN does not support separation between access, usage and communication or the separation between transport and the service architectures. To help solve this problem, the TINA-IN WG of the TINA-C have specified a TINA-IN interworking unit between the SSP and TINA[9]. This IWU, called the ITAU (IN-TINA adaptation

UNIT) acts both as a consumer, to represent an IN user for access and services sessions, and as a connectivity provider, to enable the setup of usage sessions over the IN. The principles of the ITAU have been adopted, with modifications in the platform. While the TINA-IN WG provides a mechanism to allow the setup of a TINA access session on behalf of an IN subscriber, it introduces some problems with service selection for IN services. Normally service selection in IN is done at the beginning of a call when the service code or destination address is first dialled. Later, if needed a PIN is introduced e.g. for Credit Card Calling. The PIN may vary from service to service. Additionally the calling Id (i.e. CLI) may or may not be sent. In TINA the procedure is different; first, a User name and a Password are requested and only then the access to the provider services is offered. In summary, the idea of having an access session was to allow the user to access to the Retailer domain and not to a specific service Furthermore, not all IN services require this type of Access, for many anonymous access is sufficient.

The TINA-IN WG solution proposes that the ITAU decides if a named or anonymous access session is required. This implies that, the Interworking function knows which service is to be invoked and decides what type of access is required. A better solution is to introduce additional TINA Access functionality in the consumer domain to decide whether or not to instantiate an access session depending on some criteria (e.g. type of service, terminal or network). This functionality can be kept

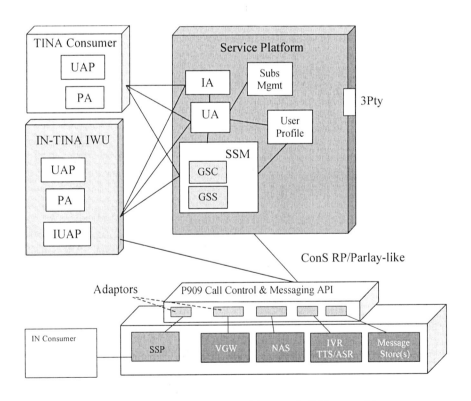

Fig. 3. TINA and the reference Architecture for IN Interworking

separate from the existing, specified TINA functionality in the consumer domain in order to maintain the service independence of the existing consumer access functions.

As with the architecture described in this paper, the TINA-IN WG have adopted Parlay for the control and usage of calls in the IN network. While the Parlay Call Control Services are used in similar ways in both TINA-IN and here, there are some differences in how they are viewed. Rather than viewing the entire IN as a single terminal and using the Parlay interlace towards that terminal, this reference architecture places Parlay at the interface between the retailer and the connectivity provider, at the ConS reference point. This greatly simplifies the setup of calls and other services across multiple network types.

Fig.3 presents a TINA view of the reference architecture. It includes TINA components to implement the functionality of the reference architecture and the interworking towards IN.

5 Products and Prototypes

In order to complete the needed architecture functionality existing products and prototypes from the telecommunications and information technology have been used, together with the TINA and Parlay based components. A survey of existing products and technologies with capabilities to enable IN-Internet integration has been performed and the results has been used in defining the prototyping and experimental phases of the P909 project. The survey also provides a snapshot of the state-of-the-art in the area of IN-Internet integration as realised using available products and technologies.

Examining the high-level architecture to be used in the project identified technology classifications. Products related to telephony networks; internet based networks; services network; service management and the DPE. Products and technologies in each of these areas have been evaluated for openness, flexibility and usefulness in the provision of advanced services.

In telephony networks, access to network functions and services through the adoption of open interfaces allows greater flexibility in the provision of advanced services, especially for third party service providers. Industry Initiatives such as JAIN[tm][10] and Parlay are driving the opening up of the PSTN network by providing open environments and interfaces for the development of advanced services. Private network interfaces, such as TAPI and JTAPI can be used for advanced service provision in PBX based networks.

Within the PSTN, SS7 and INAP remain core technologies that must be interfaced towards to insure the smooth integration of the PSTN with other network types, particularly the Internet. Companies such as Ericsson, Lucent, Radvision and Dialogic are building operational PSTN-Internet gateways to provide interworking to the PSTN at a switching infrastructure level.

Technologies for the provision of telephony and conferencing services over the Internet continue to be studied by both the ITU and the IETF. These studies have resulted in, respectively, the H.323 and SIP protocols that have been widely adopted. Currently many telecoms and IT vendors, including Dynamicsoft, Microsoft, Cisco and Radvision have products that support these protocols. H.323 Gateways are the most common means of interworking the PSTN and Internet for voice and multimedia communications. A single H.323 gateway is used for both the communications protocol (H.323) and to control the interworking between the PSTN and the Internet The IETF and ITU are jointly studying a protocol, Megacop/H.248, for the control of media gateways from the Internet, independently of any higher-level communications protocol.

Before a new service can be offered it must be designed, implemented and deployed. Object oriented technologies and methodologies, including UML and OMT, have been used in the development of sample services in the project by the use of different tools that utilise these methodologies.

The DPE is a central part of the project infrastructure. The DPE environment is seen not only as one option for a runtime environment but also as the basis for the integration of IN/PSTN and Internet within the project. DPEs based on CORBA, DCOM and Java have been utilised in the experiments.

The following table identifies the key technologies used by the partners in the project for building experimental environments and services.

Table 1. Key technologies for the provision of integrated IN-Internet services

Area	Technology
Analysis & Design	UML, OMT, SDL
Service Creation & Execution	IN platforms, DCOM Active X, JavaBeans
Public Network Interfaces	SS7/ INAP, Parlay, JAIN
Private Network Interfaces	JTAPI, TAPI, CSTA,
DPE Environments	CORBA, DCOM, RMI
Network Access	ISUP/Radius/TACACS+
Internetworking & Media Gateways	H.323, ISUP/H.323, SIP, PINT, SGCP, IPDG, MGCP, MeGaCo
Terminals	Personal JAVA Platform, Windows CE, PDAs, WAP terminals
Messaging services	X.400, X.500, SMTP, IMAP, POP, LDAP, TUP
WWW services	HTML, XML, ASP, Javascript, Java Applets

It should be noted that not all of the technologies listed in the above table have been used in operational environments. Some of them are for experimental use only; others would need some additional development and customisation to make them full products. A more detailed description of these technologies is available in [11].

6 Conclusions

Hybrid services which exploit the synergy of IN and the Internet will soon be a reality. However, as yet, there are no formal mechanisms for the development and deployment of such services. A practical approach for the specification and implementation of an integrated IN-Internet service platform will utilise existing functionality and equipment, as well as introduce advanced service concepts from the IT and Telecoms industries. In particular, products, solutions and platforms that are distributed, object oriented and support separation of service access and usage and separation of transport and service networks are desirable features in any future service platform.

This paper has described such a service platform and identified major components, technologies and initiatives needed for the development of the platform. In particular it has described call control, messaging and user interface components that are necessary to abstract services and applications away from the operation of the underlying networks. As well as these, components for user profile, service access and interworking with IN has been identified.

Assembling the service platform also requires that existing state-of-the-art products and technologies are integrated. Existing gateway solutions from IT and Telecom vendors have been used to provide access to services from PSTN and Internet terminals. Furthermore, it has been possible to reuse a wide range of, primarily Internet based products, for security, directory and user services.

The design of the service platform, including the internal and external interfaces has been validated using a set of service scenarios representing advanced services, including Click-to-Dial, Internet Call Waiting, and Unified Messaging. These services are currently being implemented for the service platform within the Eurescom P909 project.

References

1. Frost & Sullivan Report 3650 "The European Market for Internet Protocol"
2. Forrester Research; "Internet Telephony Grows Up"; March 1997
3. ITU-T Rec. H.323, "Packet based multimedia communication systems", 1998
4. Fernando Cuervo et al; "MeGaCo Protocol"; draft-ietf-megaco-protocol-03.txt; ftp://standards.nortelnetworks.com/megaco/docs/Oslo99/megacoHGCP12.pdf
5. Parlay Consortium; http://www.parlay.org
6. TINA Consortium; http://www.tinac.com
7. Parlay Consortium, Parlay API specification, version 1.2, 10 September 1999
8. TINA Consortium; TINA service architecture", Version 5.0, L. Kristiansen et al, June 1997.
9. TINA Consortium; TINA-IN Work Group RfP "IN access to TINA services & Connection management"; 14 October 1999
10. http://java.sun.com/products/jain
11. "What an IN system should be", M Barry et al., EURESCOM P909 Deliverable 1, http://www.eurescom.de

Acknowledgements

The information presented in this paper is based on the work done in the EURESCOM project P909-GI "Enabling Technologies for IN Evolution and IN-Internet Integration". This however does not imply that the paper necessarily reflects a common position of all involved EURESCOM Shareholders/Parties. The authors would like to thank their colleagues working on EURESCOM P909 for their insightful comments.

IP-based Services Boased Convergence of Fixed and Cellular Networks and Services in the Light of Liberalization

Vergados D., Vayias E., Soldatos J., Drakoulis D. and Mitrou N.

Telecommunications Laboratory
Department of Electrical Engineering & Computer Science
National Technical University of Athens
Heroon Polytechniou 9, Zographou Athens
GR 157 73, Greece
Tel: +30 1 772 2558
Fax: +30 1 772 2534
Email: {vergados; evayias}@telecom.ntua.gr

Abstract. The evolution of an "open" communications universe is now expanding to deliver new communications services, characterised by advanced and ubiquitous, multimedia capabilities. The move by telecommunication companies' corporate customers into the IP world and the need for interoperability between private and public networks will drive service providers and telecom operators to adopt IP in their core networks as a means of unifying traffic types. The particular case addressed in this paper considers a complex yet realistic scenario: the convergence of Fixed/Mobile Communications Provider and Internet Service Provider through the transition to IP technology.

Keywords: IP Network, Cellular Operators, Network Services, Convergence, Liberalization of Telecommunications, ISP, ITSP, IP Telephony, VoIP, IP Services

1. Introduction

The Internet has emerged during the 1990s as the primary force driving the expansion of demand for telecommunications companies. The growth in Internet usage has led to widespread take-up of Internet Protocol (IP) in corporate LANs, and a trend is now emerging towards IP implementation in corporate WANs as well. The move by telecommunication companies' corporate customers into the IP world, and the need for interoperability between private and public networks, will most propably drive the telecommunication companies to adopt IP in their core networks as a means of unifying traffic types [1]. This trend will be reinforced by emerging and incumbent telecommunication operators short-term need to respond to competition from low-cost IP telephony providers.

Since the IP protocol has been widely adopted for the unification of the different types of underlying physical networks, providing seamless interworking to the application layers, it is of great importance that telecom/cellular operators base their backbone networks on IP technology. Consequently, facing the liberalization of the

Telecom market, a strategic movement has to be decided towards IP technology both for access and backbone networks, that will allow the deployment of multi-media services over IP (any set of voice, fax, data) in their networks.

Our contribution will initially approach the technical standards related to IP-based control, signalling and delivery of voice, and will then proceed to attempt a technical comparison. The popular ITU-T H.323 and IETF's SIP standards are the focus of this comparison, mainly since the authors believe they are the most interesting pair especially from an academic point of view. A brief reference to QoS for IP networks and RSVP in specific follows. Finally three different scenarios are illustrated – and presented in equal sections, covering the known possible ways of establishing a call by using the current technology: PSTN/mobile terminal to PSTN/mobile terminal; PSTN/mobile terminal to IP Telephony terminal; IP Telephony Terminal to PSTN/mobile terminal.

2. Technical Approach

2.1 Introduction to Technical Approach

Today's Internet is increasingly used not only for e-mail, ftp and WWW, but also for interactive audio and video services. Historically, voice applications and data applications have required separate networks using different technologies —circuit switching for voice and packet switching for data. In the past decade, much effort has gone into finding a solution that provides satisfactory support for both transmission types on a single network. As a result there has been significant evolution of packetizing technologies for the transmission of voice traffic over data networks.

2.2 Protocols for Voice / Multimedia Provision over IP Networks

ITU-T H.323

The dominant standard for professional VoIP applications, H.323 defines in detail all the components needed for a complying Voice-over-any-packet-network implementation, and has the important advantage of currently being considered a de-facto market standard. In contradiction to popular belief H.323 does not define only voice but multimedia provision in general and implementations may be ported with moderate difficulty from one packet-based network to another. Packet-based networks defined by H.323 include IP–based (including the Internet) or Internet packet exchange (IPX)–based local-area networks (LANs), enterprise networks (ENs), metropolitan-area networks (MANs), and wide-area networks (WANs). H.323 can be applied in a variety of mechanisms—audio only (IP telephony); audio and video (videotelephony); audio and data; and audio, video and data. The standard can also be applied to multipoint-multimedia communications. H.323 is a standard that specifies the components, protocols and procedures that will enable the provision of multimedia services over packet switched networks [3]. H.323 is part of a family of

ITU–T recommendations generally referred to as H.32x, where multimedia communication services over a variety of networks are described.

The Version 1 of the H.323 recommendation —visual telephone systems and equipment for LANs that provide a non-guaranteed quality of service (QoS)— was specified by the ITU–T Study Group 16, and accepted in October 1996. Multimedia communications in a LAN environment were the main focus of this work. The emergence of voice-over–IP (VoIP) applications and the lack of a standard for voice over IP resulted in several proprietary and thus incompatible products, while additionally interworking between packet based telephony and conventional circuit-switched telephony was never defined in detail. Version 2 of H.323 emerged to accommodate these additional requirements and was formally accepted in January 1998. The standard shall evolve further and according to analysts new features should cope with performance (faster connections establishment), interworking between different media gateways and other services like fax-over-packet.

Other H.323 associated definitions

Under H.323 several other recommendations are mentioned here [8] related to

- **Audio coding**. H.323 defines the support of at least one G.711 codec (audio coding at 64 kbps quality). Additional audio CODEC recommendations such as G.722 (64, 56, and 48 kbps), G.723.1 (5.3 and 6.3 kbps), G.728 (16 kbps), and G.729 (8 kbps) may also be supported.
- **Video coding**. H.323 specifies support of video as optional, yet, any H.323 terminal providing video communications must support video encoding and decoding as specified in the ITU–T H.261 recommendation.
- **H.225 registration, admission, and status (RAS)**: Registration, admission, and status (RAS) is the protocol managing connection between any combination of endpoints (terminals and gateways) and gatekeepers. The RAS is used to perform registration, admission control, bandwidth changes, status, and disengage procedures using specific RAS (non-traffic) channels. This signaling channel is opened between an endpoint and a gatekeeper prior to the establishment of any other channels.
- **H.225 call signaling**: The H.225 call signaling is used to establish a connection between two H.323 endpoints. This is achieved by exchanging H.225 protocol messages on the call-signaling channel. The call-signaling channel is opened between two H.323 endpoints or between an endpoint and the gatekeeper.
- **H.245 control signaling**: H.245 control signaling is used to exchange end-to-end control messages governing the operation of the H.323 endpoint. These control messages carry information related to the following:
  - capabilities exchange
  - opening and closing of logical channels used to carry media streams
  - flow-control messages
  - general commands and indications
- **Real-time transfer protocol (RTP)**: Real-time transport protocol (RTP) provides end-to-end delivery services of real-time audio and video. Whereas H.323 is used to transport data over IP–based networks, RTP is typically used to transport data solely via the user datagram protocol (UDP). RTP, together with UDP, provides

transport-protocol functionality. RTP provides payload-type identification, sequence numbering, time stamping, and delivery monitoring. UDP provides multiplexing and checksum services.

- **Real-time control protocol (RTCP)**: Real-time transport control protocol (RTCP) is the counterpart of RTP that provides control services. The primary function of RTCP is to provide feedback on the quality of the data distribution. Other RTCP functions include carrying a transport-level identifier for an RTP source, called a canonical name, which is used by receivers to synchronize audio and video.

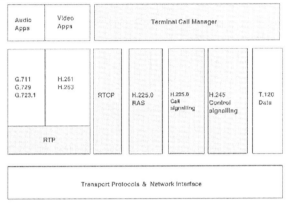

Fig. 1: H.323 protocol stack

The Session Initiation Protocol (SIP)

The Session Initiation Protocol [4], [7], runs on top of either UDP or TCP, providing its own reliability mechanisms when used with UDP. For addressing, SIP makes use of uniform resource identifiers (URIs), which are generalizations of Uniform Resource Locators (URL's), in common usage in the web. SIP defines its own URI, but its header fields can carry other URIs, such as http, mailto, or phone. SIP is a client-server protocol, similar in both syntax and structure to the well-known HyperText Transfer Protocol (HTTP). However, it defines its own methods and headers for providing the functions required in IP telephony signaling. Requests are generated by one entity (the client), and sent to a receiving entity (the server) that processes them, and then sends responses. This request - response sequence is called a transaction. The end system acting on behalf of a user in SIP is called the user agent. The client portion is called the User Agent Client (UAC) while the server portion is called User Agent Server (UAS). The UAC is used to initiate a SIP request while the UAS is used to receive requests and return responses on behalf of the user. A single host may well act as client and server for the same request. As in HTTP, the client requests invoke methods on the server. Requests and responses are textual, and contain header fields, which convey call properties and service information. SIP reuses many of the header fields used in HTTP, such as the entity headers (e.g., Content-type) and authentication headers.

Three types of servers may be found within a SIP network: Registration, Proxy and Redirect Servers. A *registration server* receives updates concerning the current locations of users. A *proxy server* upon receiving a request forwards it to the next-hop server, which has more information about the location of the called party. Thus the proxy servers may be held responsible for call routing. A *redirect server* on receiving request, determines the next-hop server and returns the address of the next-hop server to the client instead of forwarding the request. We will review the technical essence of SIP in a following section, by comparing it to the H.323.

2.3   Network Elements – Components defined by ITU

The H.323 standard specifies four kinds of network components, which provide the point-to-point and point-to-multipoint multimedia-communication services:

- **Terminals**: Used for real-time bi-directional multimedia communications, an H.323 terminal may come in many forms, e.g. as a PC application or a stand-alone device, running an H.323 and the multimedia applications. The terminal should basically support audio communications and optionally video or data communications. The primary goal of H.323 is to allow the required interworking features with other multimedia terminals. In the ITU standards world, H.323 terminals are compatible with H.324 terminals on SCN and wireless networks, H.310 terminals on B–ISDN, H.320 terminals on ISDN, H.321 terminals on B–ISDN, and H.322 terminals on guaranteed QoS LANs. H.323 terminals may be used in multi-point conferences.
- **IP Telephones:** An IP Telephone is a telephone device that is able to transports voice over an IP network using data packets instead of circuit switched connections over voice only networks. An IP Telephone consists of the following components: User Interface, Voice Interface, Network Interface (IP), and Processor Core and associated logic.
- **Gateways**: One of the primary issues the developers of the H.323 standard had to deal with, was interoperability with other multimedia-services networks. This interoperability is achieved through the use of a gateway. A gateway thus connects two dissimilar networks. This connectivity of dissimilar networks is achieved by translating protocols for call setup and release, converting media formats between different networks, and transferring information between the networks connected by the gateway. A gateway is not required, however, for communication between two terminals on an H.323 network. Also, internet telephony gateways consist of two functional parts - a dumb media gateway which converts audio data and an intelligent media gateway controller which communicates with the rest of the world over signaling protocols and controls 1-N media gateways over a gateway control protocol. ITU-T has already standardized a Gateway Control Protocol called as H.GCP protocol, which is used to control Media Gateways.
- **Gatekeepers**: A gatekeeper is a central, coordinating point within the H.323 network. Although they are not required by the standard, and are not included in some gatekeepers provide important services such as addressing, authorization and authentication of terminals and gateways; bandwidth management; accounting; billing; and charging. Gatekeepers may also provide call-routing services

- **Multipoint control units (MCUs)**: MCUs provide support for conferences of three or more H.323 terminals. All terminals participating in the conference establish a connection with the MCU. The MCU manages conference resources, negotiates between terminals for the purpose of determining the audio or video coder/decoder (CODEC) to use, and may handle the media stream.

It should be emphasized at this point that although gatekeepers, gateways, and MCUs are logically separate components of the H.323 standard, they may be implemented as a single physical device.

3.   Ease of Implementation, Compatibility and Scalability

Having briefly presented the major existing standards for VoIP, we will proceed by attempting a comparison on issues like ease of implementation, compatibility and scalability, and by referring to the engineering approaches in the implementation that result in these differences [5].

IETF SIP vs. ITU-T H.323

Ease of Implementation: H.323 is a rather complex protocol while SIP, being similar to the HTTP syntax and message exchange, is a lighter, easier to implement protocol. H.323 defines numerous (hundreds!) components while SIP defines different header types with similar structure. SIP messages are encoded as plain text, similar to HTTP. This leads to simple parsing and generation, and development can be based on languages like Perl. H.323 on the other hand uses binary representations for its messages based on ASN.1 (Abstract Syntax Notation) and the packed encoding rules (PER). Furthermore under H.323 several interactions between standard components (e.g. H.225, H.245), internally in the protocol are possible increasing the complexity.

Compatibility: SIP has built in a rich set of extensibility and compatibility functions. Compatibility is still maintained across different versions. As SIP is similar to HTTP, mechanisms being developed for HTTP extensibility can also be used in SIP. Among these is the Protocol Extensions Protocol (PEP), which contains pointers to the documentation for various features within the HTTP messages. SIP uses the Session Description Protocol (SDP) to convey the codecs supported by an endpoint in a session. H.323 provides extensibility mechanisms as well. These are generally nonstandard *Param* fields placed in various locations in the ASN.1. These params contain a vendor code, followed by an opaque value, which has meaning only for that vendor. This allows for different vendors to develop their own extensions. However extensions are limited only to those places where a non-standard parameter has been added. If a vendor wishes to add a new value to some existing parameter, and there is no placeholder for a nonstandard element, one cannot be added. Secondly, although H.323 requires full backward compatibility from each version to the next it has no mechanisms for allowing terminals to exchange information about which extensions each version supports. In H.323, each codec must be centrally registered and standardized. As many of these carry significant intellectual property, there is no free, sub-28.8 kb/s codec which can be used in an H.323 system. This presents a significant barrier to entry for small players and universities.

Scalability: SIP and H.323 differ in terms of scalability. SIP servers and gateways will need to handle many calls. For large, backbone IP telephony providers, the number of calls being handled by a large server can be significant. In SIP, a transaction through several servers and gateways can be either stateful or stateless. In the stateless model, a server receives a call request, performs some operation, forwards the request, and completely forgets about it. SIP messages contain sufficient state to allow for the response to be forwarded correctly. SIP can be carried on either TCP or UDP. In the case of UDP, no connection state is required. This means that large, backbone servers can be based on UDP and operate in a stateless fashion, reducing significantly the memory requirements and improving scalability. SIP is simpler to process than H.323; given the same processing power, SIP should allow more calls per second to be handled on particular box than H.323. As H.323 was originally conceived for use on a single LAN and thus for large numbers of domains, and complex location operations, it is normal to expect the appearance of scalability problems. As a practical example in an H.323 system, both telephony gateways and gatekeepers will be required to handle calls from a multitude of users and furthermore the connections established are TCP based, which means a gatekeeper must retain its TCP connections. H.323 will have a hard time coping with a multi-gateway situation a case where other protocols like the multi-gateway MGCP/MeGaCo shall be deployed.

4.   QOS over IP Networks – The RSVP Protocol

The IP has become a ubiquitous communications universal network that intends to incorporate traditional data-oriented services together with new multimedia services (IP telephony, videotelephony, multimedia, etc.). Within the framework of this global information infrastructure deployment, the demand for IP connectivity services suitable for business is increasing. Several QoS frameworks have been proposed in an attempt to provide IP services of a quality level suitable for multimedia and business-critical communication purposes [9],[10]. Some of these proposals have also addressed the issues of seamlessly provisioning and dynamically configuring IP services. A prominent position among these frameworks is held by the *Integrated Services* Architecture (IntServ) that works in conjunction with the *Resource reSerVation Protocol* (RSVP) towards providing QoS guarantees in a per flow basis [11]. Another important QoS framework is the *Differentiated Services* model (DiffServ). DiffServ attempts to provide scalable (i.e. no per-flow signaling or state) service differentiation in the IP Networks. Even though a lot of effort has been allocated to developing these frameworks, it seems that both present some drawbacks, which may hinder their applicability towards providing really quantitative, end-to-end QoS in a scalable manner. Specifically, the IP Network community has voiced concern over the applicability and scalability of RSVP and the Integrated Services model in the global IP Network infrastructure. The need for maintaining state and applying traffic control functions for every single connection and at every network element is considered as an obstacle to the IP Network applicability of the method. On the other hand, although the DiffServ architecture presents a great potential towards achieving scalable service provision in the next generation networks, it still lacks specific traffic control and resource management strategies for guarantying quantitative, end-to-end QoS metrics. How to control the traffic in order to achieve the advertised per-hop behavior is still not clearly defined in the DiffServ environment. Thus, the IP Networks are still restricted to offering the traditional best effort data forwarding service, which cannot meet the emerging QoS demands.

The RSVP Protocol

The network delay and Quality of Service are the most hindering factors in the voice-data convergence. The most promising solution to this problem has been developed by IETF viz., RSVP. RSVP allows prioritization and latency guaranties to specific IP traffic streams. RSVP enables a packet-switched network to emulate a more deterministic circuit switched voice network. With the advent of RSVP, VOIP has become a reality today. RSVP requests will generally result in resources being reserved in each node along the data path. RSVP requests resources in only one direction, therefore it treats a sender as logically distinct from a receiver, although the same application process may act as both a sender and a receiver at the same time. RSVP is not itself a routing protocol, it is designed to operate with current and future unicast and multicast routing protocols. RSVP interacts with entities called the packet classifier and the packet scheduler installed on the host to make quality of service decisions about the packets sent in by applications. It first queries the local decision modules to find out whether the desired QoS can be provided (this may involve resource-based decisions as well as policy-based decisions). It then sets up the required parameters in the packet classifier and the packet scheduler. The packet classifier determines the route of the packet, and the scheduler makes the forwarding decisions to achieve the desired QoS. In case the link layer at the host has its own QoS management capability, then the packet scheduler negotiates with it to obtain the QoS requested by RSVP.

The central component of the RSVP architecture, which applications interact with, is a flow specification, or flowspec, which describes both the traffic stream sent out by the source as well as the service requirements of the application. These flowspecs actually specify the desired QoS, and are used to set the parameters in the node's packet scheduler. In addition to these there are filter specs which together with the session specification define (or rather filter) the data packets which are to receive the QoS defined by the flowspecs. These are basically a set of filters, which can be used to select arbitrary subsets of packets in a given session, by selecting any fields in protocol headers, application headers etc. Data packets that are addressed to a particular session but which do not match any of the filter specs for that session are handled as best-effort traffic.

If IP Telephony were to move into the mainstream, the technology would have to deliver predictable, consistent voice quality over any network infrastructure. This meant successfully addressing a myriad of issues. Most of them are still under extensive study. The main issue is to provide QoS in every call establishment. Major standards organizations and private consortiums are putting much effort in this direction and their results should be monitored throughout the paper time period.

5. The Proposed Network Architecture

The Key Issues

This article addresses a complex scenario involving Telecom Operators of fixed network, or alternatively, Cellular Operators and Internet Service Providers. In particular, it describes in detail different types of services that should be offered by the operators and proposes various interworking schemes between the ISP network and the fixed/cellular conventional (circuit switched) network, in an IP-integrated network environment. The interworking scenario focuses both in technical aspects, and on related marketing implications based on analysis by industry experts and market analysts. In particular, opinions expressed within the paper are perfectly aligned to the following principles:

- The market for long distance services based on IP infrastructure is about to explode
- ISPs must consider the need to upgrade their infrastructure in order to turn to Internet Telephony Service Providers (ITSPs) by offering IP telephony
- By unfying voice, fax and corporate data traffic, IP-based network operators face significant profit opportunities and become a significant competitor to circuit switched telecommunication companies

ISPs may use a single infrastructure for providing both, Internet access and IP Telephony. Data-oriented switches shall be deployed for switching both data and packetized voice. IP is more fit and cost-effective for the delivery of voice services than other techniques like e.g. ATM, since ATM reserves bandwidth for Constant Bit Rate (CBR) applications like voice circuits, even when this circuit is not used [2].

The Key Areas of Application

Most of the focus on VoIP is currently centered on two key applications.

The first is private business network applications. Businesses with remotely located branch offices, which are already connected together via a corporate intranet for data services can take advantage of the existing intranet by adding voice and fax services using VoIP technologies. Businesses are driving the demand for VoIP solutions primarily because of the incredible cost savings that can be realized by reducing the operating costs of managing one network for both voice and data and by avoiding access charges and settlement fees, which are particularly expensive for corporations with multi-international sites. Managed corporate intranets do not have the QoS issues, which currently plague the Internet; thus voice quality approaches toll quality.

The second key application is VoIP over public networks. This application involves the use of voice gateway devices designed to carry voice to Internet Service Providers, who are regularly referred to as Internet Telephony Service Providers, or to the emerging Next Generation Carriers, developing IP networks specifically to carry multimedia traffic such as VoIP. ISPs are interested in VoIP as a way of offering new value-added services to increase their revenue stream and break out of the low monthly fixed fee structure currently in place for data services (a realistic example is depicted in Figure 2).

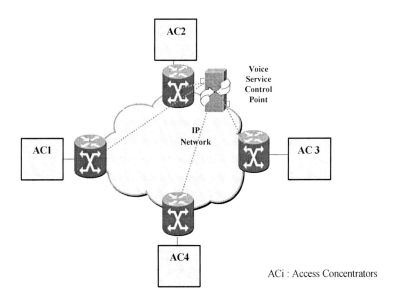

Fig. 2: The Converged Network

VoIP also allows the improvement of network utilization. These new services include voice and fax on a per-minute usage basis at rates significantly less than voice and fax rates for service through the PSTN. The sustainability of this price advantage may be short term, and is dependent on whether the regulatory agencies will require ISPs to pay the same access charges and settlement fees, PSTN carriers are obligated to pay. New carriers are interested in VoIP because data networks are more efficient than traditional voice networks. In the near term, these new carriers can avoid the access charges and settlement fees, which account a sufficient part of the cost of a long distance call. IP Telephony (IPT) networks that offer competitive voice and fax services have been the predominant new business opportunity over the last years, as vendors have introduced the equipment and functionality necessary to deploy large scale services. Now, customers may take advantage of flat Internet rating vs. hierarchical PSTN rating and save money while letting their long-distance calls be routed over Internet. The IPT users may also profit of its applications'-oriented nature: software solutions may be easily extended and integrated with other services and applications, e.g. whiteboarding, electronic calendar, or browsing.

Interoperability Scenarios

Technology in our days is mature enough to develop the right software and hardware components for supporting interoperability among different networks. The following figure (Fig. 3) illustrates the physical components with regard to the H.323 standard.

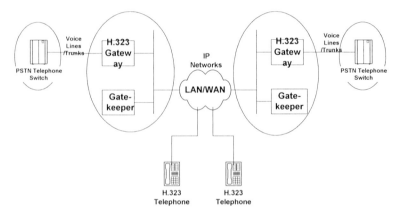

Fig. 3: A basic ITS infrastructure block diagram

As a practical example of the applicability of the presented concepts, three different scenarios of interoperability will be illustrated, covering the known possible ways of establishing a call by using the current technology. These are the following, which are going to be described in sections that follow:
- PSTN/mobile telephone to PSTN/mobile telephone
- PSTN/mobile telephone to IP Telephony terminal
- IP Telephony terminal to PSTN/mobile terminal

It is important to note before examining the scenarios that as it should be obvious by the above figure, the structure of any solution used for delivering voice traffic is only different at the end-points of the H.323 gateways. At these points only, signalling traffic from both sides is 'terminated' and specific medium-dependent hardware is installed.

PSTN/Mobile Telephone to PSTN/Mobile Telephone

1. An Internet Telephony Service subscriber is able to dial an access number provided by the Internet Telephony Service Provider.
2. The call is routed by the PSTN/Mobile to the "access" Internet telephony switch.
3. The gateway plays an announcement requesting that the subscriber enter the destination telephone number to be called.
4. The destination digit information is sent to the gatekeeper.
5. The gatekeeper determines a destination gatekeeper IP address based on the destination digit information.

6. An IP packet requesting the availability status of the destination gateway is sent to the destination gatekeeper.
7. The destination gatekeeper responds to the request by providing destination gateway availability and IP address information.
8. The originating gatekeeper then transfers this information to the originating gateway.
9. The originating gateway sets up a virtual circuit to the destination gateway.
10. This circuit is identified by a call reference variable (CRV) that will be used by both gateways for the duration of the call to identify all IP packets associated with this particular call.
11. The destination gateway selects an outgoing PSTN/Mobile voice trunk and signals to the PSTN/mobile switch to attempt to setup a call to the dialed telephone number.
12. If the PSTN/Mobile switch signals that the call setup is successful and the called party has answered, IP signaling messages are relayed to the originating gatekeeper.
13. The gatekeeper then signals the originating gateway, which in turn signals the originating PSTN/Mobile switch to indicate that the call is now completed.

In the IP network, voice packets are exchanged between the gateways for the duration of the call.

PSTN/Mobile Telephone to IP Telephone
Next we browse through a sample call flow, that originates from the PSTN/Mobile and terminates at an IP-based H.323 terminal.
1. An Internet telephony subscriber dials an access number that has been provided by the Internet telephony service provider.
2. The call is routed by the PSTN/Mobile to the "access" Internet telephony switch.
3. The gateway plays an announcement requesting that the subscriber enter the destination number to be called. (In this case, the telephone number is assigned to an H.323 terminal.)
4. The collected destination digit information is sent in a call setup request message to the gatekeeper.
5. The gatekeeper determines a destination gatekeeper IP address based on the destination digit information.
6. An IP packet requesting the availability status for the destination H.323 terminal is sent to the destination gatekeeper.
7. The destination gatekeeper responds to the request by providing the destination terminal availability status and IP address information to the originating gateway.
8. The originating gateway sets up a virtual circuit to the destination H.323 terminal. This circuit is identified by a call-reference variable that will be used by both the originating gateway and the H.323 terminal for the duration of the call to identify all IP packets associated with this particular call.
9. If the H.323 terminal indicates that call setup is successful and the called party has answered, IP signaling messages are sent to the originating gatekeeper, which then signals the originating gateway.
10. The originating gateway signals the originating PSTN/Mobile switch to indicate that the call is now completed.

The exchange of IP packets proceeds until either the calling or called party terminates the call.

IP Telephony Terminal to PSTN/Mobile Telephone

Following let us trace a sample call flow, which originates from an IP-based H.323 terminal (e.g., a personal computer) and terminates in the PSTN/Mobile.

1. An Internet telephony subscriber initiates an Internet telephony call setup procedure at an H.323 terminal (e.g., a PC).
2. The H.323 terminal sends an IP packet specifying the destination digits information to a preselected gatekeeper.

The call scenario from this point onward is similar to steps four through eight described in the PSTN/mobile to IP to PSTN/Mobile call flow (replace "originating gateway" with "originating terminal").

6. Conclusions – Expected Results

The main problem that was presented analytically in this paper, is how networks with different communication profiles will be operating under a mature and integrated way. That requires that a number of problems have to be solved. First, communication among the different nodes has to be clarified. The proposed interworking architecture explains that a signaling communication among a number of nodes needs to be established before the path is identified. In addition, there are further responsibilities that have to be affiliated, such as authentication, authorization, Quality of Service, tariffing and billing. In our proposed network, the customer management and billing system for IP telephony needs to be able to support reliable authentication mechanisms to identify the user uniquely, and be able to modify those mechanisms in real time. Also, in the converged IP telephony, the service provider may wish to check a variety of criteria before provisioning the service to the user. Customer management and billing systems used should allow the ITSP to perform real-time authorization before the service is made available to the user. The market will soon require the ability to choose their "quality of service", paying more for higher quality and reliability, or less for lower quality needs such as personal calls or one-way communication (speaker delivering a presentation). The customer management and billing system should able to support the ITSP's need to route delay-sensitive voice and video traffic. As a result, monitoring and controlling interconnection charges is key for tariffing and billing mechanisms. To maximize customer benefits, interconnection rules should be independent of technology and network topology. Authorities must ensure rules applied to the conventional Public Switched Telephone Network (PSTN) are also applied to fixed/mobile networks and IP Networks.

Consequently, the creation of the new telephony services in the proposed converged network will be must faster, cheaper and easier. In particular, an extensive technical analysis of the proposed network was presented, the open issues are identified and the problems arising are discussed in detail and proper solutions are proposed. The results of this paper together with the strategic movement to IP technology both for access and backbone networks, are of a great importance for the evolution of a new, "open" communications universe, which is now coming together harmoniously to deliver new communications services, characterized by mobility and advanced multimedia capabilities.

7. References

1. H. Schulzrinne and J. Rosenberg, "Internet telephony: Architecture and protocols -- An IETF perspective", Comp. Networks and ISDN Sys., vol. 31, Feb. 1999, pp. 237–55.
2. J. Estrin and S. Casner," Multimedia Over IP: Specs Show the Way", Data Communications, Aug. 1996.
3. ITU-T Rec. H.323, "Visual telephone systems and equipment for local area networks which provide a non-guaranteed quality of service," Geneva, Switzerland, May 1996.
4. M. Handley et al., "SIP: session initiation protocol," IETF 2543, Mar. 1999.
5. H. Schulzrinne and J. Rosenberg, "A comparison of SIP and H.323 for internet telephony," Proc. NOSSDAV, Cambridge U.K., July 1998.
6. H. Schulzrinne et al., "RTP: a transport protocol for real-time applications", IETF RFC 1889, Jan. 1996.
7. H. Schulzrinne and J. Rosenberg, "The session initiation protocol: Providing advanced telephony services across the internet," Bell Labs Tech. J., vol. 3, Oct.–Dec. 1998, pp. 144–60.
8. J. Toga and J. Ott, "ITU-T standardization activities for interactive multimedia communications on packet networks: H.323 and related recommendations," Comp. Networks, vol. 31, no. 3, 1999, pp. 205–23.
9. Busse, B. Deffner, and H. Schulzrinne, "Dynamic QoS control of multimedia applications based on RTP", Comp. Commun., vol. 19, Jan. 1996, pp. 49–58.
10. Shenker, S., C. Partridge and R. Guerin, "Specification of Guaranteed Quality of Service", RFC 2212, September 1997.
11. Wroclawski, J., "The Use of RSVP with IETF Integrated Services", RFC 2210, September 1997.
12. B. Douskalis, IP Telephony, HP Professional Books, Prentice Hall, 2000.

Development of Internet Services Based on Pure JAVA Technology

Karim Sbata[1], Pierre Vincent[2]

[1]ENIC (Villeneuve d'Ascq - France)
Morocco
[2]INT (Evry – France)
France

Abstract. This article is about the developpement of new Internet services at the Network and Computer Science Departement of the National Institute of Telecommunications (INT) of Evry. It deals mainly with a project related to an applications server, purely Java and Web (HTML) oriented, which main objective is to offer a set of online applications to any Web user (even mobile in the future), and a multi-point multi-media architecture (client/server), also purely Java, aiming to allow connnected users to exchange any type of data (text, audio, video,etc), independently from their transport protocol (TCP or UDP).

Keywords: Multi-Media, Hyper-Media, Application Server, Multi-Pointing Architecture, Multi-Terminals Clients, JAVA Technology.

1. Introduction

Traditionally, the worlds of data-processing, telecommunications and multimedia networks were regarded as distinct, even incompatible for some purists.

Indeed, in addition to the economic war that delivered (and still deliver!) the data processing and telecom lobbies, the philosophies of these various worlds were radically different: whereas the data-processing world privileges a statistical (optimal) use of the network resource, a free "best effort" service (for a right resource sharing between the users) and putting the complexity on the terminal (to sale its more and more powerful computers), the world of telecommunication prefers a deterministic reservation of resource (to ensure a quality of service), a service with variable QoS (proportionnal to the price payed by the user) and putting the complexity on the network itself, making it thus accessible from very basic terminals.

For a few years however, these two worlds have began to converge, thanks in particular to the progress made on digital technology (A/V compression, entirely digital telecom networks (e.g. GSM)) and networks performances (growing reliability and flows). It indeed proved that each one of these worlds would be soon technologically ready to ensure the services of the other, removing thus any barrier between these two fields formerly so hermetic.

The current tendency is then multimedia and multiservices networks (telephony, data transfer, A/V, etc), accessible from different terminals (PC, mobile phone, TV, and in a slightly more remote future, domestic machines (thanks to JINI technology)).

In this context, the new Internet applications will tend to be as portable as possible (100% Java) and to offer the most independent possible (from the terminal used) user interface (UI). The Network and Computer Science Departement of the ENIC (Lille) and the INT (Evry) are currently developing a server of applications purely Java, accessible from any Web navigator (ensuring then an independence from the terminal) and whose objective is to offer complex processing on simple parameters (e.g. URLs, files, etc), as well as a multimedia multipoint client/server architecture, also purely Java, which allows exchanges of all types of information (texts, images, audio, etc) between several users.

The object of this article is to detail these two projects (which can be gathered in one, the multi-point architecture being able to include the server of applications), to present their current state of advance (i.e. in July 2000), and to evoke their principal axes of evolution.

2. The server of applications

2.1. Principle

A server of applications is a Web server (thus based on HTTP) which gives access to various programs, runnable on line. Its main interest is to exempt the client to download on his terminal the executable code, to install it and to launch it. This last step is indeed very often penalising, for several reasons : first of all, the downloaded software is generally used very partially and only a small number of times; moreover, the "server" machine are often more powerful than the "client" machine and can thus be more effective for some applications (in particular in the field of multimedia); finally, it should be noticed that some terminals with very limited resources (e.g. mobile phones) can neither store nor load code.

The server of applications we are developing aims thus to provide an access to an open database of applications (i.e. allowing any programmer to contribute to it) to various users.

For that, it was first of all necessary to define a generic user interface, in order to offer the same possibilities to any client, whatever its performances. The choice of HTML language (and/or its derived languages: XML, HDML, WML [1], etc) appeared adequate.

Moreover, to be able to work with an open database of applications, it was necessary to define a standard for the applications, in particular for their input/output (to allow a single generic treatment for all the applications).

In order to avoid using directly the operating system to launch each application (from a command line or by using dynamic libraries), Java was selected like exclusive programming language. Indeed, in addition to its portability, Java offers all the advantages of an object-oriented language, in particular that of being able to instanciate and load any class from any other, to use its public fields and methods, etc.

Once these choices have been done, we had to put them into practice. The object of the following section is to give more details about this.

2.2. Implementation

The first stage was the development of the user interface. This one had to meet two conditions: first,to allow the user to choose an application and to send the associated parameters; to allow the reception of the result.

In order to have a generic interface, we standardized the form of the input parameters and the results: all the treatments are done either on files or on URLs; the results are always returned in form of files. We should notice that this standards are valid only for the traditional interface in HTML (the only one currently implemented). An interface in WML for mobile phones is under study. In this case, the input parameters will be URLs (to be treated) and an e-mail address to receive the result.

To develop such an interface in HTML was rather easy. Indeed, this language allows, thanks to its form functionalities, to send all kinds of information to the server. The result of the treatment will be simply returned in the HTTP message sent as a response to the request associated to the form.

The only difficulty encountered was the transfer of a file from the client to the server, without using FTP or any additional software on the client side (thing that would be in total contradiction with the philosophy of the project). This problem was rather quickly solved by using a relatively ignored resource of HTML's FORM tag: the input of type file (<INPUT TYPE = ' file' >), defined in [2].

This tag is recognized by the majority of recent Web navigators (from 4.x versions of Netscape and Internet Explorer). It allows the user to choose a local file (via an open file dialog box) and to send it to the Web server in a request HTTP message (using method POST), specifying a content-type "multipart/form-dated". It was thus necessary to implement within the server a module dealing with this type of messages.

Once the problem of the interface solved, we oriented our efforts to the standardization of the applications. Like it was said previously, we first decided to use exclusively the Java language for programming, in order to have a total portability, on the level of the server itself (i.e. it will runnable on any machine having a JVM) but also on the level of the applications associated. Indeed, as these last will be organized as an open database, the portability is necessary to allow all developers to contribute, whatever their programming environment (mainly their operating system).

The use of Java also has the advantage of simplifying the interactions server-applications, in particular by avoiding the use of the operating system for loading an application required by the client.

In addition to choosing the programming language, it was also necessary to define a standard for the inputs/outputs, to avoid the definition of a new treatment for each addition of a new application.

Thanks to Java, that could be done without major difficulties. Indeed, this language implements in a remarkable way the concept of streams (via the *java.io* package[3]). Thus, in the same way that CGI scripts communicate with a Web server via standard I/O streams, the applications will communicate with the server via an InputStream and an OutputStream (more exactly their generic data subclasses, the *DataInputStream* and the *DataOutputStream*).

To be integrated to the server, each application (class) will then have to contain a constructor with at least two parameters: a DataInputStream and a DataOutputStream. Other parameters are detailed in the first appendix.

We can notice that using streams of bytes allows all kinds of applications, treating any data (e.g. processing of an image sent by the client, conversion from an audio format to another, etc).

Currently, the server of applications implements only two completely operational applications: Calc, a simplified spreadsheet, which allows making arithmetic operations on numerical HTML tables, and Map, an editor of synopses, wich allows extracting a synopsis from a standard HTML document (i.e. whose titles and subtitles use headers (tag <H>)).

Others are currently under development (e.g. organization of multimedia documents form, utilities of compression for directories, etc).

We will give more details in the following section about the prospects for evolution of the server.

2.3.   Prospects for evolution

The evolution of the server concerns mainly its applications, or more exactly the fields they are associated to. Currently, the developments are done through two main axes: the installation of an online HTML oriented toolbox (including in particular the applications Calc and Map previously described) and the development of multimedia utilities (e.g. organization of multimedia documents in in a diaporama form).

The main interest of the HTML toolbox is to provide a set of online utilities which can be used either directly by the user, or by other applications. For example, in an Intranet environment, the Map application can be used by a dynamic application of type forum (internal news for example).

The development of multimedia applications, could in particular offer converters of format, compressors (audio, fixed or animated images) or applications of 3D rebuilding. Like the HTML toolbox, these tools could either be used directly, or by intermediate programs.

Other axes of development will probably appear soon. Indeed, the applications database being opened to any contribution, it will be able to evolve/move in the most various directions, with the liking of the imagination of the developers.

3. The multipoint server

3.1. Principle

There is several ways of conceiving a multipoint architecture. Indeed, according to the use it will be dedicated to, this one can vary from a centralized server model (for a reduced use, on the level of the number of potential users and on the level of the exchanged data) to a group address model (protocol IGMP [4], for newsgroups aiming to be unlimited), passing by multi-servers models.

It is noticed however that, in spite of their differences, all these architectures require the implementation of certain common functionalities: to allow the users to join the group or to leave it, to send to all the clients every new contribution, etc.

The architecture we are developing belongs to the first and last categories; it is based on a centralised server mode, which can be extended to an interconnected servers environement to increase its capacity. Indeed, being intended initially to work in an Intranet or limited Internet environment, it does not require a complex and powerful inplementation for the management of clients, broadcasting, etc. In this case, a single server is sufficient. But its extension to a larger environment needs more capacity and flexibility. In this case, a distributed model is recommended. This is why our model allows interconnecting servers (tunneling).

This architecture is based on a simple client-server model, implementing a broadcast functionality on the server. Actually, there is not only one server but three ones: two generic servers, one dedicated to UDP clients and another to TCP ones, and a specialised TCP server, implementing a basic protocol, dedicated to signalling and some specific data applications.

These three servers work mainly the same way, with few differences due to their transport protocol and their complexity.

At starting-up, they are put in a listening state on a specific port. When receiving a connection request (wich is automatically generated for TCP but not for UDP; actually, an UDP client has to send a datagram to be recognized by the server), they establishes a session with the client (via a TCP or UDP socket). Each new session is managed by a new process in the server side (the main process remains dedicated to listening). Once the communication established, the client can then send its contributions to the server. This one will broadcast them to all the connected clients (i.e. UDP or TCP clients).

As these contributions can be from very different types (text, binary files, audio, video, etc) for some applications, we developed a small applicative protocol between the client and the server and/or the clients [appendix 1].

Concerning the user interface, we wanted to preserve as possible a context of Web navigation as for the server of applications. However, in this case, the client has to be able to make a little more complex operations (to send data of different types, to list all the connected clients, etc). We were thus obliged to define more powerful interfaces than simple HTML forms.

3.2. Implementation

The first stage was the development of a server implementing broadcast functionalities, able to establish and manage several connections simultaneously. For this, Java offers very useful packages and classes.

For our network related part, we used the *java.net* [5] package. It contains in particular the *ServerSocket* (server side) and the *Socket* (client side) classes, which allow listening (ServerSocket) and establishing TCP connections. For UDP, it provides two complementary classes: the *DatagramSocket* and the *DatagramPacket*, which are used in both the server and the client sides.

To manage simultaneous communication sessions (sockets), we associated a Thread to each one of them. By using the ThreadGroup class, we can not only manage several connections (i.e. to test the alive threads, to list them, to put a threshold for the number of simultaneous connections, etc) but also to make broadcasting/multicasting with great facility. Indeed, it provide a very useful notion of group, which can be used to create users groups.

It was then necessary to develop an interface able of powerful functionalities (e.g. implementing the basic applicative protocol), without obliging the user to install any software on its machine. The choice of Java applets was then quite natural. Indeed, they allow the client to make intelligent processing from a simple Web navigator. On the other hand, this requires the use of powerful machines, excluding thus terminals like mobile phones or PDA. Other solutions could then be developed to ensure an independence of the terminal. One of them could be using gateways.

It was also necessary to develop a basic protocol of communication, to allow signalling commands and multi-media applications to recognize the type of transmitted data. This protocol is detailed in appendix 1.

However, it is still under development and will probably be improved by defining new types, in order to face the future evolutions of the model (introduction of multimedia streaming, etc).

3.3. Prospects for evolution

In its current state of development, our model allows any users, via TCP connections or UDP datagrams, to exchange any type of data (ASCII, audio/video streams or messages, etc) with a network of users groups (network of inter-connected servers)

The main objective of the project has then been reached. However, there are still several improvements that can be done.

One of them is the definition of a multi-levels hierarchy ("user ➔ user group" or "user ➔ secondary user group ➔ primary user group") to make real multicasting. Indeed, the current architecture allows multicasting based on connection points (servers) and types (TCP or UDP). To be completely multicast, the model has to permit users to define their own groups.

Another improvement can be done: it concerns the basic protocol detailed in appendix 1. In addition to the definition of new types of data, this protocol can be enriched by the possibility of conveying information concerning the users (profiles) or by increasing the set of signalling commands, to allow distant managing for example.

4. Conclusion

The two studied projects follow the current tendency of evolution of the networks and more particularly of the Internet world. Indeed, they amongst other things propose to find solutions to the problems of the independence of the terminals or resource sharing between user groups.

As they are technologically advanced, these projects are thus likely to evolve/move in directions sometimes still unknown. It is the case for example of the Web in a mobile context, which is not yet completely standardized (WAP standard).

The evolution of these projects can even become independent of the initial objectives, in particular for the server of applications. This one being based indeed on an open applications database, it will evolve in the direction the contributors will give it.

Appendices

Appendix 1: Protocol used by the multi-point architecture

The server side of our architecture is made up three twined servers: two raw servers, one for UDP and another for TCP, and a high-level TCP one. This last implements a basic protocol, specifying the type of data transmitted. It is very useful for multimedia (multi-data) applications and for signalling.

It is implemented right over TCP and has an ASCII header. A example of message can be as follows:

```
code:L CRLF
[byte1] [byte2] [byte3]... [byteL]
```

L is the length (in bytes) of the message and code identify the type of the message.

Currently, the defined codes are:

"acp",   used to notify a pseudo (user identifier) acceptance
"lcd",   for listing all the sessions and their statistics
"end",   sent by a client to end its session
"err",   for error messages
"hlo",   for identificatying a client (pseudo)
"lns",   used to update the members list after an inter-connection (request)
"a2l",   used to update the members list after an inter-connection (response)
"mbl",   for obtaining the members list
"url",   specify a URL message
"tst",   for testing loops (after an inter-connection)
"txt";   specify a text message
"ado",   specify an audio message (PCM at 8 KHz)

Remark: There is two types of codes: signalling commands and data types. As the server is not concerned by the data types, the user can define its own types.

References

1. " WAP WML ", june 1999, URL : http://www.wapforum.org/what/technical/SPEC-WML-19990616.pdf.
2. E. Nebel and L. Masinter, RFC 1867, november 1995.
3. E.R. Harold, *JAVA I/O*, O'REILLY, march 1999.
4. S.E. Deering and D.R. Cheriton, RFC 966, december 1985.
5. E.R. Harold, *JAVA Network Programming*, O'REILLY, march 1999.
6. D. Estrin, D. Farinacci, A. Helmy, RFC 2117, june 1997.

Appendix 1: Class model for the server of applications

Our server of applications is actually a server of Java classes. It can load and run any class corresponding to a defined standard, which is detailled below:

```
// A standard class

package StandardClass;

// imports
import ...;
// end imports

public class StandardClass extends WhatYouWant
                            implements SomethingElse{

  public StandardClass ( DataInputStream dis,
                         DataOutputStream dos,
                         String url,
                         String opt1, ..., optN){

  // starting treatements
  ...
  // ending treatements

  }

  // methods
  ...
  // end methods

}
```

To use this class, a user has to type the following URL in his Web navigator:

http://host:port/StandardClass:nb_of_opt:opt1:...:optN?URL   in general
http://host:port/StandardClass?URL   when there is no optional parameters

What is important to respect is the standard constructor, which has three obligatory parameters and an undefined number of optional ones (zero by default).

The optional parameters purpose is to give the developers more flexibility: they are free to use them anyway they want.

However, the obligatory ones are fixed: dis (resp. dos) is the InputStream (resp. OutputStream) associated to the navigator socket and URL is the URL to be treated (distant URL or local file). The developer can use dis and URL to get the data and to dos to send back the result to the client.

An Applicability of Transition Mechanisms for IPv6/IPv4 within the Scope of GPRS with an Internet Communication

Preetida Vinayakray-Jani, Reijo Juvonen

Nokia Research Center, P.O.Box 407, FIN-00045, NOKIA GROUP, Finland
preetida.vinayakray-jani@nokia.com

Abstract. Recent years have witnessed a new version of Internet pro-tocol and concepts of new protocol are heavily relied on for transition from the traditional IPv4-based Internet to an IPv6-based Internet. As a result it is expected that mobile node is General Packet Radio Service (GPRS) with IPv6 support likely to use IPv4 services. Therefore great concerns to transition strategy planners is how to provide connectivity between IPv6-enabled end user to IPv4. As a result many interworking techniques in terms transition mechanisms are proposed by researchers. But their applicability between GPRS and Internet influences many factors such as end-to-end integrity of data, security of communication. Therefore paper mainly focuses on the applicability of transition mechanism where these factors are main concern from GPRS user point of view.

1   Introduction

The GPRS Internet-hosted service is a TCP service that can transmit an unstructured stream between a GPRS Mobile Station (MS) and an Internet host. Thus establishing an Internet-hosted service connection involves setting up two segments, the one segment between the MS and Gateway GPRS Support Node (GGSN) and the another segment between the GGSN and Internet host.

Fundamentally GPRS inherits security features of Global System for Mobile Communication (GSM) but its access to packet data network such as, Internet brings more security threats. In other words IP vulnerabilities limit and complicate the use of GPRS networks for sensitive or secure communication. However from security point of view GPRS has quite good user authentication mechanism but confidential data transmission requires additional mechanisms.

With emerging new standard of Internet protocol, it is quite likely that the two end hosts operating over the above mentioned two segments are configured with different versions of Internet Protocol. Therefore the need of Interworking techniques which provides transition mechanisms becomes crucial, when two end hosts like to communicate with each other via different versions of the IP such as IPv4 or IPv6. Any incompetence or misconfiguration of transition mechanisms can easily amplify security threats and thereby degrading the quality of service of data communication. However an applicability of transition mechanisms requires suitable transition components such as host configuration, routers and routing protocols, domain name systems (DNS) and components dependencies. Therefore this paper focuses on proper applicability of transition mechanisms to maintain inter-operability between GPRS and external IP efficiently.

The paper is organized as follows: section 2 gives brief introduction of GPRS nodes and their functionality followed by GPRS interworking with Packet Data Network in section 3. The transition mechanisms are described in section 4, including protocol encapsulation and IP header translation. Section 5 presents the security threats in GPRS, followed by concluding remarks in section 6.

2 GPRS support Nodes and their functions

Basically GPRS is a new bearer service for GSM that has its own core network and the radio network is shared between the GPRS and GSM core networks. The core network of GPRS is attached to GSM radio network via an open interface. Thus GSM may utilize the GPRS core network to achieve more efficient performance as well as to access packet data network - Internet.

Figure 1 illustrates the basic structure of a GPRS network including possible interception points: Air interface, Base Station Subsystem (BSS), Serving GPRS Support Node (SGSN), GPRS Backbone Network, Gateway GPRS Support Node (GGSN), Border Gateway (BG) and Inter-operator Backbone Network.

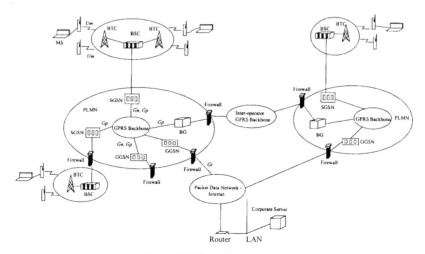

Fig. 1: GPRS Architecture

All the data and signalling is routed through at least one SGSN and at least one GGSN, both in mobile originated and mobile terminated cases. Routing depends on the subscriber IP address allocation point. The SGSN through which the messages are routed, always belongs to the network where the subscriber is currently roaming. The GGSN through which data is delivered (home network GGSN or visited network GGSN) depends on the address allocation point. Messages are routed through the GGSN from whose address pool the used address was allocated.

The SGSN is responsible for the delivery of packets to/from the MSs within its service area and communicates with the GGSN. It also keeps the tracks of the mobiles within its service area. The GGSN acts as a logical interface to external packet data network - Internet and maintains routing information to SGSN that is currently serving MS. The GPRS network can use multiple serving nodes, but requires only one gateway node for connecting to an external network - Internet.

– **Mobile Station** (MS): Generally GPRS mobile stations are classified in 3 different classes - A, B, C, depending on their configured accessibility with GSM and GPRS core. For example:
  - •Class A: mobile can have a normal GSM voice call and GPRS data transfer taking place simultaneously.
  - •Class B: mobiles are capable of using either GSM or GPRS at a time.
  - •Class C: the selection between GSM and the GPRS networks is done manually. Thus Class C mobiles can be attached to either GSM or to the GPRS network but not to both at the same time.

– **Serving GPRS Support Node (SGSN):** This is one of the main component of the GPRS network, which is responsible for handling MS registration and authentication into the GPRS network, to manage MS mobility, to relay traffic and to collect statistics and charging information.

– **Gateway GPRS Support Node (GGSN):** This is the interface between the GPRS backbone and external data networks. In conventional term it is simple IP router as it routes the data to and from external data networks to the SGSN serving the MS.

– **Border Gateway (BG):** The main function of this component is to ensure a secure connection between different GPRS networks over the inter-operator backbone network. The functionality of the BG is not defined in the GPRS specifications. It could consist firewall, security functions, and routing functions. BGs as well as their functionality are selected by the GPRS operator's manual agreement to enable roaming.

– **GPRS Backbone Networks:** There are two kinds of backbone Public Land Mobile Networks (PLMNs) as shown in Figure 1. They are called intra PLMN backbone network and inter-PLMN backbone network.
  - The intra-PLMN backbone is the IP network interconnecting GGSNs within the same PLMN. Every intra-PLMN backbone networks is a private IP network intended for GPRS data and signalling only and there must be some access control mechanism in order to achieve a required level of security. Two intra-PLMN backbone networks are connected via Gp interface using the BGs and inter-PLMN backbone networks.
  - The inter-PLMN backbone network interconnects GGSNs and intraPLMN backbone networks in different PLMNs.

3   GPRS Interworking with Packet Data Network

In GPRS network interworking is inevitable whenever PLMN is involved in communication with packet data network (PDN) to provide end-to-end communication. This interworking may be either directly with Internet or through a transit networks - intranets. Figure 2 shows the Gi reference point and protocol stack needed for GPRS interworking with IP networks [4].

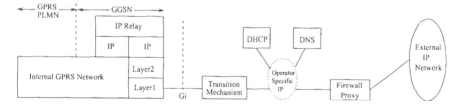

Fig. 2: Gi reference point for GPRS with IP interworking

The GGSN for interworking with the IP network is the access point of the GSM GPRS data network. In this case the GPRS network will look like any other IP network or subnetwork. Distinctively in the IP network, the interworking with subnetworks is done via IP routers. Therefore referring to Gi reference point, external IP can view, GGSN as a normal IP router. Considering generic view there are some assumptions valid between the GGSN and the external IP network:

– A firewall is configured by GPRS operators so that operators can easily make their choice for type of service they can offer.
– The DNS managed by the GPRS operator or it can be managed by the interworking with Packet Switched Data Networks and more specifically with Internet. This interworking may be either direct - Internet or through a transit network - intranet.
– The GGSN may allocate dynamic IP addresses by itself or use an external device such as DHCP server. This external server can be operated by an Internet Service Provider (ISP).
– The deployment an appropriate transition strategy that allows new IPv6 supported mobile terminals to communicate with existing IPv4 hosts or servers.

Focusing on the last assumption the transition requirements are most important for flexibility of deployment and ability of IPv6 hosts to communicate with IPv4 hosts. To place it concisely, there are 3 objectives for the proposed transition to be smooth. They are:

1. To allow IPv6 and IPv4 hosts to inter-operate.
2. To allow IM hosts to be deployed in the Internet in a highly diffuse and incremental fashion, with no interdependencies.
3. The transition should be as easy as possible for end-users, system administrators, and network operators to understand and carry out.

The mechanism of transition is a set of protocols implemented in hosts and routers, along with some operational guidelines for addressing and deployment, designed to make transition to work with as little disruption as possible. Some of the features for implementing the transition mechanism are:

1. An IPv6 addressing structure that embeds IPv4 addresses within IPv6 addresses, and encodes other information used by the transition mechanism.
2. A model of deployment where all hosts and routers upgraded to IPv6 in the early transition phase are dual capable (i.e., implement complete IPv4 and IPv6 protocol stacks).

3. The technique of encapsulating IPv6 packets within IPv4 headers to carry them over segments of the end-to-end path where the routers have not yet been upgraded to IPv6.
4. The header translation technique to allow the eventual introduction of routing topologies that route only IPv6 traffic, and the deployment of hosts that support only IM. Use of this technique is optional and would be used in the later phase of transition if it were used at all.

The proceeding discussion focuses on some of the more innovative and radical changes IPv6 brings to interworking. These interworking mechanisms are integral part of the IPv6 design effort. These techniques include dual-stack IPv4/IPv6 hosts and routers, tunneling of IPv6 via IPv4, and a number of IPv6 services, including DNS, Dynamic Host Configuration Protocol (DHCP), Application Level Gateways (ALGs), relays, proxies, caches and so on. The flexibility and usefulness of the IPv6 transition mechanisms are best gauged through scenarios that address real-world networking requirements. Therefore initial design specification of transition mechanisms specifies the use of three different types of IP nodes such as IPv4-only node, IPv6/IPv4 node (Dual stack) and IM-only node.

4   Transition Mechanisms

On the basis of above objectives and features of transition mechanisms, researchers have proposed few transition mechanisms, which are described below.

4.1   Protocol Encapsulation

When IPv6 hosts on different edges of the network are separated by IPv4 capable routers, an encapsulation mechanism is used to setup 6in4 or 6over4 tunneling [2]. Thus tunneling of IPv6 datagrams takes place by encapsulating them within IPv4 packets. This way IPv6/IPv4 hosts and routers can tunnel IPv6 traffic over regions of IPv4 routing topology which is shown in Figure 3. In simplest form the encapsulating node adds an IPv4 header to the packet and transmits it. The decapsulating node removes the IPv4 header and processes the remaining IPv6 packet as if it were normally received via IPv6 topologies. Generally tunnel can be configured manually or automatically.

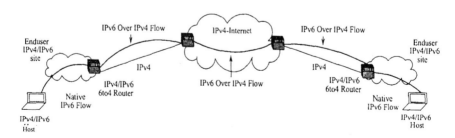

Fig. 3: 6to4 Tunnel Overview

- **Configured Tunneling (6in4):** This configuration does not require any interdependency between IPv4 and IPv6 addresses as the tunnel path endpoint IPv4 address is defined in the configuration of the encapsulating node. But this configuration requires a lot of effort to manage because:
  - Finding candidate networks when the site's choice of IPv4 service does not provide IPv6 service (either tunneling or native),
  - Determining which one are the best IPv4 path to use so that an configured tunnel doesn't inadvertently follow a very unreliable or lowperformance path,
  - Making arrangements with the desired IPv6 service provider for tunneling service, a scenario that may at times be difficult if the selected provider is not ready to provide the service or some other reasons.
- **Automatic Tunneling (6to4):** This mechanism provides a solution to the complexity problem with manually configured tunnels by advertising a site's IPv4 tunnel endpoint (to be used for dynamic tunnel) in a special external routing prefix for that site. Thus IPv6 addresses used must be IPv4compatible addresses [3]. Thus one site trying to reach another will discover the tunnel endpoints from DNS and use a dynamically built tunnel from site to site for the communication. The tunnels are transient in that there is no state maintained for them, lasting only as long as specific transaction uses the path. However this type of tunneling scenario exists when site has:
  - both 6to4 and native IPv6 connectivity, or
  - only 6to4 connectivity and trying to reach a site with both 6to4 and native IPv6 connectivity or
  - both 6to4 and native IPv6 connectivity, and trying to reach native IPv6 connectivity and vice-versa.

But most interesting when site has only 6to4 connectivity and communicating site with only native IPv6 connectivity. This will be accomplished by the use of a 6to4 relay that supports both 6to4 and native connectivity. The 6to4 relay is nothing more than an IPv4/IPv6 dual-stack router.

4.2 IP Header Translation

To enable the communication between IPv6 node and IPv4 node or vice-versa, a translator needs to perform two main functions: Address translation and Protocol translation. Address translation involves converting address for packet crossing the protocol boundary, whereas protocol translation involves mapping most of the fields from one version of IP to the other. Researchers have exploited this translation feature at network level (SIIT, NAT-PT), Transport level (SOCKS) [5] and Application level (ALGs).
- **Stateless IP/ICMP Translator (SIIT) [7]:** SIIT is the mechanism that allows IM-only nodes to talk with IPv4-only nodes by translating the packet headers between IPv6 to IPv4. Once the traffic is originated by IPv6 host and routed through SIIT node, SIIT makes simple header translation between IPv6 to IPv4 in source field of header and forwards it to IPv4 node. But SIIT proposal do not define how IPv6 host should get its temporary IPv4 address. The applicability of this mechanism is limited as
  - applications like FTP and DNS transfer where IP address is inside the payload, SIIT is unable to perform translation.

- if the arrived packet is embedded with security feature, then authentication header (AH) which is computed from IP header fields gets broken while passing through SILT. Therefore even if Encrypted payload passed through, translator partly breaks the IP security (IPsec)
- **Network Address Translator-Protocol Translator (NAT-PT):** NAT-PT also performs header translation as described in [8]. Actually there are two different methods that assigns the temporary IPv4 address to IPv6 host:
1. The translator reserves the pool of IPv4 addresses, from which it assigns IPv4 address to the IM host and caches this address mapping from IPv6-to-IPv4 for the duration of the session.
2. All IPv6 addresses are mapped to a single IPv4 address, which is the IPv4 address of the translator. This is useful when translator fall flat when the pool of IPv4 addresses assigned for translation purposes is exhausted. As a result the translator considers port number as an transport identifier - TCP/UDP port number or ICMP query identifier.

Although this translation mechanism provides stateful translation there are some limitation associated with it, such as:

- Some applications e.g., rlogind, don't accept the connections if they are not from privileged port.
- when host from outside wants to establish a connection to server which resides inside the NAPT-PT and NAPT-PT can map only one server per service.
- As translator breaks the end-to-end integrity of data it is quite likely that some information may be lost during translation.
- It also breaks IPsec partially by breaking the authentication and bypassing encryption.
- **SOCKS 64:** The SOCKS gateway - SOCKS-GATE tool is a gateway system that accepts enhanced SOCKS - SOCKS-EXT connections from IPv6 hosts and relays it to IPv4 or IPv6 hosts [5], especially for *socksified* sites, which already use SOCKS aware clients and SOCKS server. The mechanism simply replaces the standard socket and DNS resolver libraries with SOCKS versions. Besides this each *socksified* application should be configured with the IP address *of* the local SOCKS gateway. When application makes a DNS query, the SOCKS library intercepts the call and returns an arbitrary IP address for A. This IP address is never seen on the wire. The resolver library associates this address with fully qualified domain name (FQDN). When the application enables the socket call, the SOCKS library makes connection to dual stack gateway. This dual stack gateway makes standard connection to application and protocol translation. No DNS modification or address mapping is needed. Compare to previous translators SOCKS gateways are simple to maintain and configure. Thus when protocol translation is necessary, this transport level translation mechanism should be considered.
- **ALGs, relays, proxies and caches:** It is reasonable to position application level gateways, relays, proxies and caches, especially at Intranet /Internet boundary. This type of mechanism is very helpful when IP address is embedded within payload. Although ALGs provides transparency, they do so in precise way, correctly terminating network, transport and application protocols on both sides. They can however exhibits some shortfalls in ease of configuration and fail-over. However,

in some application level mechanism such as proxies or relays they grab and modify traffic in an inappropriate way and generate totally unforeseeable side effects.

- **Temporary Address Allocation:** Knowing the limitations of protocol translator some proposals such as Allocation of IPv4 address to IPv6 nodes (AIIH), Realm Specific IP (RSIP) [1, 6] are made to provide publicly routable IPv4 address so that dual stack host machine can communicate with IPv4 server.

  - The AIIH is essentially a DHCPv6 server, which allocates the temporary IPv4 address to dual stack host using global address extension. When connection is initiated by only IPv4 host then AIIH server also needs to contain DNS server, so that IPv4 host can make DNS query about dual stack host node. The AIIH server will respond this query by assigning temporary IPv4 address and simultaneously sending 'DHCP reconfigure' message to dual stack host so that it can update its interface with new IP address. A dual stack host acknowledges this by sending confirmation to AIIH server which then finally updates DNS record.

  - With RSIP proposal the basic idea is very similar to AIIH but currently existing network stack of client requires modifications. Thus RSIP client makes the request for temporary public IPv4 address from local RSIP server which is located at the boundary between two routing realms. This protocol standard is till evolving but current proposal requires that RSIP clients use a tunnel to the RSIP server, which matches very well with GPRS architecture.

5   Security threats in GPRS systems

Here we will examine security threats originating from external IP and leading to different nodes in GPRS.

5.1   Denial of Service (DOS)

This type of threat occurs when malicious party manages to do a bogus registration of a new care-of-address for a particular mobile host within GPRS. Such a bogus registration gives rise to two problems:
1. The particular victim mobile host gets terminated.
2. The malicious party gets to see all traffic directed to victimized mobile host.
   The usual protection again DOS is a firewall, but it is not effective in all cases, e.g., protection against viruses is always one step behind the sources.

5.2   Session Stealing/ Spoofing

This is one of the active form of information theft. An attacker performs following steps once the mobile host gets registered with its home agent:
1. An attacker eavesdrops and floods the mobile host with bogus packets, thus putting it out of the action.

2. Thus attacker steals the session by originating packets that seem to have come from the mobile host and at the same time intercepting the packets destined for the mobile host.

Such type of attack could occur either at the foreign link or at some other point between foreign agent and mobile node where foreign agent is not colocated. This can be avoided by applying strong encryption so that even if session gets stolen the attacker cannot get the actual data.

5.3   Incompetent translator

The major drawback of protocol translator is that it breaks the end-to-end integrity of data as it needs to change IP headers including upper layer header and simultaneously destroying authenticity of data, which shown in Figure 4.

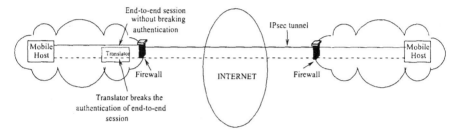

Fig. 4: Translator Applicability

Once the attacker has gained the 'physical' access via unattended network socket, by following first three steps he breaks the mobile host from the network:

1. The attacker figures out a network prefix to use. This can be done by listening for mobile IP agent advertisement, by examining IP address in packets flowing around on the network segment, or even by just doing a DHCP configuration request.
2. Then attacker guesses an available host number to use. This can be done by listening for a while and just picking one that does no appear to be used. Then by making ARP request for the resulting IP address and see if it goes unanswered otherwise again a DHCP request.
3. Once the above steps succeeded the attacker gets access to IP hosts and floods it bogus packets.
4. Another attack path is the external net to backbone through GGSN. Since the GGSN is an IP router, it will handle any packet addressed to a backbone node as any other packet unless firewall at Gi prevents it.
5. A simple attack is from one intranet to another through through GGSN Gi. Since the GGSN is an IP router, it can mediate packets between the two, or even the Internet as mentioned in the above step.
6. In GPRS Tunneling Protocol (GTP) specification are public, as they can be implemented by anyone. The GGSN unwraps the only the outer GTP envelope, supposedly wrapped by the SGSN. If the remaining packet is addressed to a backbone node, it will be forwarded by to that node. This opens attack possibilities via all interfaces. Thus any Internet host can attack the backbone nodes by sending valid GTP packets and target the billing system.

6 Concluding Remarks

The recommended suggestions are not comprehensive, but they do provide the cornerstone for proper applicability of interworking techniques within the scope of GPRS and external IP. The main problem of security is not the available tech nology, but its ease to use it. Concluding above discussion here we suggest some key points to consider so that end-to-end integrity of data can be maintained.

– Applicability of transition mechanism:
  - Viewing carefully transition mechanisms mentioned above 6to4 encapsulating mechanism shows more competitive than others as it allows the isolated IPv6 routing domains to communicate with other IPv6 routing domains even in the total absence of native IPv6 service provider. It is a powerful IPv6 transition mechanism that will allow both traditional IPv4-based Internet sites to utilize IPv6 and operate successfully over existing IPv4/based Internet routing infrastructure.
  - Upgrade the existing IPv4 servers with dual stack support, so that both IPv4 and IPv6 host can communicate easily
  - In case of limited public IPv4 addresses, currently the use of RSIP is more preferred choice as it provides temporary assignment of IPv4 addresses and simultaneously preserving the transparency of data.
– Security Consideration: The loss of transparency of end-to-end data is one of the main concern from security point of view. If network level translator is in the path, then the best that can be done to is to decrypt and re/encrypt IP traffic in the translator. This traffic will therefore be momentarily in clear text and potentially vulnerable. In the environment where this is unacceptable, the encryption must be applied above network layer instead. Anyhow this break in security provides well defined point at which to apply the restriction by using firewalls or Active filters. In case when protocol translation one should use SOCKS or ALGs rather than SIIT or NAT-PT.

In an ideal transition concept the pushing force for transition mechanisms would be the technical advantage and exploit the new features offered by the new version of protocol. However here we try to make the transition concept feasible with GPRS where coexistence of both IPv4 and IPv6 can be arranged in a more practical and efficient way.

References

1. Borella, M et al.: Realm Specific IP: Protocol Specification. Internet Draft, draft-ietf-nat-rsip-framework-01.txt. (1999)
2. Carpenter, B., Moore, K.: Connection of IPv6 Domains via IPv4 Clouds without Explicit Tunnels. Internet Draft, draft-ietf-ngtrans-6to4-02.txt.(1999)
3. Deering, S., Hinden, B.: Internet Protocol, Version 6 (IPv6) Architecture. RFC 2460.(1998)
4. ETSI TS 101 348 V7.1.0.: GSM 09.61: Digital Cellular telecommunications system (Phase 2+); Genreal Pacekt Radio Service (GPRS); Interworking between the Public Land Mobile Network (PLMN) supporting GPRS and Packet Data Networks (PDN). (1999-07)
5. Kitamura, H.: SOCKSv5 Protocol Extensions for IPv6/IPv4 Communication Environment. Internet Draft, draft-ietf-ngtrans-socks-gateway-02.

6. Montengero, G., Borella, M.: RSIP Support for End-to-End IPsec. Internet Draft, draft-ietf-nat-rsip-ipsec-02.txt. (2000)
7. Nordmark, E.: Stateless IP/ICMP Translator (SIIT). Internet Draft, draft—ietf-ngtrans-siit-06. txt. (1999)
8. Tsirtsis, G, Srisuresh, P.: Network Address Translation - Protocol Translation (NAT-PT). Internet Draft, draft-ietf-ngtrans-natpt-06.txt.(1999)

Implementing the Integrated Services QoS Model with IPv6 Over ATM Networks

David Fernández[1], David Larrabeiti[2], Ana B. García[1], Arturo Azcorra[2], Luis Bellido[1], Julio Berrocal[1]

[1] Dpto. Ingeniería de Sistemas Telemáticos, ETSI Telecomunicación
Universidad Politécnica de Madrid, Ciudad Universitaria s/n
E-28040 Madrid, Spain
{david, abgarcia, lbt, berrocal}@dit.upm.es
[2] Área de Ingeniería Telemática
Universidad Carlos III de Madrid, Avda. Universidad, 30
E-28913 Madrid, Spain
{dlarra, azcorra}@it.uc3m.es

Abstract. This paper describes the experience gained and the results achieved developing and integrating a protocol stack for IPv6 over ATM networks with Quality of Service (QoS) support based on the Integrated Services model defined by IETF. This work, which constitutes one of the first implementations of such protocol stack over Windows NT operating system, was developed as part of the authors' contribution to European funded Broadband Trial Integration (BTI) ACTS Project[1]. The paper focuses on the description of the technologies selected in BTI to provide QoS over a residential access network, and on the main characteristics of the protocol stack developed as well as the problems faced and the technical solutions applied. Finally, some conclusions about the experience are presented.

1   Introduction

In order to cover the general demand for improved quality of Internet Services, the BTI [1] project pointed its objective to develop and demonstrate a concept for improved QoS based on the integration of IPv6 and ATM. The project focused on implementing QoS in an ATM based Passive Optical Network (APON) extended to the user by means of VDSL technology, supporting unicast and multicast with well-defined QoS control in terms of the controlled load service of Integrated Services Internet approach (IntServ from now on). Fig. 1 presents the detailed architecture of the access network used in BTI.

To explore QoS features offered by the integrated IPv6/ATM networking environment, a set of distant education applications were selected in the project, including tools to access digital video servers or to co-operate synchronously by means of virtual workspaces and video/audio conferencing tools. All these

[1] This work has been partially supported by the EU Commission under the ACTS project 362.

applications were migrated to work over IPv6 and modified to include QoS support, in order to allow the users to demand the network a QoS level through a simple user interface.

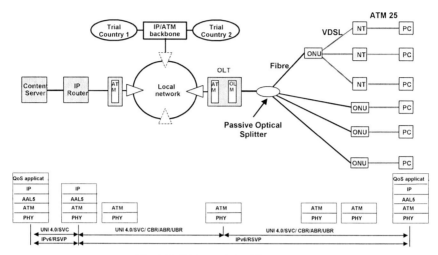

Fig. 1. BTI Network Architecture

Apart from the technical evaluation of the solutions used in BTI network, a program of structured usability testing was performed to evaluate the user perception of the QoS control and user interface. Students and teachers at universities and schools in Denmark, Poland, and Portugal participated in trial experiments over real access networks.

One of the main problems faced by the project was the unavailability of a complete protocol stack running over the target end-user's stations operating system (Windows NT) that implemented and integrated all the protocols used in BTI access networks: IPv6 over ATM with full multicast support based on multipoint ATM SVC's, and RSVP over IPv6 with traffic control functionality over ATM networks.

In principle, BTI planed to reuse some existing solution available in the market or, more likely, in the research arena; with the aim of concentrating the project effort on the adaptation of applications. However, as time progressed and no complete solution was available, the project decided to develop and integrate its own protocol stack, starting from existing protocol blocks available on the network (mainly RSVP and IPv6 implementations with source code available) and adding the missing parts (mainly IPv6 over ATM drivers and all QoS traffic control functionality). All the components, existing and new, had to be integrated to offer applications the QoS demanded through standard programming interfaces.

The rest of the paper is organized as follows. Section 2 describes the main technologies involved in BTI access network, focusing on how IPv6 works over ATM and how IntServ QoS approach is implemented over this scenario. Section 3 briefly presents the three applications selected in BTI, emphasizing on the requirements they impose on the network. Later, Section 4 describes the integrated protocol stack developed, bringing out all the difficulties found during the design, implementation and testing phases, and presenting the main technical solutions adopted. Finally, section 5 summarizes some conclusions about the work presented.

2 IntServ over ATM Networks with IPv6

This section presents how the main technologies used in BTI access network -IPv6, ATM and IntServ- were integrated to provide QoS support. First, we discuss how IPv6 works over ATM networks for *Best Effort* traffic; then we present how IntServ model has been implemented over ATM networks to provide end-to-end QoS support.

2.1 IPv6 over ATM Networks

Most ATM-based solutions implemented today in IP access networks make use of permanent virtual connections (PVC). In these scenarios, IP-over-ATM devices at customer premises communicate with routers at provider premises by means of pre-configured PVC's, either directly from hosts equipped with ATM cards or through a local router or bridge. This setup guarantees an agreed quality of service –ATM connections used are typically CBR or VBR- for the whole traffic coming out or going to the client, but does neither allow any adaptation to dynamic usage requirements nor any differentiation among data flows sent by clients. Furthermore, this approach does not take full advantage of subnetwork facilities like direct ATM connections between clients, or the ATM multipoint service, in the case of IP multicast traffic.

Carrying IP packets over ATM networks using PVCs is relatively simple; the only decision to make is the adaptation layer and encapsulation to be used -by default AAL5 and LLC/SNAP, respectively [2]- and configuring IP routing to work over these static point-to-point channels.

However, the situation is much more complex if switched virtual circuits (SVC) are used. For example, the lack of a mechanism to broadcast packets to all nodes belonging to a logical IP Subnet (LIS) on ATM prevents the usage of simple protocols, like ARP, for link-level address resolution.

Moreover, IP multicast support over ATM is rather complicated. Although multicast services exist in ATM networks -in the way of multipoint SVC's-, it is not possible to directly map IP multicast addresses to multicast subnet addresses, as it is done in LAN. Multicast model in ATM networks does not include multicast addresses at all, so an external entity to manage the association between IP multicast addresses and the set of ATM unicast addresses of hosts that belong to the group is needed. In the data plane, a multipoint ATM connection must be established to all those endpoints that have joined the group prior to sending any multicast IP packet.

In the case of multicast, the connectionless (IP) to connection-oriented (ATM) gap is greatly enlarged by the highly dynamic nature of IP multicast membership. This problem is partially simplified if UNI 4.0 ATM signaling version is used, thanks to the new leaf-initiated join capability.

In IPv4 two new protocol entities were defined when the Classical IP over ATM model (CLIP) was designed: the ATM ARP server initially defined in RFC1577 [3] and the MARS (Multicast Address Resolution Server) defined in RFC2022 [4].

The ATM ARP server is in charge of address resolution function over the ATM subnet. It is basically a client-server version of ARP that, given a destination IP address, supplies the corresponding ATM endpoint address. Likewise, MARS is a client-server protocol that resolves IP multicast addresses to the set of ATM addresses of the nodes in the LIS that have joined the target group.

The usage of these protocols is conceptually simple. Nodes in a LIS register themselves in the servers. When a node needs to forward a datagram, firstly computes the next-hop address and issues a request to the server to find out the ATM address (*addresses* in the multicast case) associated to that next-hop if it is not available in its neighbors cache. Once the ATM address is known, the source node can proceed to set up an ATM connection to the target node and transmit the datagrams over the SVC to that next-hop.

In the case of MARS, clients connect to the MARS server to register themselves and to issue multicast group joining/leaving updates. Once registered, nodes become leaves of a cluster control multipoint VC rooted at the MARS, which is used to multicast membership updates to all registered LIS nodes and usually also to respond to IP-multicast address to ATM-addresses resolution requests. Knowing of this information is required by clients to keep up the ATM multipoint SVC to the right IP group members, issuing the necessary ATM leaf add/drops.

At the time of sending multicast data, MARS allows two possible scenarios:

- *VC mesh.* Every sender to a group issues a request to the MARS to find out the ATM addresses of the group members and, once known, places a multipoint call to all of them as leaves.
- MCS (Multicast Server). In this case, the MARS instructs its sender clients to setup the connection to a single leaf -the MCS entity-, where a multipoint connection to all group members is rooted and shared by all senders to the group. The MCS serializes and forwards all packets received from group senders to group members via the multipoint connection.

MARS RFC [4] discusses advantages and disadvantages of each approach. Just let us outline that, although both configurations were feasible in BTI scenario, in principle the MCS solution was more suitable, due to the tree topology of the access network as well as for scalability reasons (less signaling overhead and less number of SVCs).

However, VC mesh option was chosen instead because MCS does not provide the QoS support. In particular, it is not possible to delegate on the MCS a multipoint connection to a subset of group members, as it is an essential requirement in the IntServ scenario to differentiate users who has made a reservation from the rest. Therefore, the choice for the VC mesh for the IntServ scenario implemented was a must, as our design principle was just using standardized solutions.

The way IPv6 works over ATM networks is similar to the solutions described for IPv4, but includes some important differences that make it simpler. For example, IPv4 stacks that work over LAN and ATM networks duplicate at some extend the address resolution functions, as they are both included in LAN and ATM drivers (Fig. 2).

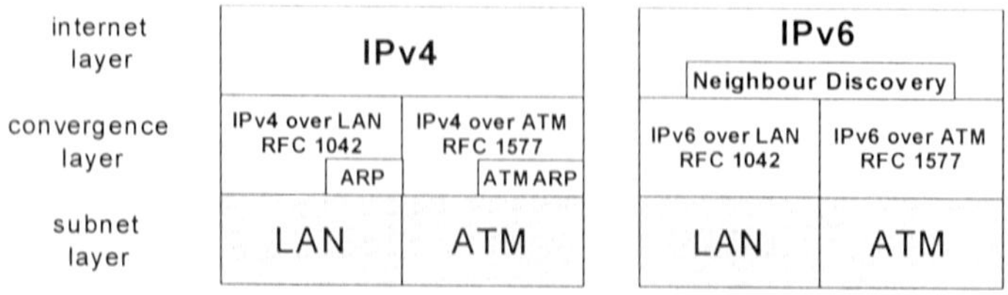

internet layer	IPv4		IPv6 Neighbour Discovery	
convergence layer	IPv4 over LAN RFC 1042 ARP	IPv4 over ATM RFC 1577 ATMARP	IPv6 over LAN RFC 1042	IPv6 over ATM RFC 1577
subnet layer	LAN	ATM	LAN	ATM

Fig. 2. IPv6 vs. IPv4 Address Resolution Architecture

IPv6 architectural model solves this problem in a simple an elegant way. In IPv6, Neighbour Discovery protocol is defined independently of the underlying subnet, and it assumes that all subnets provide multicast link-layer services to allow the multicast of Neighbour Discovery messages. In this way, address resolution function is common to all subnets (Fig. 2), and the ATMARP server is not needed. The disadvantage of this approach comes from the fact that multicast must be always supported, and therefore a MARS server must be available in the ATM subnet when using IPv6.

2.2. IntServ over ATM

The potential interest of implementing IntServ over ATM is obvious, specially if we think that flows requesting a given quality of service can take advantage of ATM short-cuts by-passing and unloading routers from packet scheduling and signaling, delegating real QoS implementation and even flow aggregation, on ATM. Nevertheless, implementing IntServ QoS approach over ATM networks it is not easy at all, due to the important differences between QoS support models of RSVP and ATM networks. To be mentioned:

- RSVP is receiver-based and ATM is sender-based. This can be partially solved using signaling at layers above ATM and, for the multicast case, by the leaf-initiated functions included in UNI 4.0.
- Resource allocation in RSVP is unidirectional. However, in ATM it is bidirectional for unicast circuits (although reservations can be asymmetric and even void in one direction), and unidirectional for multicast ones.
- RSVP allows dynamic changes of QoS reservations; however, ATM obliges to close a circuit and reestablish it again to change QoS parameters.
- RSVP allows receiver heterogeneity in multicast reservations; however, ATM imposes uniform QoS for all receivers of a multipoint connection. This implies that multicast can be performed very efficiently at the ATM layer with a guaranteed QoS only if reservations are coordinated at the application layer in such a way that a single mutipoint connection can be shared by many receivers.

In the past years a lot of effort have been invested to have a set of standards that define how to map RSVP reservation requests into ATM connections. In outline, the alternatives for each service category, as given by RFC2381 [5], are the following: the Guaranteed Service should be implemented as a CBR or rt-VBR connection, Controlled Load reservations as CBR, nrt-VBR or ABR including a minimum cell rate, and Best Effort as UBR or ABR connections.

In the BTI scenario, the Controlled Load Service was realized by means of CBR (BCOB-A) connections in order to guarantee the portability to any ATM device, since this service is always present in all implementations. However, the best theoretical match, corresponds to nrt-VBR where the mapping to ATM traffic descriptors is as simple as mapping the receiver TSpec's peak rate to PCR (Peak Cell Rate), the token bucket rate to SCR (Sustained Cell Rate), and the bucket size to MBS (Maximum Burst Size). Usually, these traffic translations must be complemented with buffering edge devices to achieve a seamless coupling between the two models.

It is important to note in this context that IP nodes linked by ATM networks must establish bidirectional best-effort connections to exchange RSVP messages. These connections, together with the necessary connections for MARS signaling and the fact that currently no standard way of aggregating sessions over a single ATM connection exist, add up complexity to this solution in terms of number of VCs.

3. Application Scenarios

In order to explore QoS features offered by the integrated IPv6/ATM networking environment, a set of distant education applications was specified including Digital Video Library, Virtual Workspace and Video/Audio Conferencing tools [6].

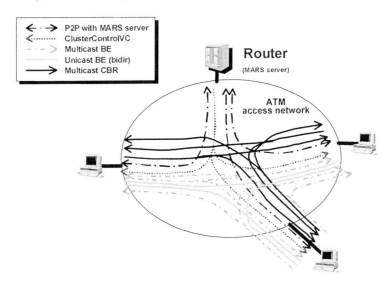

Fig. 3. A multipoint to multipoint videoconference

The Virtual Workspace is made of a set of integrated data-conferencing applications that support collaborative education. It was basically reused from the environment developed in LEVERAGE EU funded ACTS 109 project [7], and it allows users to create virtual meetings and to interact in real time by sharing documents. The Digital Video Library consists of a video-streaming engine, database, and content manager server. User interfaces for video retrieval, playback, uploading, and QoS control are embedded in a standard web browser.

The Video/Audio Conferencing tool allows the users to establish point to point and multipoint videoconferences. It was based on well-known VIC (Videoconference) and RAT (Robust Audio Tool) multicast tools used in Internet Mbone. All these applications were adapted to work over IPv6 and modified to include a simple and friendly interface to allow users to control the QoS, basically by selecting the desired quality level.

All this target applications were supported by BTI platform. However, each application had different requirements from the network, in terms of resources and functionality demanded. For example, Virtual Workspace did not required multicast, as it was implemented at application level; Digital Video Library did not required multicast services from clients, only from video server.

As expected, the most complex and resource-consuming[2] application was multipoint videoconferencing. To illustrate that, Fig. 3 shows the scenario of an audioconference with tree terminals using multicast. In this "simple" scenario each client has to maintain 10 different ATM circuits, in order to send and receive all traffic types: control traffic with the MARS server, unicast and multicast best effort traffic and multicast traffic with guaranteed QoS. In addition, if the conference uses video as well (sent to a different multicast address than the one used for audio), 6 more circuits are added to the list. As already mentioned, a centralized approach (MCS like) would have minimized the amount of SVCs, but this is not currently feasible in a fully standard way.

4. Integrated Protocol Stack

DIT-UPM and IT-UC3M main effort in BTI project was devoted to develop an integrated protocol stack named PATAM (IPv6 over ATm Adaptation Module with RSVP support, [8]). PATAM is a Winsock2 compatible protocol stack running on Windows NT, the operating system chosen by BTI application developers and trial organizations.

PATAM includes an IPv6 stack able to run over Ethernet and ATM subnetworks with full multicast support, as well as an RSVP over IPv6 implementation with traffic control support over ATM interfaces.

IPv6 support of PATAM was based on a modified version of Microsoft Research's IPv6 implementation for Windows NT [9] -that only supports IPv6 over Ethernet interfaces- with the addition of a completely new IPv6 over ATM driver developed by the authors. RSVP for IPv6 support was based on the well-known ISI's RSVP implementation for UNIX operating system [10], which has been migrated to Windows NT and adapted to offer a Winsock2 interface and to interact with MSR's IPv6   stack.

---

[2] From the point of view of control resources (number of circuits to establish, signaling messages, etc). In terms of bandwidth, the Digital Video Library is the most resource-consuming application clearly.

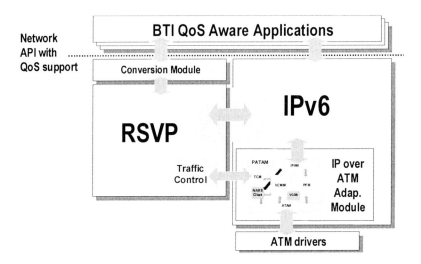

Fig. 4. Protocol Stack Architecture

The whole stack integrates under the Winsock2 based networking architecture of Windows NT. QoS aware applications use the standard Winsock2 API to access either IPv6 or RSVP services offered by the stack. Fig. 4 depicts the general architecture of PATAM stack. The next two subsections describe with more detail the two main parts of our contribution: the IPv6 over ATM driver (PATAM) and the RSVP protocol daemon.

4.1. IPv6 over ATM driver

Fig. 5 shows the detailed architecture of the IPv6 over ATM driver, including its relation with other modules of the integrated stack. PATAM is a user-mode multithreaded driver that implements all the necessary functions to carry IPv6 packets over ATM networks using dynamic circuits (SVCs) with full multicast support. The driver is made of several components:

- *Flows* Database, which manages all the information about the Best Effort (BE) and Controled Load (CL) active IPv6 flows. Each time a new IP flow is created, either unicast or multicast; BE or CL, a new entry in the Flows Database is created, storing all the information necessary to later classify and schedule the sending of packets belonging to that flow.
- IPv6 Access Module (IPAM), which manages the communications between the driver and the IPv6 stack. Each time an IPv6 packet is received through any of the ATM circuits, IPAM passes it to the IPv6 stack; and each time the IPv6 stack has a packet directed to the ATM interface, it is received by IPAM, that delivers it to the classifier and scheduler modules.
- *Packet Forwarding* Module, which is in charge of sending and receiving IPv6 packets to and from ATM network. It includes all the functions needed to classify

IPv6 packets according to the different flows in the database and schedule their transmission according to QoS reservations.

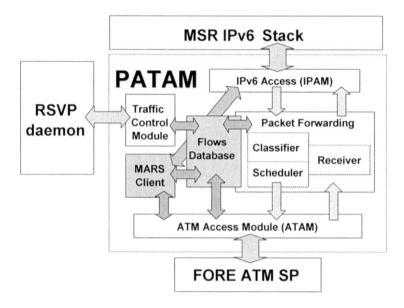

Fig. 5. IPv6 over ATM driver Architecture

- *ATM* Access Module (ATAM), which manages all the ATM circuits associated with IPv6 flows. It is in charge of creating and releasing SVCs, adding or deleting leafs to multipoint circuits and, in general, reporting other modules about any event related to ATM circuits. It accesses ATM network services using the standard Winsock2 API defined for ATM. This interface allows the creation of UBR and CBR point-to-point and multipoint circuits, however, no support for ABR was available.
- MARS Client Module, which implements MARS Client functionality according to [4]. All requests to send and receive to or from IPv6 multicast addresses and all the communications with the MARS server are managed by this module. This module was developed starting from a public LINUX implementation developed by NIST. This implementation was modified to support IPv6 and later migrated to Windows NT and adapted to work over PATAM architecture.
- Traffic *Control Module,* which manages the communications with the RSVP daemon for the creation and release of Controlled Load flows and their correspondent CBR ATM multipoint circuits. It is described with more detail below.

4.2. RSVP daemon

The RSVP functionality developed by the authors for BTI project includes a complete RSVP engine according to current standards [11] [12]. The main features are:
- Standard Winsock2 API.
- IPv6 support (no IPv4 support is provided), using native IPv6 encapsulation.
- Support for both Ethernet and ATM interfaces.
- Interaction with PATAM driver to offer Traffic Control (TC) implementation over ATM subnetworks, supporting FF and SE styles, and IntServ's Controlled Load reservations.
- Host (not router) implementation.

As already mentioned, the development was started from the well-known RSVP implementation of ISI [10], which works on UNIX platforms. This daemon was migrated to Windows NT and adapted to the Windows-specific asynchronous event notification, and later was completed in order to provide actual Traffic control support for ATM subnetworks. Figure 6 shows the specific architecture of the RSVP module.

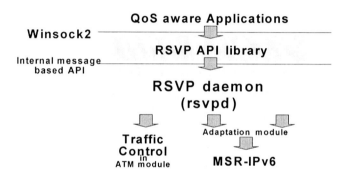

Fig. 6. RSVP Architecture in PATAM

The main tasks carried out in the developing of RSVP module are briefly discussed:
- Adaptation to Winsock2 architecture. To offer the standard Winsock2 RSVP API to applications, a standard WS2 Service Provider (SP) Library was developed. This library performs the translation between WS2 interface offered to applications and an internal interface with the core RSVP processing, which has been kept the same as it is in ISI's implementation.
- Interfaces with the IPv6 stack: routing and I/O. RSVP module has to access IP functionality at a lower level than a standard application. Regarding pure input/output (I/O), at least raw IPv6 access must be provided. It is also necessary for RSVP to know, for instance, what interfaces the system has, or through what interfaces a PATH should be forwarded according to IP routing information. Although advanced APIs [13] are being defined for this purpose, they were not available in the IPv6 protocol stack used, so operating system specific interfaces were used instead. This fact implied some important modifications to the MSR's IPv6 stack.

- *Traffic Control for ATM subnetworks.* ISI's implementation introduces an intermediate layer between core RSVP processing and the actual Traffic Control module: the Link Layer Dependent Adaptation Layer (or LLDAL). It also provides a LLDAL and an (almost) empty TC implementation suitable for Ethernet interfaces. In BTI new ATM-specific LLDAL (pertaining to the daemon) and ATM TC module (actually implemented within PATAM) have been developed. The interface offered by LLDAL has been maintained, making it possible for the daemon to work both over Ethernet and ATM interfaces applying the correct Traffic Control to each one.

The ATM TC offers to the ATM LLDAL a simplified and adapted-to-ATM version of the TC interface specified in RFC 2205 [11]. This simplified interface basically allows the daemon to open and close CBR circuits (or leafs of circuits) when reservations must be placed or torn down. Fig. 7 summarizes the modules involved in the TC for the two types of interfaces supported. BTI work has been focused on the remarked modules.

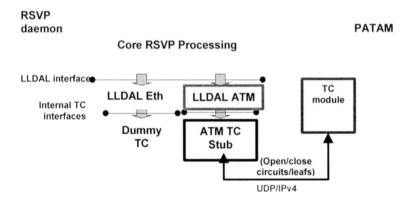

Fig. 7.: Traffic Control Architecture in PATAM

5. Conclusions

Although the timeframe for the development and integration of the protocol stack was very tight (in fact, the whole development was made in less than 8 months), the effort dedicated to the task concluded with a successfully running solution. That solution fulfilled the requirements imposed by BTI network scenario and applications running on Windows NT (videoserver, videoconference and data applications), and made possible the experimentation with the whole BTI system during usability testing phase.

With respect to the practical problems found during protocol development, in summary, most of them were due to the use of state-of-the-art implementations of the technologies being integrated in BTI project. Even some technologies we thought at the beginning that should be stable enough were found to be unstable.

Many conclusions can be drawn from the experience acquired by the development of the protocol stack and the usage trials over the broadband access scenario. Let us outline the most important ones. On the positive side:

- It has been demonstrated the feasibility of implementing dynamically provided end-to-end QoS using IPv6 over ATM, with the added complexity of an underlying VDSL access network. This solution provides a tight way of controlling network resources in an access network.
- IPv6 implementations for Windows NT (from Microsoft Research), Ericsson-Telebit and Solaris proved to be stable and interworked properly. Basic IPv6 functionality could be used in production networks. However, the lack of advanced functions made the integration of the ATM driver and the RSVP daemon difficult and costly. Advance functionalities need to be work out and stabilized before it can be used in production environments.
- The way Neighbor Discovery is organized in IPv6 compared to how it is done in IPv4 has simplified driver development. Instead of having to write an ATM ARP client for address resolution, as it should have been the case for IPv4, we only have to add multicast support to the driver in order to have ND functions working. Some small modifications were needed to the IPv6 protocol stack, for example, to cope with bigger subnet addresses (20 octets for ATM addresses compared to 6 octets MAC addresses). But basically, as all ND functions were located inside the protocol stack, the driver development was simplified.

On the negative side, let us address the following problems and proposed solutions:

- Too many ATM circuits must be setup in order to support IPv6 multicast over ATM using multipoint switched QoS VCs in a full standard way. Solutions to this problem could be: multiplexing different flows over the same ATM circuit, improving multicast signaling over ATM by using only MARS signaling and not the ICMPv6 Multicast Listener Discovery procedure (for example, to avoid the multipoint circuits used only for sending MLD report packets in receiver-only clients), and finally the possibility to order multipoint connections with a given QoS at the MCS over a given subset of group subscribers.
- As stated above, the way IPv6 is conceived streamlines the development of new network drivers for new media if multicast support is one of the targets. Otherwise, for a non-broadcast medium, the implementation of neighbor discovery is costly and treating ATM as a static point to point single LIS link can be worth the extra effort required to simulate multicast at the link layer.
- Lack of integrated QoS APIs. The protocol stack provides two different interfaces: one is used to access IPv6 stack to send and receive data, and the other to access RSVP services to make reservations. Although the interfaces are not complex, the applications are in charge of coordinating the activity between them, and that has been demonstrated problematic for application developers. The use of an integrated network API with QoS support, like the one defined for Winsock2 in [14], will simplify greatly the development or adaptation of QoS aware applications.

Acknowledgements

This work was been partly supported by the EU Commission under the ACTS project 362 BTI. DIT-UPM and IT-UC3M would like to thank all the partners involved in BTI project and also the ACTS Project Manager for their positive collaboration and contribution to the project success.

References

1. Andersen, N., Azcorra, A., Bertelsen, E., Carapinha, J., Dittmann, L., Fernandez, D., Kjaergaard, J., McKay, I., Maliszewski, J., Papir, Z.: Applying QoS Control through Integration of IP and ATM. IEEE Communications Magazine, July 2000.
2. Ginsburg, D.: ATM. Solutions for Enterprise Internetworking. Second Edition. Addison Wesley, 1999.
3. Laubach, M.: Classical IP and ARP over ATM. Request for Comments 1577. IETF Proposed Standard, January 1994.
4. Armitage, G.: Support for Multicast over UNI 3.0/3.1 based ATM Networks. Request for Comments 2022. IETF Proposed Standard, November 1996.
5. Garrett, M., Borden, M.: Interoperation of Controlled-Load Service and Guaranteed Service with ATM. Request for Comments 2381. IETF Proposed Standard, August 1998.
6. Fernández, D., García, A. B., Larrabeiti, D., Azcorra, A., Pacyna, P., Papir, Z.: Bouquet of Multimedia Services for Distant Work & Education in IP/ATM Environment. Pending publication, June 2000.
7. Fernández, D., Bellido, L., Pastor, E.: Session Management and Collaboration in LEVERAGE. First LEVERAGE Conference on Broadband Communications in Education and Training, Cambridge, January 1998. Available at: http://www.dit.upm.es/leverage.
8. PATAM Protocol Stack v0.9, March 2000. Available at: http://www.dit.upm.es/bti.
9. IPv6 implementation for Windows NT. Microsoft Research. Available at: http://research.microsoft.com/msripv6.
10. RSVP daemon. Information Science Institute. University of Southern California. Available at: http://www.isi.edu/rsvp.
11. Braden, R., (ed.): Resource ReSerVation Protocol (RSVP) -- Version 1 Functional Specification. Request for Comments 2205. IETF Proposed Standard, September 1997.
12. Wroclawski, J.: The Use of RSVP with IETF Integrated Services. Request for Comments 2210. IETF Proposed Standard, September 1997.
13. Stevens, W., Thomas, M.: Advanced Sockets API for IPv6. Request for Comments 2292. February 1998.
14. Bernet, Y., Stewart, J., Yavatkar, R., Andersen, D., Tai, C., Quinn, B., Lee, K.: Winsock Generic QoS Mapping. Windows Networking Group.

Transmission of DVB Service
Information via Internet

Artur Lugmayr[1], Seppo Kalli[1]

[1] Technical University of Tampere, Digital Media Institute, P.O. Box 553,
FIN-33101 Tampere, Finland
{lartur, skalli}@cs.tut.fi

Abstract. It is a fact, that more and more applications make use of networks. One of those applications will be digital interactive television, relying on a generic protocol solution to provide broadcasting and interaction over network dependent protocols. A digital television transmission, based on a MPEG2 transport stream consists of multiplexed video, audio and Service Information (SI) streams. Service information is important for describing the content of a transport stream and encapsulating various protocols. Because plenty of research work is done on pure MPEG2 audio and video transmissions, this paper deals with service information transmissions, its mechanisms and network requirements if broadcasted over the Internet only. For this reason a DVB-SI broadcast protocol stack will be introduced. UDP and RTP with its packetization schemes will be evaluated and bandwidth consumptions, boundaries and requirements in a LAN/WAN environment evaluated. Finally a software framework for transmission and decoding SI will be presented.

1 Introduction

Digital television is going to be realized in the next few years and seems to be the biggest revolution in broadcasting since the introduction of color television [4].

The provided standards are (almost) complete and ready to be implemented. For Europe the key standard will be the *Multimedia Home Platform (MHP)*, published by the *European Broadcasting Union (EBU)*. It is relying on appropriate *Digital Video Broadcasting (DVB)* specifications for digital video broadcast and associated interactive services. The MHP is applicable to all DVB defined transmission media and networks such as satellite, cable, terrestrial and microwave [5].

Several topics of and standards of digital television are covered by the "umbrella-name" DVB: interfaces visible to applications, integrated TV sets, multimedia computers, communication protocols, recommendations, interaction, broadcasting, etc.

Figure (1) presents a generic system reference model for interactive DVB systems, consisting of broadcast service provider, interactive service provider, interaction network, broadcasting network, several interfaces and end user. Both, broadcast and interaction channels - established between user and service provider - are presented.

The broadcast channel represents a unidirectional broadband broadcast channel including video, audio and service information. A bi-directional interaction channel is established between the user and the service provider for interaction purposes. It is formed by a return interaction path and a forward interaction path [6]. The channel specific protocols may be split between network dependent and network independent protocols.

DVB relies on MPEG2 for encoding, decoding and transmission of video, audio and services. This section contains a brief description of MPEG2 basics and focuses on the DVB-SI especially.

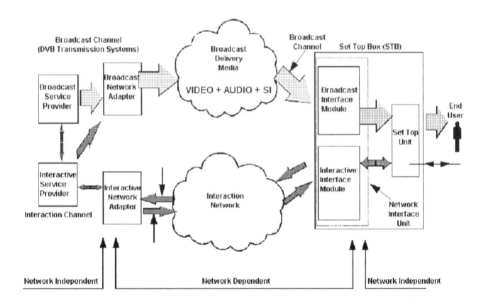

Fig. 1. Generic System Reference Model for DVB [6]

Moving Picture Experts Group (MPEG) is one of the most popular audio/video compression techniques because it is not just a single standard. Instead it is a range of standards suitable for different applications but based on similar principles [7]. Multiple audio, video and service streams, thus program streams, are multiplexed within a Transport Stream (TS). For each program a different compression factor and a dynamic bit rate can be used. This behavior allows dynamical bandwidth handling and is called statistical multiplexing. Besides video and audio a TS contains service information concerning PSI and DVB-SI. Concerning digital video broadcasts service information describes the delivery system, signal content and scheduling/timing of broadcast data streams etc. and includes MPEG2-PSI together with independently defined extension [8].

As mentioned above a TS is a multiplex of video, audio and data. The raw output of an encoder, containing only data to approximate the original video and audio, is called *Elementary Stream (ES)*. Note that the continuous elementary stream contains raw data for one audio/video/data stream only and no error recovery is provided. For practical purposes it is essential to break this continuous elementary stream into packets. Each packet in the *Packetized Elementary Stream (PES)* consists of a header and a payload. The header contains synchronization time stamps and is used for identifying PES packet. The size of each packet varies with the application. Those PES packets are multiplexed within TS packets, which are uniquely identified by its PID.

DVB Service Information (DVB-SI) describes the delivery system, content and scheduling/timing of broadcast data streams etc. DVB-SI is multiplexed in the MPEG2 transport stream and is based on *Program Specific Information (PSI)* and several extensions.

2. Related Work

The MHP is specified in [5] and is a good starting point to get an overview over digital broadcasting standards and specifications. Furthermore this specification defines the interface visible to applications. The MPEG2 based transport stream multiplex is specified by [9]. An excellent technical description covering MPEG2 compression, streaming and testing can be found in [7]. Network dependent protocols in the DVB context address the interaction channels and are described in [6] and [12]. Network independent protocol specifications can be found in [10], [13], [15], [16], [17] and [19], issuing DVB encapsulation in existing networking protocols. A detailed description of DVB-SI can be found in [8] and implementation issues are covered by [18]. *Basso, Cash and Civanlar* addresses the transmission of MPEG2 streams over non-guaranteed quality of service networks. They discuss issues related to the transport of MPEG2 streams over such network by the *Real-Time Transport Protocol (RTP)* [2].

3. DVB Protocol Layers

The DVB protocol stack facilitates the OSI reference model up to the fourth layer. The application layer is not covered by any standards, and applications are left open to competitive market forces [6]. Basically the DVB protocol stack identifies a network dependent (up to the 4th OSI layer) and a network independent protocol (upper layers). The purpose of the different DVB protocol stacks is acting as interface between upper and lower layers and to maintain a logical connection between user and broadcast/interactive server provider, as presented in Figure (1).

It is important to distinguish between different kinds of connections that can be established between broadcast service provider, interactive service provider and end user. The broadcast channel is used as downstream channel for video, audio and data. The interaction network addresses application control data, application communication data and/or data download control [6]. Depending on the transport medium, various protocols for transmission are provided, e.g. [16] addresses the interaction channel through GSM.

Basically two different protocol stacks are specified for combining network independent and network dependent protocol suites: broadcast channel protocol stacks and interaction channel protocol stacks. Figure (2) shows all layers involved in encapsulating a DVB stream within a broadcast medium. The lower four layers are exactly the same as in the ISO-7-layer reference model and consist of physical, data-link, network, transport and application layers. Each layer has its ISO predefined purpose and protocols, depending on the kind of connection established between broadcast service provider, interactive service provider and the end user. A description of ideas, underlying protocols and encapsulation methods of each layer can be found in the following sections.

3.1 Broadcast Channel Protocols

The purpose of the broadcast channel protocol stack is to encapsulate the MPEG2-TS to the broadcast medium. The responsibility of the interaction channel protocol stack is to carry information from the user to the provider and vice.

3.1.1 Broadcast Delivery Layers

The responsibility of the broadcast delivery layer is to establish a point-to-multipoint connection for the delivery of a MPEG2 transport stream between the broadcast service provider and the end-user.

There are many possible network configurations covering the currently specified DVB broadcast options including satellite, terrestrial, cable, SMATV and MMDS in conjunction with PSTN, ISDN, cable and other interactive channel options [5]. As this paper focuses on the transmission of DVB-SI over the Internet, only possible Internet protocols are considered and presented.

The content of the stream may be video, audio and/or data. The main application area and profiles addresses *data piping, data streaming, multiprotocol encapsulation, data carousels and object carousels* [12]. Data pipes support asynchronous, end-to-end delivery of data and is carried directly in the payload of a MPEG2 TS packet. Data streaming supports a streaming-oriented, end-to-end delivery of data in an asynchronous, synchronous or synchronized way. This addresses the ability to broadcast streaming media, which contain video and/or audio. Within the TS multiplex such data would be the PES.

Multiprotocol encapsulation is used to map higher-level protocols in the data portion of lower layer protocols. It is essential for carrying transport units of some protocol within another type of protocol. Especially mutliprotocol encapsulation is essential for two reasons: Firstly it allows to broadcast services that require the transmission of IP-datagrams of various communication protocols. Secondly multiprotocol encapsulation guarantees the transmission of a DVB stream over another protocol type, thus broadcasting over the Internet for example.

A point-to-multipoint transmission of a very large amount of data has to be established between the broadcaster and the end-user. The considered protocol has to be able to establish a point-to-multipoint connection. Thus manly the *User Datagram Protocol (UDP)* and the *Real Time Transfer Protocol (RTP)* can be considered.

The UDP provides an unreliable connectionless delivery service using IP to transport messages between machines. It uses IP to carry messages, but adds the ability to distinguish among multiple destinations within a given host computer [1]. UDP utilizes the networking services provided by the underlying IP layer in addition allows multiplexing, checksum services and distinguishing among multiple processes within the operating system by assigning port numbers to each message.

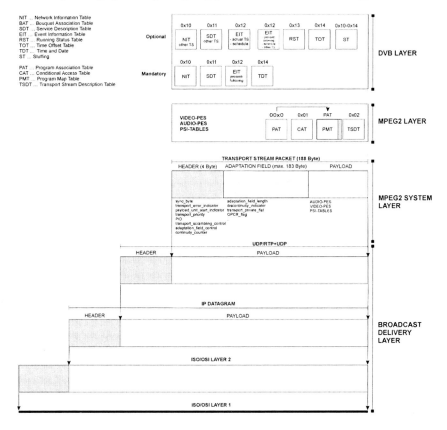

Fig. 2. DVB stream encapsulation by using the Internet as broadcast delivery medium

The *Real Time Transfer Protocol (RTP)* usually relies on UDP and provides payload type identification, sequence numbering, time stamping and delivery monitoring [2]. RTP provides a time stamping service and a synchronization mechanism for jitter measurement. Time stamping is derived from a 90 KHz clock and carried within a 32 bit field. Currently three different types of payload are defined to carry a MPEG2 stream: Firstly the payload based on encapsulated MPEG2 PS and TS as described in [21]. This approach relies on the MPEG2 timing model as based on MPEG2 Program Clock Reference (PCR), Decoding Time Stamps (DTS) and Presentation Time Stamps (PTS) [9], the RTP timestamp mechanism is not used. TS packets are packetized that each RTP packet contains multiple of MPEG2 transport packets. Secondly the payload based on separate packetization of MPEG2 audio and video elementary streams [21]. This approach reduces packet dependencies, maximizes throughput and uses the RTP timing mechanisms through synchronization provided by the *Network Time Protocol (NTP)* [22].

The third approach uses one transmission and one reception port per bundled A/V and services, as audio/video and relating service information are inseparateable and mostly used together. This approach provides an implicit synchronization of audio/video and service information, reduces the header overhead and reduces overall receiver buffer size allowing a better bandwidth control [2]. This should reduce any overheads through redundant information in various layers and provides easier error recovery. According to [2] the overhead can be up to 2-3% of the overall bandwidth.

3.1.2   The MPEG System Layer

Each transport unit within a MPEG2 transport stream is called a *transport stream packet*. Conceptually a transport stream packet consists of three parts: 4-byte header, adaptation field and payload area. The whole length of a transport stream packet is exactly 188 bytes. The length of the payload depends on the length of the adaptation field, the length of which is limited to 183 bytes. Figure (2) shows the header fields of a TS packet. The adaptation field is optional and need not be used at all. It is used if more header information is essential and the header is larger than usually. To maintain the packet length of 188 bytes, the payload is reduced. The extended header information includes priority and clock reference.

3.1.3   MPEG and DVB Layer

The MPEG layer contains demultiplexed audio and video streams as *Packetized Elementary Streams (PES)*. Service information is represented in PSI tables. This layer describes and holds information of only one transport stream. The description of more transport streams can be found in the DVB layer. The DVB layer contains not only SI tables of one TS, but it contains various other tables containing information of a whole program bouquet, networks, satellite transmission, program events and service descriptions. The table and service information mechanism is described in further detail in the next sections.

The raw output of a MPEG encoder is an ES and contains any information needed to approximate the original video/audio. No fault tolerance and no error recovery are provided. The continuous ES needs to be broken in packets out of practical purposes. This stream is called PES.

3.2   Interaction Channel Protocols

For the interaction channel the whole bunch of Internet protocols can be used. The MHP standard allows access to IP, TCP, UDP and HTTP. But there is no limitation in using other protocols that can be encapsulated within IP. MPH also provides a registry mechanism for new broadcast protocols for data broadcast services.

4. DVB Service Information (DVB-SI)

Various tables, defined by ISO/IEC or ETSI, describe DVB-SI. Each DVB-SI carried in transport stream packets has unique PIDs, whether standardized or assigned by the *Program Association Table (PAT)* or *Conditional Access Table (CAT)* described in this section. Each packet carrying DVB-SI must be sent periodically in every TS. ISO/IEC 13818 [9] specifies service information, which is referred to as PSI. The PSI data provides information to enable automatic configuration of the receiver to demutliplex and decode the various streams of programs within the multiplex. For identifying services and events for the user, additional information is needed. This includes information on services and events carried by other transport streams or networks. The coding of this data is defined by ETSI in ETS 300 468 [8].

In the following the most important tables are mentioned:
- *Program Association Table (PAT):* The PAT provides the correspondence between a program number and the PID value of the TS packets that carry the program definition. The program number is the numeric label associated with a program [9]. It specifies the PIDs of all Program Map Tables (PMT) packets.
- *Conditional Access Table (CAT):* The CAT table provides the association between one or more CA systems, their streams and any special parameters associated with them [9]. The CAT always has a PID of 1 and must be multiplexed in every TS.
- *Program Map Table (PMT):* The PMT provides the mappings between program numbers and the program elements that compromise them. Thus, the PMT is the complete collection of all program definitions for a simple TS [9].
- *Network Information Table (NIT):* The NIT conveys information relating to the physical organization of the multiplexes/TSs carried via a given network and the characteristics of the network itself. Through the combination of two fields in the NIT section (original_network_id and transport_stream_id) each TS can be uniquely identified [8].
- *Service Description Table (SDT):* Each sub-table of the SDT describes services that are contained within a particular TS. The services may be part of the actual TS or part of other TSs, these being identified by means of the table_id [8].
- *Event Information Table (EIT):* The EIT provides information in chronological order regarding the events contained within each service [8]. The EIT contains information about the current TS or other TSs. It covers present/following event information or event schedule information.

Sections defined by the broadcaster can be sent via private sections. They can occur in transport streams, that are exclusively assigned for sending private sections, or in various other streams, who's PID is defined in the PMT.

4.1 Decoding of DVB-SI

The first issue in demultiplexing a transport stream is to look for PIDs 0 and 1 in the multiplex. Packets with the PID 0 contain the PAT, whereas the PID 1 is carrying the

CAT table. Upon reading the PAT, the PIDs of the NIT and each PMT can be found. By decoding the PMT, the decoder is able to obtain each PID of each elementary stream. If programmes are encrypted, the CAT has to be accessed; therefore the CAT section at PID 1 has to be read. For the whole DVB-SI more tables have to be decoded and read. The previous four tables describe only information to demutliplex various streams of programs within the multiplex and therefore called program specific information. Figure (2) shows the process of decoding PSI data from a transport stream. More detailed information for identifying services and events for the user can be found in other DVB-SI tables.

Each DVB defined service information table might consist of a descriptor loop. The descriptors identify the content of the current table and are defined in various ETSI standards. Each descriptor consists of a descriptor_tag for identifying the descriptor, the descriptor_length and the specific descriptor data – for example the title of the current movie or the classification information for a current event is classified by its descriptor_loop.[1]

4.2   Mapping of Sections in Transport Stream Packets

It is essential to distinguish between tables and sections. A section is a syntactic structure used for mapping all predefined service information into transport stream packets [8]. The length of one section is variable, but does not extend 1024 bytes, except the EIT, whose maximum length limit is 4096. Whereas a table consists of one section minimum and is uniquely identified by the table_id. The content of the table depends on its use and its definition. A table provides information about a certain matter, addressing various needs of a DVB transmission. Each service information section is uniquely identified by table_id and table_id_extension. All sections are directly mapped into transport stream packet, without any mapping into PES. The start of a section in the payload of a transport stream packet is pointed to by the only one pointer_field or by default at the beginning of the payload. One pointer field is enough, as the length of each section can be estimated by the length field of each section and as one restriction of the MPEG standard issues, that no gaps between sections within a transport stream packet are allowed. Each section has to be finished, before the next one can start in packets with single PIDs, as otherwise it is not possible to identify to which section the data belongs.

In case a section does not completely fill up one transport stream packet and it is not advised to start a new section, stuffing has to be used to fill up the free space. There are to ways to perform stuffing: Firstly through the adaptation field mechanism, secondly by filling remaining bytes with the value 0xFF. The decoder can discard the bytes with the value 0xFF. Therefore the table_id 0xFF is prohibited.

---

[1] Example: The EIT consists of a content_descriptor to provide classification information for an event. If the current stream carries a romance, this is classified by its content_descriptor.

5. Requirements in Broadcasting a SI Stream

Network requirements in broadcasting a DVB stream are related to signal acquisition time, bandwidth utilization, table repetition rate and protocol performance. These concepts are presented more detailed in the following:

Signal Acquisition Time: Signal acquisition time is the time needed to decode a particular table as part of the SI. It is assumed in [9], that the worst-case contribution to signal acquisition time for retrieving PSI is approx. 80ms based on a 25Hz PSI frequency. Signal acquisition time depends on protocol performance, implementation, hardware configuration and error recovery.

Repetition Rate: The repetition rate addresses the frequency at which each service information table is sent. MPEG2 defined PSI tables should be sent with a frequency of 25Hz. For DVB rates only minimum repetition rates are defined. The NIT, BAT, SDT for other TSs, EIT for other TSs and the EIT for the first 8 days ahead shall be transmitted every 10 seconds. Whereas SDT and EIT for the actual TS should be transmitted at least every 2 seconds. TDT, TOT and EIT for further than 8 days need only to be transmitted every 30 seconds.

	1			10			128		
	PSI	**DVB**	**Sum**	**PSI**	**DVB**	**Sum**	**PSI**	**DVB**	**Sum**
1	3008	155063	158071	4512	234692	239204	33088	18305184	18338272
*	75200	73397	148597	112800	101104	213904	827200	9132741	9959941

Table 1. Comparing bandwidth usage (in bps) of DVB SI with frequency table sent and number of programs per TS. "*" frequency marks the minimum frequency all tables must be sent according the standards

Bandwidth Utilization: Bandwidth utilization varies in number of programs, MPEG bandwidth for audio/video, frequency of sending service information tables, size of tables and number of encapsulated IP datagrams and private sections. Previous experiments already evaluated the performance issues of different MPEG2 packetization schemes for video/audio [2]. The following bandwidth calculations assume that the transmission of minimum SI without any protocol overhead, conditional access, scrambling and stuffing tables. The following are based on following assumptions: Average descriptor length of 20 bytes, 10 descriptors per descriptor loop and minimum SI information sent. This means minimum additional information in various table sections; furthermore no information about other TSs or networks, no user defined tables and encapsulated IP datagrams are included. Each section is fitted into a single TS packet, which is filled up with essential stuffing bytes. The number of TS packets needed to transmit one-table results in the bandwidth usage (bps). Only one section can be fitted into each packet.

Table (1) shows the MPEG and DVB related bandwidth usage. Generally the SDT and the EIT are the largest tables in the multiplex. Assuming ten 8Mbps programs (MPEG video/audio at high quality), the total SI bandwidth would be 213,904bps. This leads to an SI overhead of about 0.27%.

Protocol performance can mostly be evaluated on practical tests as it highly depends on the implementation of the protocol stack, packetization scheme and network load. One way of measurement is to set up criterion addressing the whole bandwidth of disturbance that might occur in transferring service information. This includes packet loss, counting the number of transmission disturbances at the client side and packet acquisition time delay in a LAN and WAN environment.

6. Implementation Issues

The implementation platform is a Intel Pentium II with Windows98 as operating System. For retrieving a real MPEG2 TS a COCOM CC1016 DVB-S data receiver connected to a satellite dish was used.

6.1 Software APIs

The whole implementation is based on Java with native interfaces to the SDK to the DVB-S receiver card. It can be divided in four packages: *COCOM Drivers/SDK, CC1016 API, MHP SI Access API and Application API.*

The COCOM Drivers/SDK API provides a wide range of functionality: selection channel parameters and frequencies, open/close a TS packet stream, add/remove PIDs to each stream, stop/start data reception, read data, get statistics and various system status calls. This API represents the lowest layer of the implementation and allows access to the data received by the COCOM DVB-S receiver card.

The objective of the CC1016 API is to provide access to data in MPEG sections. It should allow a convenient application-level access to data transported in MPEG sections and filtering mechanisms. Furthermore it provides mechanisms to retrieve pure PES packets, TS packets and channel navigation possibilities. Through the implementation is purely in Java and to provide fully functionality for upper software layers several access methods of the COCOM SDK are included by the use of Java native interfaces.

The MHP SI Access API is completely MHP (Revision 16/Section M) compliant. It makes use of the CC1016 API as well as the COCOM SDK. Its objective is to provide an accurate SI access. This includes parsing of SI tables, managing and monitoring SI, handling SI requests and initializing the whole filtering process.

The Application API is multifunctional. Its responsibility is to provide an accurate user interface, controlling the initialization and configuration process, managing network connections, accessing lower layer APIs and retrieving SI out of a file or from the DVB-S adapter. Currently there are three major possibilities in retrieving SI: from a file, server or directly from the DVB-S adapter. SI can be visualized and accessed by sending requests to the MHP SI Access API - independent by which way SI is obtained - by a very simple user-interface.

The generation of SI service out of a file bases on minimum repetition rates and table sizes as defined in the file. This included the sending of PAT, PMT, NIT, SDT and EIT per TS. The length of the tables was restricted to their minimum lengths[2].

---

[2] 2 byte for program associations; a PMT size of 26 bytes; an EIT size of 4048 bytes, assuming average descriptor length; NIT, TDT and SDT of their minimum size;

The applied encapsulation schemes have been described in previous sections. UDP and the three different packetization schemes of RTP have been used. In one UDP packet one TS packet got encapsulated.

6.2   Process Flow

Figure (3) shows the data flow diagram of the whole system. Grey boxes illustrate SI retrieving elements, which represent source nodes for SI information. Setting initialization and configuration options start the whole process. This information defines which TS is obtained by defining its frequency and channel parameters. These parameters are static during the whole process. For the file version and the client version these information does not have to be set at all, as the server or the content of the file specifies it. The same applies for navigation information, which is also static during the whole process. Navigation information includes mainly filter options.

Filter options are *pure PID filtering, PID filtering with a positive filter* and *PID filtering with a negative and positive filter.* The first method filters only objects, which match a specific PID. It also allows getting raw stream of TS packets with this certain PID, if no table identifier is specified. The other two methods allow section filtering by specifying filter patterns. Positive filtering means, that a certain filter gets only triggered if the defined values match the specified bytes in the section. The third filter does completely the opposite. It gets triggered if the defined pattern does not match. This can be used to retrieve a section when its header changes (e.g. version_number).

The filtered SI stream can be treated in several ways: A UDP/RTP server can forward encapsulated SI by using mutliprotocol encapsulation or it can be passed directly to the MHP SI Access API. Encapsulating raw stream of TS packets is the simplest method: One UDP/RTP packet contains one or more TS packets. In the current application one TS packet can be sent in one UDP packet. If section filtering is applied, one specific section has to be encapsulated in one UDP/RTP packet. The maximum payload length is limited to 4048 bytes, as the EIT section has this maximum length. The client side retrieves the encapsulated sections or TS packets and has to unpack the payload of RTP and UDP packets. If the payload contains whole sections they can be passed easily to the MHP SI Access API. Otherwise, if the payload contains TS packets, they have to be parsed to obtain whole sections, which can be passed to the MHP SI Access API.

The MHP SI Access API represents a lightweight database, which stores the whole SI. Applications have the possibility to request information by several criteria (e.g. table_id, transport_stream_id, PID, description tags).

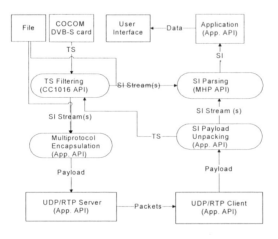

Fig. 3. Data Flow Diagram of the whole System

7. Results

Various test constellations have been applied. The first was based on a generation of SI by a file and forwarding them to LAN/WAN clients. The second test consisted of a forwarding of a whole feed of SI from a dedicated satellite card. Each test was carried out under consideration of UDP and the different packetization schemes of RTP. Preliminary assumptions were that no CAT access, no transmission of private sections and encapsulated IP datagrams were included. The criterion to analyze protocol performance was to measure the number of disturbances at the client side. This was done by setting up the client in a LAN and WAN environment under usual network load. Bandwidth measurements were performed under typical internet/intranet network load by using a network sniffer tool. The repetition rate depended on the kind of experiment: The generation of a service information stream out of a file assumed minimum rates, whereas the forwarding of service information form the COCOM card relied on the transmission rate broadcasted by the satellite. As estimated, UDP encapsulation showed the best performance values, through its less header overhead. But RTP offered a more robust behavior concerning reliability and synchronization possibilities with payload-based packetization and joint packetization. The whole service information could be retrieved in a satisfying amount of time, even through some packet losses and re-readings of information tables. The main problem that occurred was in network environments with high packet losses, where lost packets lead to a high amount of table re-readings. But as the DVB environment is based on a not reliable transmission media packet losses are always expected. Packet losses lead to longer acquisition times of the tables and therefore the frequency in sending service information is assumed to be high enough. The encapsulation of one TS packet per

UDP/RTP packet caused a big header overhead and is avoidable. As suggested in [24] more TS packets could be encapsulated in one UDP/RTP packet. Very surprisingly Java was fast enough to generate service information in appropriate time and also the forwarding time delay caused by the server was negligible. The reason for this is the native interface to C++ source code in an implementation.

8. Conclusions and Future Work

This paper considered the bandwidth utilization of a DVB service information stream in a LAN and WAN environment. The utilization highly depends on the number of programs and information that is sent to the end-user. It showed that not only the video/audio stream needs its bandwidth - also the service information transmissions demand its part. Future work will focus on the demand of multiprotocol encapsulated IP over a broadcast network and its demand in bandwidth, signal acquisition time and reliability of transmission of different protocol types. Especially multiprotocol encapsulation issues this matter. Reliability is essential because a broadcast media does not allow "conventional" acknowledgement schemes. Therefore a reliable transmission has to be established on the basis of other mechanisms. Their exploration and practical studies is another major subject[3].

References

1. Comer, E.D.: Internetworking with TCP/IP; Vol. I: Principles, Protocols, and Architecture. Prentice-Hall Inc. 1995.
2. Basso A.; Cash G.L.; Civanlar M.R.: Transmission of MPEG-2 Streams over Non-Guaranteed Quality of Service Networks. Proceedings of Picture Coding Symposium: Berlin Germany 1997.
3. Krupczak, B.; Calvert, K.; Ammar, M.: Protocol Portability through Module Encapsulation. Network Protocols, IEEE Proceedings: pp. 56-63, 1996.
4. Fox, B.: Digital TV comes down to earth. IEEE Spectrum: October, 1998.
5. TAM232r16: DVB: Multimedia Home Platform; Revision 16. European Broadcasting Union: February 2000".
6. ETS 300 802: DVB: Network-Independent Protocols for DVB Interactive Systems. European Telecommunications Standards Institute: November 1997.
7. Tektronix: A Guide to MPEG Fundamentals and Protocol Analysis. Tektronix Inc.: 1997. http://www.tektronix.com.
8. ETS 300 468: DVB: Specification for Service Information (SI) in DVB systems. European Telecommunications Standards Institute: January 1997.

[3] *Acknowledgements:* The FutureTV project is funded by the National Technology Agency of Finland together with major Finish television, telecommunications and digital media companies, especially YLE. We would like to thank especially Teemu Lukkarinen for all his help and the whole FurtureTV tream: Stephane Palomba, Florina Tico, Jukka Rakkola, Olli Savia, Olli Stroem and Chengyuan Peng.

9. ISO/IEC 13818-1: Generic Coding of Moving Pictures and Associated Audio: Systems. International Organization for Standardization: November 1994. Recommendation H.222.0.

10. ETS 300 800: DVB: Interaction Channel for Cable TV Distribution Systems. European Telecommunications Standards Institute: January 1997.

11. Veeraraghavan, Malathi; Karol, Mark: Internetworking Connectionless and Connection-Oriented Networks. IEEE Communications Magazine: Vol. 37, Issue 12; pp. 130-138; December 1999.

12. EN 301 192: DVB: Specification for Data Broadcasting. European Telecommunications Standards Institute: December 1997.

13. ETS 300 801: DVB: Interaction Channel through Public Switched Telecommunications Network (PSTN)/Integrated Services Digital Networks (ISDN). European Telecommunications Standards Institute: August 1997.

14. ETS 300 468: DVB: Specification for Service Information (SI) in DVB systems. European Telecommunications Standards Institute: January 1997.

15. EN 301 193: DVB: Interaction Channel through DECT. European Telecommunications Standards Institute.

16. EN 301 195: DVB: DVB Interaction Channel through GSM. European Telecommunications Standards Institute.

17. EN 301 199: DVB: DVB Interaction Channel for LMS Distribution Systems. European Telecommunications Standards Institute.

18. ES201218: DVB: DVB Interaction Channel for LMS Distribution Systems. European Telecommunications Standards Institute.

19. TR 101 201: DVB: DVB Interaction Channel for SMATV Systems. European Telecommunications Standards Institute.

20. Civanlar, M. Reha; Glenn, L.; Cash; Barry, G.: RTP Payload Format for Bundled MPEG. draft-civanlar-bmpeg-01, Internet draft. February, 1997.

21. Hoffman, D.; Fernando, G.; Goyal, V.; Civanlar, M.R.: RTP Payload Format for MPEG1/MPEG2 Video. draft-ietf-avt-mpeg-new-01, Internet draft. June, 1997.

22. RFC1305: Network Time Protocol (Version 3); Specification, Implementation, Analysis. David L. Mills: March, 1992.

23. Sarginson, P. A.: MPEG-2: A Tutorial Introduction to the Systems Layer. The Institution of Electrical Engineers: pp. 4/1-4/13; 1995.

24. Chengyuan Peng; Petri Vuorimaa.: A Digital Television Navigator. ACM Multimedia 2000: March 20th, 2000.

Measurement of the Performance of IP Packets over ATM Environment via Multiprotocol-over-ATM (MPOA)

Dipl.-Ing. Kai-Oliver Detken
Director *wwl network*, Bremen

WWL Internet AG,
Goebelstr. 46, D-28865 Lilienthal/Bremen, Germany
kai.detken@wwl.de,
Business: http://wwl.de
Private: http://kai.nord.de

Abstract. Multiprotocol-over-ATM (MPOA) is a service which supports layer-3 internetworking for hosts attached to ELANs (running a LAN Emulation Client-LEC), hosts attached to ATM networks, and hosts attached to legacy LANs. MPOA provides and delivers the functions of a router and takes advantage of the underlying ATM network as much as possible. That means, MPOA works as a virtual router with real QoS features of ATM. Furthermore, the integration of existing network with reduced protocol overhead and reduced equipment is a crucial point in terms of network performance cost efficiency. But MPOA has already some disadvantages, because MPOA is a very complex technology and the success of MPOA is doubtful. IP might be worth the complexity because it is so widely used, but it can be doubted if this holds true for other layer 3 protocols as well. Nevertheless, the MPOA model is a very promising technology which has also negative aspects like very complex software implementation (hard to handle) and QoS can only be supported if you use ATM network interface cards (NICs) directly without Ethernet. This paper will discuss the measurement results (latency, throughput) from the Internet Protocol (IP) in ATM and Ethernet environment via MPOA. Furthermore this paper will give an overview about available ATM components regarding MPOA and the further development of this technology.

1. Introduction

Witnessing this latest round of ATM switch tests at the WWL, it is difficult to compare ATM core switch products against those from the Gigabit-Ethernet camp, because of the fast, simple and efficient technology from Gigabit-Ethernet. However, ATM is not a dead issue; it's just not in favor anymore. The reason is, ATM is too hard to use. A big disappointment, however, was support for the Multiprotocol-over-ATM (MPOA) specification. Originally, the WWL wanted to measure several components from different manufactures and vendors. But out of more than eight invites sent out, only two vendors was ready to come into the lab and demonstrate an MPOA solution. These vendors were Cisco Systems and Cabeltron Systems (today: Enterasys). What made MPOA so important to companies and customers is that this was the great hope of the ATM Forum to integrate legacy network protocols with ATM's cell traffic structure. Using an MPOA server, network managers would in theory be able to use all the Layer 3 switching features common to Layer-3-Ethernet switches, plus a few other goodies.

For example, MPOA finally made it feasible to route ATM and other protocols into the same VLAN. If this seems like the MPOA server would become a traffic bottle-neck in large networks, it would be wrong. MPOA also makes use of PNNI to provide wire-speed routing across the entire ATM segment simply by establishing direct cli-ent-to-client connections when congestion gets ugly. The WWL used for the tests different packet sizes, because of exact measuring of performance and latency. Frag-mentation cause by to big packets, influence the measuring results dramatically. Typi-cal IP packet sizes are 64 byte, 512 byte, and 1518 byte. Small packets are usual for client-server communication regarding databases. This load is normal for a network and decribe the daily operation.

On the other hand, packets with a size of 512 byte are typical for the smallest IP packet size. Additionally, 1518 byte fits exactly in the maximum size of an Ethernet frame. The packets have not to fragment in smaller packets. For automatically connec-tions in LANE/MPOA scenarios, the LEC/MPOA client must communicate with the LES/MPOA server. Between this communication an additional measurement equip-ment (in our case: WinPharaoh from GN Nettest) was used. Additionally, the traffic was produce from an analyser (in our case: Smartbits 2000 (SMB-2000) from Net-Com) at the edge of the network on the Ethernet site. Thereby we was able to measure the traffic and the performance end-to-end.

By the test scenario in Figure 1 the WWL was able to test the following parame-ters:
- Performance Ethernet-Ethernet
- Latency
- Frame loss

MPOA was used with its specific protocols like P-NNI, LANEv2, NHRP. For the test of the shortcut functionality the MPC devices got different values for the Shortcut Frame Count, especially at the Cabletron equipment. This parameter is the threshold for the number of frames per second and is responsible for the establishing of short-cuts. We recommend an value of 10 for a fast establishing of shortcuts. Normally the value has the number 65000.

This paper will show by the use of two different manufacturers that MPOA is work-ing, fast, sufficient, and efficient. But also, the paper will show the weakness of the products or the MPOA approach. Additionally, we tested quality-of-service (QoS) features with NICs from Fore Systems and Olicom. Therefore, mainly two vendors Cisco and Cabletron was tested at the WWL for our customer and a evaluation of this technologies, which has an deep impact of the development of QoS and Multi-Protocol-Label-Switching (MPLS).

2. Comparison of LANE and MPOA

For the adaptation or integration of IP into ATM, there are different methods existing developed by the ATM-Forum (LANE and MPOA) and IETF (CLIP and Multiprotocol Label Switching). For this paper, LANE and MPOA have been tested compared with CLIP, because they are based on each other, they are in ATM products currently available and support legacy LANs such as Ethernet and Token Ring.

LAN Emulation (LANE) is after CLIP the second solution for IP-over-ATM and can be best characterized as a service developed by the ATM Forum that will enable existing LAN applications to be run over an ATM network. In order to do so, this service has to amalgamated the characteristics and behaviors of traditional Ethernet, Token Ring and FDDI networks. Moreover, it has to support a connectionless service as current LAN stations send data without establishing a connection before the operation takes place. Therefore, LANE must support broadcast and multicast traffic such as the kind of traffic allowed over shared media LANs. It has to allow the interconnection of traditional LANs with the amalgamated LAN and at the same time to maintain the MAC address identity associated with each individual device that is attached to a LAN. Finally, it has to protect the vast installed basis of existing LAN applications and enable them to work in the same way as before over an ATM network. This may be put into practise over the OSI layer 2, whereas the LANE technology can be understood as a bridging technology.

MPOA clients established VCCs with the MPOA server components in order to forward data packets or to request information. Using these components, the client is able to establish a more direct path. MPOA supports various kinds of routable protocols (IP, IPX, AppleTalk, etc.), integrates existing internetworking protocols (RFC-1577, RFC-1483, NHRP, MARS, RSVP) and the IETF and ATM Forum solutions (LANE, P-NNI) into a virtual router environment. Yet actually, MPOA supports only IP and based on LANE, RFC-1483, and NHRP. MPOA is a service with layer 3 internetworking support for hosts attached to ELANs, ATM networks and legacy LANs. Thus, the real premise behind MPOA is to provide and deliver the functionality of a router and to take as much advantage of the underlying ATM network as possible. MPOA works as a virtual router on the OSI layer 3. [1]

2.1 TCP/IP-over-ATM

IP is an important protocol used to achieve interoperability in a heterogeneous network. In contrast, the effectiveness of IP in high speed networks is not widely known. There is much doubt about the TCP performance, in particular, as the acknowledgement mechanisms, overhead size and parameter setting are considered to be obstacles. UDP is more appropriate for real-time data streams, but does not have any security mechanisms.

The TCP/IP protocol stacks were not designed for high speed performance networks in the first place. Several extensions of TCP protocols have been suggested in order to achieve higher performance over these networks and to improve the performance of connections with a high bandwidth delay. New bandwidth-intensive applications such as multimedia conferencing systems and characteristics of high speed networks have triggered research on advanced transport protocols. Therefore, the discovery of additional error sources is not surprising. Following, the reader will find factors identified by WWL that are responsible for inefficiencies of TCP protocols over ATM: [3]

- Send and receive socket buffer size
- Network: Maximum Transport Unit (MTU)
- Protocol: Maximum Segment Size (MSS)
- Transmitter: use of Nagle's algorithm
- Round Trip Time (RTT)
- Receiver: delayed acknowledgement mechanisms
- Transmitter: Silly Window Syndrome (SWS)
- Copy strategy at the socket interface
- Network congestion and lost notice

2.2   Scenarios for MPOA

The following ATM devices were used during the tests in order to measure the performance of MPOA and LANE:

1. Smart Switch 2200 (MPOA Client – MPC): Firmware Release No.: 40726; 24 x 10/100 Mbps Ethernet Ports (not modular); 1 x FE-100TX(10/100 Mbps Ethernet Module); 1 x APIM-21R (ATM-Module)
2. Smart Switch 2000 (MPOA Server – MPS): Firmware Release No.: 40771; 1 x FE-100TX(10/100 Mbps Ethernet Module); 1 x APIM-21R(ATM-Module); Device Hardware Revision: 00C; Device Firmware Revision: MPS_1.0.1; Device BOOTPROM Revision: 01.05.02
3. Smart Switch Router 2000 (MPOA Router): Router Device connected to MPS via IEEE 802.1Q; Firmware Release : ssr22a2; 2 x 8 10/100 Mbps Ethernet Ports (not modular); Software Version: 2.2.A.2; Boot Prom Version: prom-1.1.0.5; System Type: SSR 2000, Rev. 0; CPU Module Type: CPU-SSR2, Rev. 0; Processor: R5000, Rev 2.1, 159.99 MHz
4. Smart Switch 2500: ATM Switch with LECS, LES/BUS device; Firmware Release: 02.03(35); 3 x 4 OC-3 Ports (ZX-IOM-21-4); CPU Model: i960 CX; CPU Speed : 33 MHz
5. Catalyst 5506 (LEC ELAN 1, MPC-A): 100BaseTX Supervisor (Hardware 2.3; Software 4.5(3)); 10/100BaseTX Ethernet (Hardware 1.2; Software 4.5(3); MM OC-3 Dual-Phy ATM (Hardware 3.0; Software 11.3(5))
6. Catalyst 5500 Inge (LEC ELAN 2, MPC-B): 10/100BaseTX Supervis (Hardware 3.1; Software 4.5(2)); 10/100BaseTX Ethernet (Hardware 1.2; Software 4.5(2)); MM OC-3 Dual-Phy ATM (Hardware 2.0; Software 11.3(8)); ASP/SRP
7. Catalyst 5500 "C50-01" (LECS, LES, BUS, MPS, Router): 100BaseFX MM Supervis (Hardware 2.2; Software 4.5(3)); 10/100BaseTX Ethernet (Hardware 1.2; Software 4.5(3)); Route Switch Ext Port; Route Switch (Hardware 7.0; Software 12.0(6.5)); MM OC-3 Dual-Phy ATM (Hardware 2.0; Software 11.3(10); ASP/SRP

8. ATM NIC from ForeRunnerLE series with throughput of 155 Mbps, bus architecture PCI 2.0/2.1, 32-Bit, 33 MHz PCI Bus and Windows95/NT 3.51/4.0, NDIS 4.0 und NDIS 5.0
9. ATM NIC from Olicom Rapid Fire OC-6162: 155 Mbps over MMF (SC Connector), NDIS 3.0 NIC driver

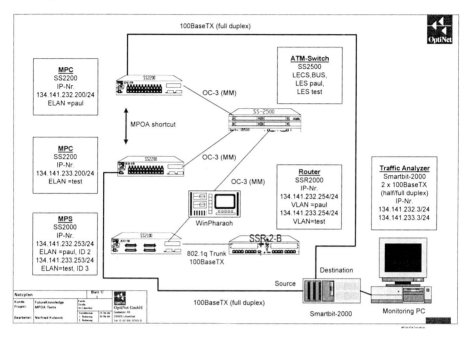

Fig. 1. MPOA scenario from Cabletron Systems (Enterasys): For the measurements with and without MPOA shortcuts, we used different shortcut frame counter for the MPOA devices. This value is responsible for the number of frames per second (fps) for the threshold, which establish a shortcut. During the tests the shortcut threshold was 10. This scenario makes it possible to establish the shortcut directly between MPC 1 and 2. On the other hand, the traffic between the SSR and ATM switch has been controlled by the Win Pharaoh. The Smartbit equipment was responsible for the traffic generation and the analysis of the test information.

As measurement equipment we used the following software and hardware from GN Nettest and NetCom:
1. Smartbits 2000 (SMB-2000): Firmware 6.220012; 2 x ML-7710 (Ethernet); 2 x AT-9155C(ATM); 1 x AT-9622(ATM); Software: SmartApplications ver. 2.22; Smart Flow 1.00.010 Beta; Library: 3.07-48; SmartSignaling 2.1(Missing authorization file, will run in demo mode); SmartWindow 6.51
2. Win Pharaoh: Hardware: LAN, WAN and ATM Line Interfaces, ISA-Bus; Prozessor: 10 MIPS RISC Prozessor; 16 MByte On-Board RAM; Software: based on Windows; works on a laptop, PC or Rack-Mount-PC; ATM Remote Software; ATM Site License; ATM Corporate License; Adapter: LAN: Fast-Ethernet, Token Ring and FDDI; WAN: RS-232, RS-422, RS-449, RS-530, V.35, X.21, V.10, V.11, Basic Rate ISDN, Primary Rate ISDN/T1 and Primary Rate ISDN/E1; ATM:

155 MBit/s OC3c/STM-1 single mode and multi mode, 155 MBit/s UTP-5, DS3/DS1, E3/E1

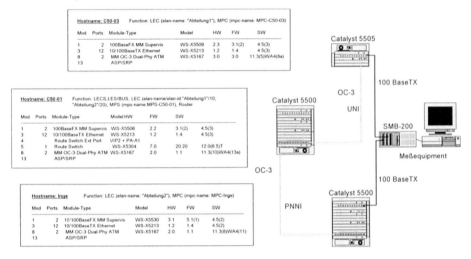

Fig. 2. MPOA scenario from Cisco Systems: The measurements with the components from Cisco Systems were realised with the Catalyst5xxx series. This are the only switches from Cisco, which support MPOA. Additionally, the router Cisco 4500, 4700, 7200, and 7500 have been integrated MPOA, too. For the tests two Catalysts worked as MPC, LEC, and Ethernet access switch. Furthermore two devices worked as MPS, LECS, LES/BUS, and router. Because of the close time frame, there is only a MPOA shortcut within the ATM switch possible. For the preparation of a MPS a RSM with VIP2 (Versatile Interface Processor) module with one ATM NIC (PA-Ax) is necessary. Two modules were devided one slot and were connected via the backplane with the Supervisor Engine and the RSM.

2.3 TCP/IP-over-ATM scenario

The tests were carried out over a 155 Mbps connection between two separate Pentium -II/350MHz and Pentium-II/266MHz workstations with operating systems Windows98 and Windows NT 4.0. Both clients had an ATM Network Interface Card (NIC) from Fore Systems (ForeRunnerLE155) and also the RapidFire6162 ATM 155 PCI Adapter from Olicom. The central ATM switch was Cabletron's SmartSwitch2000 with a single ATM module and four OC-3 interfaces. AAL-5 was used for all measurements to encapsulated IP packets. The sender buffer varied between 16-64 kbyte, whilst the receiver buffer was always of the same value. Classical IP (CLIP) was used for the scenario without any routing. Therefore, it was a point-to-point connection between both workstations under optimal conditions.

Netperf, originally developed by Hewlett Packard, was the software used for test-purposes on both clients. It is a benchmark program that to measure the performance of networks. Measurement opportunities for burst traffic and request/response performance are the main tasks of the program. Optional extensions of the software are available. In contrast to the full Netperf version that is only available for Unix operation systems, the used test-version was especially designed for Windows end systems.

2.5.1 MPOA Measurement results

Cabletron Systems (Enterasys) latency test without MPOA: The measurements were done by increasing of the throughput by 10 Mbps steps up to 100 Mbps. The duration of one test run was 10 sec, because of the total time of the measurements. In this diagram we used unidirectional connections established by the Smartbit.

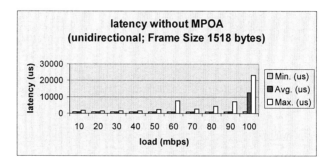

Fig. 3. The frame size differ between 64 to 1518 byte. Here you see only the 1518 byte test results, because is the significant size in the network area. Appropriate without MPOA there should not appear high latency. This is the fact, because only at 90-100% load there were higher values. In this case, the test results differ between zero and approx. 23000 μsec.

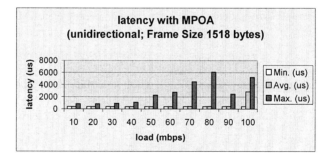

Fig. 4. Cabletron Systems (Enterasys) latency test with MPOA: The direct connection between virtual clients is different, established by a shortcut with MPOA. Here were only approx. 6000 μsec latency measured as maximum delay time. This was noticed at a load of 80%. Therefore, the delay has been minimised by MPOA from approx. 23000 to 6000 μsec.

117

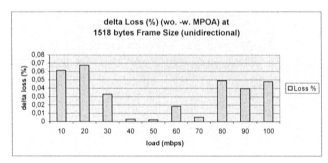

Fig. 5. Cabletron Systems (Enterasys) packet loss test with/without MPOA: If you measure the delay, you have also to keep in mind the frame losses. Here you can see the test results for Cabletron's components for MPOA (w. = with) and without MPOA (wo. = without). For packet sizes of 64 byte, there was no packet losses recorded. If there is a load of 70% there was in every case no packet losses. The value of packet losses was under 1% in every case of this measurement.

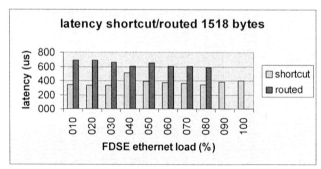

Fig. 6. Cisco Systems latency test with/without MPOA: In the first step, the latency was measured with MPOA and a load of 10-100%. Here we divided the packet sizes in 64, 128, 256, 512, 1024, 1280, and 1518 byte. So-called Under-runs (frame sizes under 64 byte) were rejected, while Giants (frame sizes over 1518 byte) were ignored by the LANE module. In the second step, the latency of the routed path was measured. Here, the router showed after a frame frequency of approx. 65000 fps, that packets were deleted. This was the reason, why no test results are there. According to Cisco Systems, this behaviour was caused by the ATM module PA-A1, which has not sufficient performance.

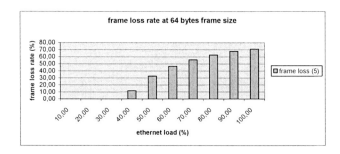

Fig. 7. Cisco Systems packet loss test with/without MPOA: In the direct comparison between with and without MPOA, you can see at 64 byte, that the routed packets need essential more time than the shortcut packets. After a load of 40% there were not measured any values, because of the frame losses.

2.6   TCP/IP-over-ATM performance

The measurement of TCP/IP-over-ATM was carried out by using various packet sizes in order to represent the effectiveness of IP-over-ATM. The throughput alone was not really interesting, because of the different bottlenecks that were already mentioned before. The fluctuations and throughput breaks were more important. Classical IP (CLIP) has been used for this measurements point-to-point, because of the highest effectiveness. The test phase had a time duration of 60 seconds for each measurement via Netperf.

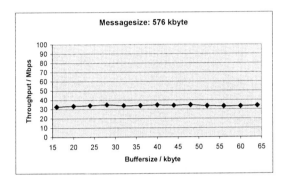

Fig. 8. IP-over-ATM, 576 byte: This figure shows the TCP/IP measurements with a packet size of 576 byte via CLIP. The buffer varies from 15-64 kbyte. The packet size of 576 byte represents a normal datagram in the Internet environment. Figure 8 demonstrates the effectiveness of the achieved throughput during a normal point-to-point session with minimal overhead. By the less datagram size the effectiveness went down to approx. 35 Mbps.

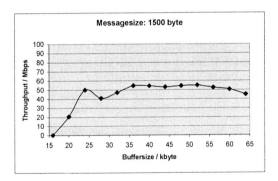

Fig. 9. IP-over-ATM, 1500 byte: Additional breaks and interruptions happened with small and big packet sizes. The MTU of 1.500 byte required a fragmentation of the packets and this has to be considered the reason for it. If the fragmentation was increased the performance went down. Figure 9 shows a differing result. In this case, the packet size was 1.500 byte which allows higher data rates then before. The results are only limited to approx. 60 Mbps. The measurements show some fluctuations, because of the different buffer sizes, which was used.

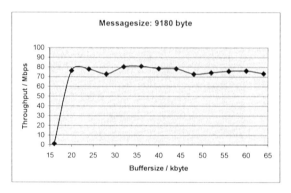

Fig. 10. Fluctuations happened, especially at a buffer size of 55 kbyte. The best results were achieved at 40 and 65 kbyte. A fundamentally improved performance was achieved with a packet size of 9.180 byte. The throughput increased up to 80 Mbps. Fluctuations happened, again (especially on less buffer sizes), like during the other measurements. The small fragmentation of the packets has to be considered the reason as the MTU size was 9.180 byte.

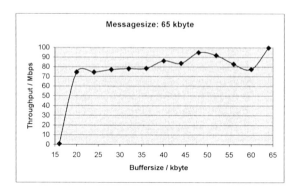

Fig. 11. IP-over-ATM, 65 kbyte: IP-over-ATM, 9180 byte: For the last measurement a packet size of 65 kbyte was chosen. It triggered several fluctuations as figure 11 shows. At first the throughput went down to approx. 5 Mbps. This performance has to be considered poor and was not efficient enough to be used in practice. Otherwise this results show the best data rates till 100 Mbps. Concluding, the big packets are useful for data networks with big buffer sizes and the maximum transport unit (MTU) is responsible for an effective data throughput.

The NIC RapidFire6162 from Olicom together with the LE155 from Fore Systems have been an efficient throughput and high Switched Virtual Circuit (SVC) performance. The LE155 has been also an active processor for direct traffic support without influence the CPU of the workstation. Both NICs need 128 Mbyte for a efficient operating. Additionally, both cards support in MPOA mode different service classes as Constant Bit Rate (CBR), Variable Bit Rate (VBR), and Unspecified Bit Rate (UBR). Available Bit Rate (ABR) was only available at the RapidFire6162. But, for the support of ABR you need this service inside the ATM switch, which was not integrated in every component yet. [3]

3. Conclusions

Original, more vendors like Olicom, IBM, 3Com, and Nortel Networks promised to take part on the tests of the WWL. But on the one hand, some manufacturers developed at the time of the test MPOA, other companies go to MPLS instead of the MPOA approach. The focus of Cisco & Co. has been changed, because the switches will be developed for frame switching and not for cell switching. [4]

Cabletron's components worked sufficient in the test environment. The shortcut functionality has been established after five packets. Therefore after 10 sec there were a shortcut established, which work with MPOA. By the measurement equipment of GN Nettest, the Win Pharaoh controlled the main connection in real-time. Netcom's Smartbits simulated and analyzed MPOA and turn over the advantages of MPOA in the LAN area. Fast layer-3-switching was possible and more effective for different packet sizes than without MPOA. The solution from Cabletron includes one more feature: the router works as layer-3-switch with IEEE802.1p and adapt the different approaches of QoS and CoS. Therefore, you can work with Gigabit-Ethernet and ATM networks.

The components of Cisco Systems established a shortcut after 10 packets. Additional, the shortcut functionality worked also efficient and sufficient. Only the blockade of the router if the system was overloaded prevented to get the test results. But the other test results were in the foreseen parameters. Therefore, the tests of the components of Cisco and the MPOA approach were successful and manageable.

TCP has been extended and further developed for better and more efficient mechanisms in high-speed networks. Yet in practice, TCP quite often does not have the optimal throughput. Several parameters, devices and mechanisms were responsible for this effect. This caused fluctuations and delays during the transmission of data traffic. Mainly TCP's own mechanisms were responsible for the small throughput such as acknowledgements, small window sizes and sequence number overloads. Additionally, the performance of the ATM boards and the end stations were important. The adapter boards from Fore Systems and Olicom showed only a maximum throughput of approx. 100 Mbps, if you use CLIP. Therefore, there are three main issues regarding latency in Internet protocols, which are able to improve the performance:

1. The default "window size" in many TCP/IP protocol implementations acts as a bottleneck on communications over high-latency links. On many implementations, the standard window prevents sending enough data to fill a high-latency connection.

2. TCP includes two essential congestion control mechanisms called "slow start" and "congestion avoidance". These mean that all Internet connections (such as viewing web pages and sending e-mail) start out at lower bandwidth and then throttle up to higher speed if no congestion is encountered. The problem is that each cycle of speed increase requires a full round-trip communication between sender and receiver, and dozens of such round-trips can be necessary to reach the full potential of a link.

3. There are research efforts to look at increasing the performance of TCP by trick-ing the connection on the other side into believing it is communicating over a low-latency link. Unfortunately, these schemes fundamentally alter the semantics of TCP communications, introducing the possibility of data corruption. Moreover, they are incompatible with the IP security protocols (IPsec), which promise to bring an unprecedented and badly needed degree of security to the Internet

In order to implement IP-over-ATM effectively, it is necessary to assign enough time to the configuration of the participating devices and software. ATM is a new technol-ogy which considerably extents the usable bandwidth with regard to QoS parameters. Nowadays, workstations are designed for work in traditional networks such as Ethernet or Token Ring. These deadlocks must be compensated in order to use effi-ciently IP-over-ATM. [2, 3]

ATM is stuck in the data center, since deploying ATM to either the wiring closet or the desktop is more trouble than its worth. Some ATM vendors are trying to help this issue by building their switches to require only software upgrades for new improve-ments, but certain upgrades-like moving from OC-12 to OC-48 and beyond-will al-ways require additional hardware. But you can't count ATM out completely. It still retains the strengths its always had, namely fault tolerance and reliability. ATM's cell-based traffic and dynamic routing capabilities make it a much more resilient backbone technology than Gigabit Ethernet and considerably faster than FDDI. And, it has little if any distance limitations, which makes it a great WAN connector. Obviously then, we will continue to see ATM in the service provider arena, both in telecommunica-tions as well as ASPs, and even large ISPs. If your corporate network especially re-quires these strengths, then ATM is probably in the best shape of its life as far as the enterprise is concerned. But the ATM solution will be build without MPOA and with MPLS to establish QoS mechanisms, because MPOA works but is not widespread. [4]

References

1. Detken, K.-O.: Interworking in heterogeneous environment: Multiprotocol over ATM (MPOA); European conference Interworking98; Ottawa 1998
2. Detken, K.-O.: ATM Handbook; publishing house Hüthig; Heidelberg 1999
3. Detken, K.-O.: Quality-of-Service (QoS) in heterogeneous networks: CLIP, LANE , and MPOA performance test; Networking 2000; Paris 2000
4. Rist, Oliver: ATM Still Around, But Struggling; Internet Week 10/99; USA 1999

Evaluation of a New Resource Reservation Scheme for MPEG Transmission in Highly Available Real-Time Channels

Enrique Hernández Orallo[1,] Joan Vila i Carbó[1]

[1] Departamento de Informática de Sistemas y Computadores (DISCA)
Universidad Politécnica de Valencia,
Camino de Vera s/n, 46022 Valencia, SPAIN
ehernandez@bcj.gbancaja.com, jvila@disca.upv.es

Abstract[1]. Backup real-time channels is a technique to provide uninterrupted service in the presence of router failures during real-time transmissions. This technique adds the notion of availability to the concept of QoS (Quality of Service), usually expressed in terms of guaranteed throughput and maximum delays only. Availability comes at the cost of increasing the required resource (bandwidth, buffers) reservations, due to the needs of a backup channel. However, this extra resources reservation is potentially wasted, since fault rates are very low. This paper proposes a systematic method for estimating and optimizing resource reservations. This approach is based on inaccurate failure detection in order to reduce latency is proposed. The cost of inaccurate failure detection is that the backup channel will be activated and utilized unnecessarily upon detection of "false failures". However, the paper shows, through simulations and using MPEG transmission traces, that the percentage of false failures is almost negligible.

1 Introduction

QoS is a key issue for using computer networks for multimedia transmissions. Although there is not a clear and standardized approach of how this requirement will be provided, it seems that it will be based on establishing a set of contractual clauses between the client and the provider of the service in order to guarantee some performance requirements. The actual proposals for guaranteeing QoS mainly rely on network resource reservation schemes.

QoS requirements usually include parameters like throughput, maximum delays or jitter. However some critical real-time applications, like scientific and medical monitoring, demand not only transmission performance and predictability, but also availability: maintaining the QoS service in spite of router or link failures.

---

[1] This work has been supported by the Spanish Government Research Office (CYCYT) under grant TIC99-1043-C03-02.

The availability requirement can be dealt with through the use of redundant channels. Redundancy can be managed using two extreme approaches. In the *Multiple Copy Approach* [1] all redundant channels are simultaneously active, while in the *Backup Channels Approach* [2] a backup is only activated upon failure detection. The backup approach is, in principle, more efficient, since it does not waste bandwidth in the absence of failures, but this paper shows that its performance strongly relies on the failure detection latency and channel setup time. Backup channels have the drawback of keeping some bandwidth reserved without being used. This resource waste can be improved by using the technique by Shin [3] based on reserving spare resources. However, two important issues concerning resource reservation remain unsolved: how to estimate the bandwidth reservation for the primary and backup channels and how does failure latency affects these estimations.

The resource reservation for a given workload on a real-time channel can be estimated from its real-time requirements by properly defining that workload under the flow model of a particular resource reservation protocol. In this sense, it is worth noting that every protocol defines its own different flow model. This way, the IETF standard proposal for resource reservation, known as RSVP (*Reservation Protocol*) [4,5], defines a flow for a guaranteed service [6] based on the token bucket scheme while other proposals, like the *Tenet Suite* [7] or ATM [8] use different flow models.

On the other hand, the real-time requirements for a particular workload under the backup channels scheme is determined by the failure detection scheme. This is due to the following observation: the longer is the failure latency time, the more demanding the delay requirements of the backup channel are, in order to maintain the desired QoS. According to this, the paper introduces a scheme for fault detection whose goal is to reduce failure latency (and, thus, resource reservations). The proposed technique reduces failure latency by "suspecting" a failure before it occurs. This yields unnecessary activations of the backup channel when a failure is not finally confirmed. This technique can be considered a trade off between failure detection latency and the utilization of the backup channel. The demand of resources of this technique is compared to previously proposed schemes under the RSVP protocol using MPEG workloads. The percentage of "unnecessary activations" of the backup channel is evaluated using a simulator. Results show that this rate is negligible.

The mechanisms for establishing disjoint paths, failure detection, recovery and reporting and channel reconfiguration are beyond the scope of this paper and there exist well-known solutions that can be easily introduced to the proposed scheme. However, in section 6, a simple way to implement this protocol is sketched.

2.   Failure Detection Scheme

There are several schemes to provide fault-tolerance for real-time channels [1,3] in multi-hop networks with several alternative routes between the sender and receiver:
- **Multiple active channels**: several redundant real-time channels are setup through different routes; packets are sent simultaneously through all channels. The scheme works as long as the receiver gets at least one of the replicated packets. This is a costly technique but it provides uninterrupted service in spite of failed routers.

- **Backup channels**: there is a primary channel and a backup channel but this channel is only setup upon a failure of the primary channel. This technique saves unnecessary bandwidth utilization but has the drawback of requiring failure detection and the delay introduced for setting up the backup channel. Besides, it may occur that there exist no enough resources when trying to setup the backup channel.

Reserving resources in advance for the backup channel can solve some drawbacks of the second approach: that would improve the setup time and avoid the lack of enough resources. A key aspect of the backup channel approach is failure detection. Proposals for failure detection rely on behavior-based techniques for detecting communication failures. More precisely, heartbeats ("I am alive" messages) are used to detect failures of neighbor nodes, or in an end-to-end fashion. An important problem pointed out by Shin [11] is failure detection latencies, especially in the end-to-end detection technique. This latency can be a serious obstacle for applying this technique, since it makes the delay requirements of the backup channel very demanding in order to keep the service uninterrupted, and that causes the corresponding resource reservations to be sometimes unacceptable.

The scheme for failure detection proposed in this paper is aimed to achieve an efficient channel resource reservation. It is based on "suspecting" a failure when the packet delay through the primary channel is close to the maximum guaranteed delay. Whenever a failure is suspected, the backup channel is activated. This yields unnecessary backup channels activations due to inaccurate failure detection (false failures), but it has the advantage that the delay requirements of the backup channel are not so restrictive. That allows reducing the required bandwidth for the backup channel. In summary, the proposed trade-off is to allow inaccurate failure detection in order to minimize latency optimizing, thus, resource reservations. The only situation where the inaccuracy of the failure detector could be considered non-acceptable would be when the sender has periods of relatively long inactivity. This would imply either to inject heartbeats or to modify the traffic to ensure that the connection is not used for some (relatively) long time. In practice, it has been proved that using MPEG video there are not periods of inactivity, due to the MPEG encoding scheme.

Let d_f be the failure detection time. It is assumed that d_f is that it is strictly less than the maximum delay of the primary channel, denoted as d_A. The problem consists in how to choose d_f in order to trade-off failure detection latency and resource reservation. The maximum delay experienced by the first packet retransmitted through the backup channel upon a failure, denoted as d_{total}, can be expressed as $d_{total}=d_A+d_s+d_B$ where d_A and d_B are the maximum delays of the primary and the backup channel respectively and d_s is the setup time for the backup channel. If d_{total} has to be less than the maximum allowed delay for the real-time transmission, then d_A and d_B will have to be more demanding. The more demanding is the maximum delay for a channel, the more resource reservation is required.

According to the above proposed technique, the idea is to choose a failure detection threshold d_f such that $d_f<d_A$ in order to allow a larger d_B. When a packet delay exceeds d_f then the backup channel would be activated, although the primary channel would not be discarded, since it could be due to a "false failure". Using d_f as a failure detection threshold, d_{total} can be now expressed as $d_{total}=max(d_A, d_f+d_s+d_B)$. This way, the upper bound for d_A could be d_{total}, while d_B could be chosen as $d_{total}-d_f$

d_s. The fact that d_f is always less than d_A implies an important reduction in resource reservation for the backup channel. In addition, resources reserved in routers for the backup channel could be available for non-prioritary packets, so this technique would not imply resource wasting. The only situations where resources are wasted are upon detection of "false failures".

3. A Method for Optimal Network Resource Reservations

The goal of this section is to introduce a method for obtaining optimal resource reservations for a given workload in an accurate way, so admission control algorithms do not over-estimate the resources required by a connection.

The end-to-end delay bound is a function of the reserved bandwidth in the links and it is usually calculated using a model for the traffic and a network model. The IETF specification for a *guaranteed Service*[6] uses the *token bucket (b, r, p)* traffic model where the number of bits that the source transmits is less than $b + pt$, for any interval of time t. The maximum end-to-end queuing delay bound can be calculated using these equations:

$$Q_{delay} = \frac{(b-M)(p-R)}{R(p-r)} + \frac{M + C_{tot}}{R} + D_{tot} \quad (p > R \geq r) \tag{1}$$

$$Q_{delay} = \frac{M + C_{tot}}{R} + D_{tot} \qquad (R \geq p \geq r) \tag{2}$$

where C_{tot}, D_{tot} are the parameters defining the network, p, r, b, M are the traffic flow parameters, R is the bandwidth reservation in the network and Q_{delay} the end-to-end delay. Note that equation (2) does not depend on the flow parameters, since the reserved bandwidth R is greater than the peak rate p, so there are no packets queued from this flow. For instance, with a WFQ (*Weighted Fair Queueing*) scheduling algorithm C_i and D_i can be calculated as follows: D_i is equal to the *MTU* (Maximum Transmission Unit) of the link divided by the link bandwidth B_i, with the condition that M must be smaller than the minimal MTU of the path. The value C_i is assumed M in order to consider packet fragmentation. The buffer size needed in the nodes is $b + C_{sum} + D_{sum}r$ where C_{sum} y D_{sum} are the sum of all the previous C_i y D_i parameters.

A simpler equation that not depends on the peak rate to obtain the end-to-end delay is the one based in the Parekh's work [16] that is usually used for nodes with WFQ schedulers (some other schedulers use this equation too: Virtual Clock, WF^{2}Q, FFQ, SPFQ). If the source traffic is constraint by a *leaky bucket* (σ, ρ) flow then the maximum end-to-end delay is:

$$D_i = \frac{\sigma + nL_i}{R} + \sum_{j=1}^{n} \frac{L\max_j}{C_j} = \frac{\sigma + C_{tot}}{R} + D_{tot} \tag{3}$$

where L_i is the maximum packet size for session i, $Lmax_j$ is the maximum packet size in node j, C_j the bandwidth of the link j and n the number of nodes. This equation is equivalent to (1) when the peak rate p is infinite (b and r parameters are equivalent to σ, ρ, $L_i = M$ and $Lmax_j = MTU_j$). The buffer size in the j node is $\sigma + jL_i$. It is important to remind that the network minimal latency must be added to the described equations (1)(2)(3) in order to obtain the complete delay. This latency is a fixed value and depends mainly on the physical transmission, which is usually negligible compared to the network delay.

Shortly, for calculating the reservation of networks resources it is necessary: (1) a network model, to obtain the C_{tot}, D_{tot}, M parameters already determined, and (2) the calculation of the flow parameters (b,r,p) or (σ,ρ). Of these parameters, the peak rate is obtained directly. The problem is how to calculate the σ and ρ parameters to optimise network resources, that is, to minimise the value of R in equations (1)(2) and (3). To find the optimal bandwidth reservation, a costly and not bounded iterative method on traffic traces can be used. However, a much more convenient and fast method based on the concept of empirical envelope is introduced in [12].

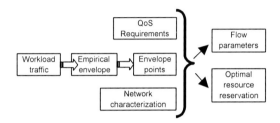

Fig. 1: Process for obtaining optimal reservation

The concept of the empirical envelope was introduced in [7,9] by the Tenet group. The empirical envelope $E(t)$ gives a time invariant bound of the traffic. Function $A(t)$ is defined as the number of bits transmitted in a given time t. This function can be used to characterise the traffic, but it is time dependent, i.e, its value depends on the origin for measuring the time interval t. So $E(t)$ is defined as the most accurate constraint function that bounds input traffic and is time independent. Using the empirical $E(t)$, a set of points are calculated, the *envelope points*, that give a condensed description of the traffic flow. The number of points obtained have been proved to be very low (always less than 100 points for 30000 frames of MPEG videos). Using these points the optimal bandwidth reservation of the nodes can be obtained in a fast and bounded way.

In short, the process is as follows (figure 1): from the workload traffic the empirical envelope is calculated and then the envelope points. With the QoS requirements (maximum delay) and network characterization, the σ and ρ parameters that optimise network resources are calculated. Another conclusion of this work [12] is that the flow parameters that make optimal the reservation are the same for the Parekh's delay equations (3) and IETF/RSVP guaranteed service (1)(2) using a WFQ class scheduler. This implies a simplification of the calculation for IETF/RSVP and an extension of the results to both models.

4. Description of the Network Model and MPEG Workload

Figure (2) describes the networks used in this paper. Network 1 is a typical IP network with 3 nodes and bandwidths over 50 Mb. Network 2 has 9 nodes and it is an example of a multi-hop long path network, and Network 3 is an ATM like with fixed packet size (53 bytes) with high bandwidths. The scheduling of the nodes is the well-known WFQ [14] algorithm. In the table M is the maximum packet size and m is the minimum policed unit. The total delay of the network is the sum of the queue delay (Q_{delay}) and the minimum path latency. Without loss of generality, for all cases the setup time d_s will be considered 0. In the last point of this article it can be shown that this value does not affect the results.

Nowadays, the foremost objective of real-time networks is the transmission of video. This is usually transmitted using the MPEG compression algorithm. Therefore, it is important to evaluate our new scheme with these kinds of traffics. In this paper, we use some well-known Rose's MPEG-1 video traces [13]. The original sequences have 30000 frames but in this paper only the first 1500 frames are used (about a minute of video) due to the large simulations times. However, using the complete sequences has been proved to produce similar results.

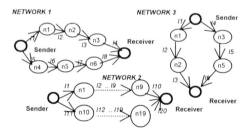

Network 1			Network 2			Network 3		
Link	BW (Mbps)	MTU (bits)	Link	BW (Mbps)	MTU (bits)	Link	BW (Mbps)	MTU (bits)
l1,l5	50	10000	l1,l5,l9,l11,l15,l19	50	25000	l1,l4	100	428
l2,l6	35	12000	l2,l6,l10,l12,l6,l20	100	20000	l2,l5	52	428
l3,l7	100	15000	l3.l7,l13,l17	52	15000	l3.l6	155	428
l4,l8	50	20000	l4,l8,l14,l18	50	20000			
M = 1Kbyte m = 1000 bits			M = 1Kbyte m = 500 bits			M = 428 bits		

Fig. 2: Sample Network parameters

The workloads used are different types of traffic, which are representative of the most usual ones. The first one, the **LAMBS** traffic, is part of the movie "The Silence of the Lambs". It is a bursty traffic as can be shown in figure 3(a), with big rate variation between the movie sequences. Figure 3(b) shows the corresponding cumulative arrivals and envelope functions. The **SOCCER** traffic is recorded from the Soccer World Cup 1994 Final: Brazil vs. Italy. It is a bursty traffic (see figure 3(c)) with a big peak over the frame 1250. Figure 3(d) shows the corresponding cumulative arrivals and envelope functions. The **NEWS** traffic is a German news show (see figure 3(e) and 3(f)). It has a very regular pattern, with very low variations.

5. Analytical Evaluation

The goal of this section is to compare the resource reservations of the proposed backup channels scheme with the other two schemes introduced in section 2. These results will be used in the next section, to evaluate the goodness of this new scheme simulating the same scenarios. The only parameter that has to be specified is the total delay d_{total} and the primary and backup channel delays (d_A, d_B).

- **A&B Scheme**: The sender always sends packets simultaneously through channels A and B. The maximum delay for both channels is d_{total}. This is an expensive technique because the channel B is always in use.

- **A|B Scheme**: In this scheme, channel B is only activated upon detection of a failure in channel A. Failures are detected when the delay of a packet exceeds $d_{total}/2$. Therefore, the maximum delay for both channels is $d_{total}/2$. Channel B is not used until a failure is detected. This allows routers to use the resources of channel B to transmit non-priority traffic. This avoids wasting assigned network resources.

- **A^B Scheme**: The proposed scheme still has a maximum delay of $d_{total}/2$ for channel B, but allows channel A to be less restrictive by doubling its delay to d_{total}. Like in the A|B scheme, channel B is not used until the delay of channel A is more than d_f. Recall that this can generate false failures, since the expected packet could arrive later through channel A, but still on time to meet its deadline d_{total}.

In all evaluations a minimum value of d_{total} is selected in order to reserve the maximum resources the network allows. From this value, another two bigger multiple values of d_{total} are selected. From the traffic description, it is possible to calculate the False Failure Rate (FFR). This FFR is an estimation of how many times the backup channel will be activated due to "false failures". The goal is to obtain the portion of traffic that causes a delay bigger than $d_{total}/2$. Consider a new reduced traffic obtained by eliminating the peaks from a given traffic. This reduced traffic gives less delay than the original one. Figure (4) shows the original SOCCER traffic. and a reduced SOCCER traffic in a 1% . The new reduced traffic has a peak rate of 2464167b/s and a mean rate of 584922b/s.

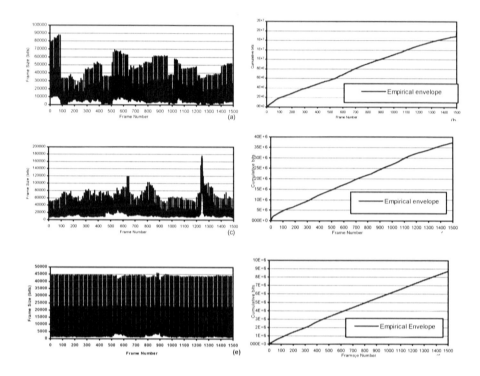

Fig. 3 (a) Lambs traffic, (b) cumulative and envelope functions (c) Soccer traffic, (d) cumulative and envelope functions (e) News traffic (f) cumulative and envelope functions.

The False Failure Rate (FFR) can be thought of as the rate of traffic reduction that gives a delay of d_{total} /2. For example, if the SOCCER traffic is reduced in 1% by eliminating the peak rates, then thie reduced traffic has a lower delay d_x of 0.016s using the same network parameters. Therefore, the goal is to obtain the reduction rate that yields a delay of d_{total} /2. This portion of the traffic produces the difference of delay between d_{total} and d_x. In order to calculate the delay, the traffic is first reduced in a rate r and then, b and r parameters are calculated. With these parameters and the reservation of the original traffic, the delay can be calculated using equations (1) and (2). Thus, an approximation of the False Failure Rate (FFR) can be calculated in an iterative way.

NETWORK 1 : Table (1) shows the results for all three traffics in Network 1. Three total delays (d_{total}) are used in these comparisons. The delays for channels A and B (d_A and d_B) and their respective reservations are obtained from d_{total}. R_A and R_B are the reservations for channel A and B respectively, and R_{total} is the total reservation for both channels.

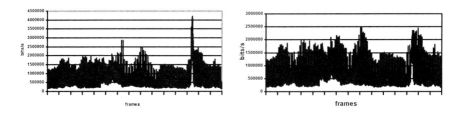

Fig. 4: (a)Original traffic,(b) Reduced traffic

Table 1: Reservations for Network

	A&B			A|B			A^B		
$d_{total}(s)$	0.1	0.05	0.02	0.1	0.05	0.02	0.1	0.05	0.02
$d_A(s)$	0.1	0.05	0.02	0.05	0.025	0.01	0.1	0.05	0.02
$d_B(s)$	0.1	0.05	0.02	0.05	0.025	0.01	0.05	0.025	0.01
LAMBS									
$R_A(Mb/s)$	0.87	1.36	2.05	1.36	1.89	4.6	0.87	1.36	2.16
$R_B(Mb/s)$	0.87	1.36	2.05	1.36	1.89	4.6	1.36	1.89	4.6
$R_{total}(Mb/s)$	**1.74**	**2.72**	**4.1**	**2.72**	**3.78**	**9.2**	**2.23**	**3.25**	**6.76**
% A^B	-/50	-/50	-/50	18/36	14/28	27/53			
FFR							0.055	0.023	0.04
SOCCER									
$R_A(Mb/s)$	2.43	2.74	3.53	2.74	3.25	4.25	2.43	2.74	3.53
$R_B(Mb/s)$	2.43	2.74	3.53	2.74	3.25	4.25	2.74	3.25	4.25
$R_{total}(Mb/s)$	**4.86**	**5.48**	**7.06**	**5.48**	**6.5**	**8.5**	**5.17**	**5.99**	**7.78**
% A^B	-/50	-/50	-/50	5.6/11	7.8/16	8.4/17			
FFR							0.004	0.004	0.004
NEWS									
$R_A(Mb/s)$	0.57	0.89	2.17	0.89	1.73	4.6	0.57	0.89	2.17
$R_B(Mb/s)$	0.57	0.89	2.17	0.89	1.73	4.6	0.89	1.73	4.6
$R_{total}(Mb/s)$	**1.14**	**1.78**	**4.34**	**1.78**	**3.46**	**9.2**	**1.46**	**2.62**	**6.77**
% A^B	-/50	-/50	-/50	18/28	24/49	26/53			
FFR							0.084	0.039	

In general, resource savings are more important as the total deadline becomes more restrictive. This way, in the LAMBS workload for $d_{total} = 0.02$s it saves about 2.5Mb/s of bandwidth over the A|B scheme (27% off) but, for 0.05s the savings are only of 0.53Mb/s (14%). In the SOCCER workload, the savings are of 0.7Mb/s over A|B (8.4%) for the more restrictive deadline of 0.02s. In the News traffic, the savings are of 1.8Mb/s over A|B (24%) for a deadline of 0.05s . These results hold for every link in the path, so in the first example the total resource savings in the network are 9.5Mb/s. Assuming that the resources of channel B could be used by the router when the channel is not in use (that is, considering only the resources of channel A), results are substantially better. This way, in the NEWS workload, savings are of 0.89Mb over A|B (49% off) and 0.89Mb over A&B (50%) for a $d_{total} = 0.05s$.

The estimated false failure rates are very low for all calculated deadlines. Taking into account the cost that supposes the false failures in the use of the channel B in the last example, the equivalent reservation is (0.039 * 0.89 = 0.034), the total reservation is 0.93, that is still lower than the other two schemes.

The resource gain depends mainly of the difference between the bandwidth reservation obtained for d_{total} (R value) delay and d_{total} (R' value). Figure (5a) shows graphically this difference for the LAMBS traffic with deadlines between 0 to 1s. It shows that the peak of resource savings is for a deadline of 0.2s. From 0.2 to 0.4s the savings decay greatly to later increase very slowly. For the SOCCER traffic (see figure 5b) the curves are very different. The gain is high for very low deadlines, but decreases very quickly until 0.7s to later increase lineally. The NEWS traffic (figure 5c) is very similar to the LAMBS traffic with a peak gain of 42%. Figure (5d) shows the total gain for all three traffics in Network 1.

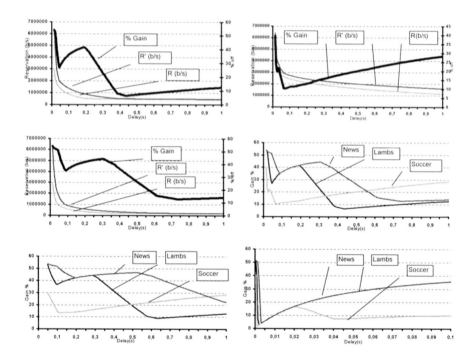

Fig. 5: (a)(b)(c) Reservations for LAMBS, SOCCER and NEWS in network 1. (d)(e)(f) Total percentage gain in network 1, 2 and 3.

In Network 2 an evaluation of the reservations over a long path is done. Figure (5d) shows the gain from 0 to 1s, which is very similar to network 1. The FFR is also very similar.

In Network 3 the delays are very small due to the high bandwidth of the links and little size packets. In this network the percentage gain curves differ from the last networks. It has a peak gain at very low delays and falls very quickly to approximately 0 at 0,005 seconds to later increase exponentially.

In summary, the proposed A^B scheme reserves a total of resources that is in between the other two schemes. In general, the greatest savings are obtained with the

LAMBS traffic, but with a false fail rate of about 10%. This is reasonably because it has a regular pattern. On the other hand the SOCCER traffic has lesser gains, due to its high variability. Despite this, it has the best FFR. The resource savings are between 10 to 30% over the A|B protocol, and considering only the primary channel, the savings are between 10 to 50% over the A&B and A|B protocols.

6. Network Simulations

The goal of this section is to validate the new introduced scheme by evaluating how many times the backup channel will be activated due to a "false failure". This rate will be shown that it is very low. This is mainly due to the fact that the equations used to guarantee the maximum delay of a packet (as the ones used in the last section) give a very pessimistic bound of the traffic, and that only happens when the network is fully loaded. Normally, packets arrive much sooner than their deadlines.

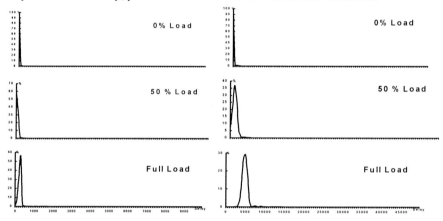

Fig. 6 : Density functions for packet arrival. SOCCER traffic in Network 1 and 2.

For this study, the ONetSim [15] network simulator has been developed. ONetSim is a C++ object oriented discrete-event simulation program specifically designed to test traffic workloads and scheduling algorithms. ONetSim allows to specify the network in a simply description language and it can use a deterministic flow as the MPEG traces seen in former sections. The traffic is introduced in the network following the token bucket flow. The WFQ scheduler is used for all simulations. One of the problems that arises in the simulations is the one of how to load the network in order to compare schemes involving guaranteed performance connections. To solve this problem, 10 channels are created in each node with a load index that ranges from 0% (no load) to 100% (full load). The load index is the percentage of the rest of the free bandwidth of the link.

Figure (6) represents the density functions for packet arrival in the simulations with **SOCCER** traffic. It is clearly shown that all packets usually arrive much sooner than their nominal deadlines and the more loaded the network is, the more the packets are delayed. These two obvious facts, is the basis for the efficiency of the proposed scheme. For example, in Network 1 with full load the maximum delay is *0.024s* seconds, that is below the threshold (*0.025s*). In Network 2 only with full load, the delay of some packets goes beyond the threshold for activating the backup channel: *0.025s*. This implies that all the nodes are full loaded (a very improbable condition).

Table 2: Max. packet delay (µs) in Network 1.

Delay / Load	0	25	50	75	100
LAMBS Traffic					
100000	643	1601	3191	5003	8929
50000	643	1520	3223	4671	6903
20000	643	1447	2793	4056	5816
10000	78	470	1084	1584	2166
SOCCER Traffic					
100000	643	1574	3137	5143	78470(*0.52%*)
50000	643	1479	3000	3926	24178
20000	643	1275	2348	3478	10895(*0.05%*)
10000	643	1008	1956	2603	4114
NEWS Traffic					
100000	613	1420	2676	3963	7047
50000	576	1481	2997	3736	6693
20000	78	493	1048	1671	2136
10000	78	422	1015	1580	2095

Table (2) shows the results of the simulation of the maximum delays for the traffics in Network 1. It can be observed that only for the SOCCER traffic the failure detection threshold d_f is surpassed. This condition only happens when the network is full loaded (For example *0.52%* FFR for full load for 0.1s). The simulation results of the maximum delays in Network 2 are very similar to Network 1. Only for soccer traffic the detection threshold d_f is surpassed. And finally, in table (3) are the simulation results of the maximum delays for the traffics of Network 3. The delays are very low due to the high bandwidth and little (and fixed) packet size. In this ATM like network, the obtained delays obtained are very close to the calculated deadlines. The reason is that the little packet size makes more fluent the traffic so the network behaves more like a network flow and the deadline equations yield better estimations.

In general, the packet delays of the simulations are similar to those of other papers [18]. The first question that arises is why the packet delays are so much lower than their deadlines. The main reason is the bursty characteristics of the traffic. Some other reasons are that traffic characterization (the flow model) makes a very coarse approximation of the traffic dynamics and that the delay bound equations are very pessimistic.

With these experimental results, the resources in the backup channel can be reduced even more using a greater delay. For example, in Network 1, for the NEWS traffic with $d_{total}=0.05s$ a failure detection threshold $d_f=0.01s$ can be used, since there is no packet that surpasses this value (see table (1)). That gives for backup channel a maximum deadline $d_B=0.4s$. The reservation for this deadline is about *0.99Mb/s* (the original deadline was *0.025s* with *1.73Mb/s* reservation). That gives a total reservation of *1.89Mb/s* that saves *1.57Mb/s* over the A|B schemes (*46%* off). For $d_{total}=0.01s$ the savings can be up to 60%.

Table 3 :Maximum packet delay in Network 3. All values are in µs.

Delay / Load	0	25	50	75	100
LAMBS Traffic					
100000	15.27	93.06	101.85	103.74	47859
50000	15.27	99.25	101.86	103.04	26735(0.42%)
10000	15.27	92.33	96.41	103.20	3297
5000	15.27	93.54	99.54	100.53	373
SOCCER Traffic					
100000	15.27	70.11	104.6	105.46	96091(21.05%)
50000	15.27	71.21	104.17	104.76	46569(13.71%)
10000	15.27	70.43	102.72	103.35	8201(0.36%)
5000	15.27	69.96	103.48	104.53	3392(0.87%)
NEWS Traffic					
100000	15.27	98.33	98.19	100.05	5424
50000	15.27	82.20	103.55	102.66	1730
10000	15.27	97.24	100.12	96.95	307.58
5000	15.27	92.35	94.15	99.59	173.2

7. Implementation

The proposed scheme for backup channels can be implemented as a part of a transport protocol on top of all other network protocols, as shown in figure (8a). The FSM (Failure Suspect Module) module will provide the mechanisms to calculate the primary and backup delays, activate the backup channel when a packet delays more than d_f, and inject heartbeats in the network when there is not traffic. However, for MPEG video it has been proved that the injection of heartbeat packets is not necessary because there is always traffic to send. The receiver starts counting the delay of the next packet when receives the previous one. Then, the receiver activates the backup channel when a packet is delayed more than d_f, by sending a system message (using the reserved backup channel) to the sender (Figure 8b). Simulations prove that the delay between the received packets are very uniform in time, i.e., if a packet P1 takes the minimum delay to arrive to the receiver, then the next packet P2 will take a very similar time. It is not possible that this packet takes a very long delay (but less than d_f), forcing the receiver to activate the backup channel, because the difference of time between P2 and P1 is greater than d_f.

When the sender receives the backup activation message, it starts transmitting through the backup channel. The setup time d_s for this new channel is the time needed for this high priority message to reach the sender. In the simulations this time is very low (it ranges from $15\mu s$ for the ATM like network to $187\mu s$ for Network 2). When the sender starts transmitting through the backup channel, it begins with the first packet so the buffers are empty for this channel. This implies that the firsts packets are delayed below its calculated deadline d_B, only depending on the network load; so we can assume that the time d_s will not influence in the total delay d_{total}.

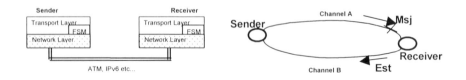

Fig. 8 : Implementation of the FSM.

The network or transport layer must provide some services: some way to create disjoints channels, a reservation mechanism to ensure delays (like ATM, RSVP, etc) and a time-bounded and robust transmission of failure report messages and channel activation messages (from receiver to sender).

8.   Conclusions

This paper has introduced a new scheme for fault detection in highly available real-time channels that provides an efficient resource reservation. The idea behind this scheme is that the delay bounds in real-time network are very pessimistic. Normally packets arrive far behind from its deadline as it has been shown. This characteristic can be used to reduce the reservations in the network, ensuring uninterrupted service.

The scheme has been compared to the resource reservations of other fault detection alternatives using MPEG traffics traces. From these results it can be concluded that the proposed scheme is a trade-off between the Multiple Copy Approach and the Backup Approach: it reserves more resources than Multiple Copy and less than Backup, but the penalty is that backup channel unnecessarily used even when "false failures" occur. Simulations show that for token bucket flow model the activation of the backup channel happens very seldom: in the majority of simulations this rate is 0 and always remain below 5%. False failures are only detected with fully loaded networks (a very improbable condition) and very low deadlines. The total saving in the network ranges from 10% to 35% over the Backup Channels scheme. Considering that the router may use the resources of the backup channel for congested traffic when the channel is not in use, savings range from 10% to 55%. The fact that simulations shows that packets are much lower than their deadlines can be used to increment the failure detection threshold d_f. This enables higher deadline for the backup channel and higher savings in resources reservations: up to 60%.

References

1. P.Ramanathan and K.G.Shin, Delivery of Time-Critical Messages Using a Multiple Copy Approach. *ACM Transaction on Computer Systems*, Vol. 10, No. 2, May 1992, 144-166.

2. A. Banerjea, C. Parris and D. Ferrari, Recovering guaranteed performance service connections from single and multiple faults, *Technical Report TR-93-66*, Int. Computer Science Institute, Berkeley, CA.

3. S.Han and K.G.Shin, Eficcient Spare-Resource Allocation for Fast Restoration of Real-Time Channels from Network Component Failures, *IEEE 1997*,99-108.

4. P.P.White, RSVP and Integrated Services in the Internet : A tutorial, *IEEE Communications Magazine*, May 1997,100-106

5. IETF:Internet Draft: Resource Reservation Protocol (RSVP). Functional Specification.

6. S. Schenker, C.Partridge, R.Guerinm, Specification of Guaranteed Quality of Service, RFC 2212

7. A. Banerjea, D. Ferrari, B. Mah, M. Moran, D. Verma, and H. Zhang, The Tenet Real-Time Protocol Suite: Design, Implementation, and Experiences, *IEEE/ACM Transactions on Networking*, vol. 4, n. 1, 1-10, February 1996.

8. J. Crowcroft, Z. Wang, A. Smith, and J. Adams. A Rough Comparison of the IETF and ATM Service Models. *IEEE Network, Dec. 95*

9. E.W.Knightly, H. Zhang, Traffic Characterization and switch Utilization using a Deterministic Bounding Interval Dependent Traffic Model, *In Proceedings of IEEE INFOCOM'95*

10. E.W.Knightly, R.F.Mines, H. Zhang, Deterministic Characterization and Network Utilizations for Several Distributed Real-time Applications, *In Prodings of IEEE WORDS'94*

11. S. Han and K.G. Shin. Experimental Evaluation of Behavior based Failure-Detection Schemes in Real-Time Communication Networks, To appear in *IEEE Trans. In Parallel and Distributed Computing.*

12. E. Hernández and J. Vila, "A Fast method to optimise Network resources in Video-on-demand transmission" *Proceedings Euromicro'2000. Maastricht, Holland. Sep. 2000.*

13. O. Rose, "Statistical properties of MPEG video traffic and their impact on traffic modeling in ATM systems", *Proc. of the 20th Annual Conference on Local Computer Networks, Minneapolis, MN, 1995, pp. 397-406.*

14. A. Demers, S. Keshav, and S. Shenkar "Analysis and Simulation of a fair Queueing Algorithm", *Internet Res. and Exper. vol. 1,1990.*

15. E.Hernández "ONetSim:Simulador de redes orientado a objetos"(Object Oriented Network Simulator).*PhD report*, 1998, *DISCA .Universidad Politécnica de Valencia.*

16. H.Zhang "Service Disciplines For Guaranteed Performance Service in Packet-Switching Networks", *Proceedings of the IEEE, Vol. 83, N° 10*, 1374-1396

17. E. Hernández, J.Vila , "An efficient Resource Reservation Scheme for Highly available real-time Channels", *Proceeding of PDCS'99, Nov. 1999,* M.I.T., Cambridge, USA.

18. D. Clark, S. Shenker, and L. Zhang, "Supporting Real-Time Applications in an Integrated Services Packet Network : Architecture and Mechanisms", In *Proceedings of ACM SIGCOMM'92*, pag. 14-26, Baltimore, Maryland, August 1992.

Performance Evaluation of Diffserv Driven HFC System

Giannis Pikrammenos, Helen-Catherine. Leligou

National Technical University of Athens
Greece

Abstract. Shared access networks such as hybrid fiber/coaxial and passive optical networks have emerged as promising ways to reduce the cost of the transition to a broadband access infrastructure and provide a graceful upgrade path towards the photonization of the local loop. The MAC protocol as the only arbiter of the upstream bandwidth directly affects the Quality of Service (QoS) provided to each upstream traffic flow and must meet several constraints. Such constraints include the adequate speed of operation, flexibility to support efficiently the largest number of services and applications offering an adequate number of QoS classes, and independence of higher layers, protocols and future extensions to traffic management specifications. The implementation of a MAC mechanism targeting these goals and aligned to the emerging Differentiated Services Internet strategy is evaluated using computer simulation results in this paper. The benefits of the adopted prioritization scheme to provide and guarantee different QoS levels are illustrated.

Keywords: Interworking, HFC, ATM, IP, DiffServ, QoS, Access Network.

1. Introduction

Re-use of the existing infrastructure greatly reduces the initial investment outlay and provides a graceful upgrade path in step with service demand. Typical examples are Hybrid Fiber Coaxial (HFC) that use legacy CATV systems and their coaxial medium beyond the fiber node and Passive Optical Network (PON) systems that use the twisted pair based telephone networks enhanced with xDSL beyond the Optical Network Unit (ONU). The legacy topology is that of a broadcast tree-shaped access network. The attractiveness of HFC systems for the delivery of broadband services lies in the high reuse of existing infrastructure and a sound gradual upgrade strategy. The initial investment mainly consists in the incorporation of a return channel and fiber feeders to customer clusters of a few hundred homes. Sharing the common feeder and cable in the upstream direction (from stations to head-end) requires a MAC protocol to allocate slots to stations in a TDMA fashion [5], [6].

This sharing results in a distributed queuing system characterized by the long time required to pass control information from the queuing points to the service controller residing at the head-end. The allocation of upstream slots in a tree-topology access system is based on a reservation method which allows to dynamically adapt the bandwidth distribution to traffic fluctuations. The MAC controller works on the basis of collected access requests to allocate the upstream slots by sending access permits. It is important to note that the service policy of the MAC governs the distributed multiplexing from a central point situated at the head-end. This has the important implication that, as regards delay control, the acquisition of arrival information and buffer fill levels includes a considerable delay element not found in the centralized multiplexing typically operating in a switch queuing point. In contrast, drop policies must be distributed over all network terminations (cable modems) where the flow identity and fill levels are known. Because of the much larger reservation delay and statistical behaviour of the aggregations from many customers, special care must be taken to safeguard QoS to sensitive traffic. This requires a prioritization scheme and the differentiated services strategy of IETF [1] is a very suitable approach to handle the problem.

The paper is partitioned in two sections. The concept of operation of the MAC protocol and the actual implementation is presented in section 1, while in section 2 the performance is evaluated using computer simulation.

2. Presentation of the MAC concept – integration aspects

Tree-shaped access system exhibits a very different behaviour between the upstream and the downstream direction. In the downstream, replicas of the signal are created due to the broadcast nature of the medium, giving rise to privacy and security issues, which are typically dealt with by means of encryption. In the upstream, the MAC control function implies the multiplexing and concentration of the traffic from all active modems. However, in the upstream case the MAC is characterized by the distributed nature of the queuing points and the additional difficulties in the exchange of control information. To apply any scheduling or priority discipline requires the correlation of the traffic from all multiplexed sources going to the one common egress point of the system.

The Differentiated Services (DS) strategy, recently adopted by IETF as a scalable and relatively simple methodology towards enriching with QoS the IP services, is applicable and quite appropriate in the case of such tree-shaped access systems, like HFC, where IP services are or could be dominant. To align such an access system to the concept of differentiated services requires the incorporation of priorities in the MAC function, for the appropriate handling of each flow aggregation with respect to its requirements. The strength of DS lies in the slow and graceful introduction of such complexities in line with revenues from a previous stage of introduction of such mechanisms. Dealing with behaviour aggregates and starting with static management based Service Level Agreements (SLAs) executed at slow time scales while keeping traffic conditioning at the edges of the network enables a low cost starting phase while smaller granularity levels can be sought out at later stages of deployment. Non-compliant intermediate nodes can be transparent but at risk of reduced overall performance should they become the bottleneck in the route of the flow. Slow distributed access multiplexer such as an HFC system residing at the network edge can not be relied upon to operate in a transparent fashion as regards the DS strategy since the MAC directly affects the temporal properties of the egress stream. This can not be realized in an HFC system without embedding suitable differentiation mechanisms into the MAC control function.

The basis of the approach is the use of access priorities in the reservation system, which can be programmed to fit with required PHBs by means of the mapping of flows to priorities. [1] Logically separate queues for each priority are necessary for the proper operation of the prioritization scheme.

The characteristics of the four aggregation levels/priorities are the following:

1. The high priority is devoted to services with very strict delay requirements, which undergo strict traffic profile control (traffic conditioning) such as the EF (Expedited Forwarding) service [2].
2. The second priority level is devoted to real-time variable rate flows, such as video services or VoIP and it is provided with peak rate policing for guaranteed QoS. In the DS context it could be used for the top AF (Assured Forwarding) class [1].
3. The third priority is devoted to data services with higher requirements than best-effort. The traffic profile control assumed for this class aims at minimizing the loss of packets and the disturbance to other traffic. The 3rd priority mechanism is suited to the support of all four or the lower three AF classes [3]. (Drop policies can be independently applied at the modem queuing points).
4. The fourth priority is reserved for plain best-effort services which employ loss based flow control at the TCP level and can be very disruptive to the other classes when sharing the same queue.

The implemented system employs a TDMA slotting designed for ATM cells. We will simply consider the ATM size slots as a quantum of MAC assigned bandwidth allocation. The last three priorities employ reservation while the first unsolicited permits.

The reservations in the HFC MAC use a request field piggy-backed in the upstream slot and are usually assisted by contention on special reservation mini-slots [4]. The latter is needed for the announcement of the first arrival of a burst since the piggy-back mechanism is not self-starting but relies on the existence of previous traffic for the announcement of new arrivals [6]. However, the implemented MAC of the under discussion system departs from this approach in two points: first it provides a three simultaneous request field – one for each queue – which is embedded in every upstream slot irrespective of which queue provided the cell. Second, it employs round robin polling instead of contention. That was chosen for its simplicity and came without serious drawbacks in the targeted environment. In the AROMA system the one byte request field uses the place of the HEC field of the ATM cell which is not needed for cell delineation since an additional synchronization preamble is employed because of the burst mode operation. The multiple requests on the other hand are the strong point of the AROMA system and the tool for the higher QoS capabilities of this system. They are necessary if higher priority traffic is to be quickly made known to the head-end. This feature enables the algorithm in the MAC to offer precedence to the high priority cells when issuing access permits on a global basis and not just among the cells of the same termination which is of a very limited value.

The described MAC function was implemented on the Access Network Adaptation board of the under discussion system. Since the MAC considered a sub-layer of the physical layer and is closely related to the physical layer functions (TDMA, framing, etc.) the permit generation function was selected to be integrated with the downstream framer component, which was responsible for the construction of the downstream frame including the transmission permits. The whole design was programmed and placed on a Field Programmable Gate Array (FPGA) chip enabling the re-design of the implementation to adjust to modified frame structures, rates etc. An embedded processor (On Board Controller) was used to calculate and modify all non-real time parameters such as the programmed (pre-allocated or provisioned) bandwidth distribution, which may vary with time due to the switch on/off of the modems. An external static RAM chip was used for buffering necessary protocol information in addition to the available on-chip memory provided by the FPGA, which was exploited to keep state information and speed-up the permit generation process.

In the MAC operation algorithm, a list of 512 permits is employed, which has been prepared by the OBC on the basis of subscription data for the scheduling of all pre-arbitrated permits. The OBC executes the Call Acceptance Control (CAC) scheme (or Service Level Agreement in the DS terminology) and writes periodically permits for the 1st and 2nd priority connections at the peak and the minimum guaranteed rate accordingly (to support the expedited and assured forwarding classes). Permits are also inserted for polling requests by modems that have not established a 1st or 2nd priority connection. This list is cyclically read out by the MAC controller H/W and the permits are sent downstream. Permits are spaced in the list according with techniques given in [6]. This list is stored on the external SRAM chip. At the end of each cycle, the embedded controller updates also a list of credits for the 2nd and 3rd priority connections. These credits are used for policing 2nd priority connections at the contracted peak rate and guaranteeing the minimum calculated bandwidth of 3rd priority connections as described above. Since ATM signaling is also supported in AROMA, the permit and credit lists can be updated dynamically to add new connections using a second copy.

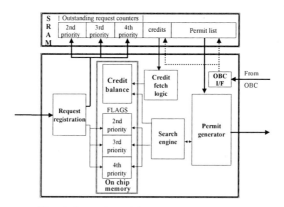

Fig. 1. Operation of the permit scheduler

The three lower priorities are serviced dynamically, on the basis of requests, by filling in empty locations in the permit list (representing unallocated bandwidth). The requests per queue, for each modem, arrive at the MAC controller in the head-end in a format of one byte, with every cell transmitted from the modem (as mentioned above), partitioned into 2, 3 and 3 bit areas for the 2nd, 3rd and 4th priorities respectively. This information is used to update the 3 outstanding request counters per modem, one for each priority. These counters are holding the total request number for each modem, and are incremented by the priority relevant part of the request byte content for each new cell arrival.

At the same time, to replace the unallocated permits, a search engine scan in a round robin fashion the outstanding request counters and reduces them by one for each permit scheduled. The higher priority counters are inspected first and only if all are empty the same process is repeated for the immediately lower priority. To expedite the process, flags are used to quickly detect and skip all empty locations. The 2nd and 3rd priority dynamically assigned permits are issued provided that the credits are not exhausted. The MAC H/W subtracts the credits as it issues permits and stops serving any modem queue that exceeded its allocated apportionment. This policing action guarantees that 2nd priority malicious users can not disturb complying traffic. As far as for the 3rd priority, the credits guarantee that this class will achieve a better than best-effort service. When credits are exhausted, 3rd priority left-over requests are added to the 4th priority ones, thus a minimum rate is guaranteed but any excess is considered plain best-effort, in accordance with Assured Forwarding rules. The operation of the permit scheduler and the execution of the MAC algorithm are depicted in Figure 1. In the modem side, arrive the permit the cell waiting in the higher priority queue is transmitted.

3. Evaluation of the MAC algorithm

The behaviour of the above MAC mechanisms was studied with the help of a computer simulation model created with the PTOLEMY tool. PTOLEMY is a graphical simulation environment, where different software entities are represented with graphical objects. The derived topology of the AROMA system is further processed in order to generate executable code, which is to be finally simulated.

The scenarios used 10 cable modems loaded with uniform traffic for each priority. Each source used a common ON-OFF model, generating traffic at the slot/cell level, with on and off periods exponentially distributed. The peak rate was equal to the system rate. For the simulations, IP packets with geometrically distributed lengths were considered and were segmented to fit into slots. There was no limit set to credits, since the policing function would only be meaningful to demonstrate with malicious sources that was out of the scope of these tests, since the results are determinist in this case. The duration of the runs was 1.5 million slots because the PTOLEMY model was heavy and slow to run in the Solaris OS available. The upstream slot duration was used as a time unit for the simulation (170.6 μs).

Figure 2 depicts the probability distribution function (pdf) of the access delay for the three lower priorities (2nd, 3rd, and 4th) under a total load of 85%. Only the variable part of the delay is shown, as the fixed round trip time (about 4 slots) required for the modem request (upstream) to reach the head-end and the permit response (downstream) to reach the modem is not counted. The 10% of the load were devoted to CBR traffic (or virtual leased lines by EF in the DS model of usage) through the 1st priority queue of the modems, while the other 75% were equally distributed among the other 3 priorities. The 1st priority is not presented since it exhibits deterministic behaviour, as the transmission permits are pre-programmed in the list [7], and thus, the delay never exceeds the fixed permit distance.

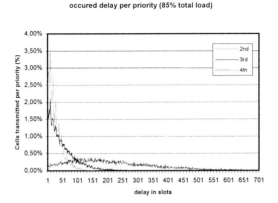

Fig. 2. Pdf of access delay

The most important result caused by the prioritization scheme is the delay advantage provided to the higher priorities. The transmission of almost all the 2nd priority cells delay is less than 250 slots (i.e. 43ms) while those of the 3rd less than 350 slots (60ms). Of course there is no bound for the 4th priority which can exceed any limit depending on the total loading.

Under a higher offered load of 110%, the probability distribution function of delay is shown in Figure 3. The 1st priority occupied 10% of the load while the 2nd and 3rd priority sources were offered a load equal to 30% of capacity each, and 40% for the 4th. As it can be observed from the figure, 2nd and 3rd priorities do not exhibit any significant behaviour difference with the previous scenario, since their service up to the contracted rate is guaranteed. The effect of prioritization is exactly to hide the

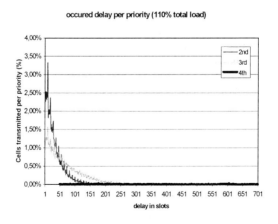

Fig. 3. Pdf of access delay

presence of any lower priority traffic, which is prevented from competing against the protected sensitive traffic. The effect on the 4th priority is very strong since it is throttled down to the 30% of capacity left over from the other three. The occurred delay distribution is thinly spread till theoretically infinite delay. This can not be shown in the simulation results since buffers overflow no matter how much memory is allocated. The average delay is not bounded for the 4th priority, as shown if Figure 4.

However in Figure 3 only the cells that managed to get through are included. Those values are showing a low probability to depart at any delay value as they reflect the seeking from the 4th priority traffic of "holes" left out in the permit generation of the higher priority traffic. In addition to this phenomenon, Figure 5, the line for the 4th priority extends, theoretically, up to infinity, but only a small section of the very large values that occurred in the simulation run is included in the figure. This explosion of delay values is better illustrated in Figure 4, where the buffer fill levels for each priority are shown. The tendency to an almost linear long-term increase of the 4th buffer is clear due to the steady long-term average cell birth generation rate by the ON-OFF source model. The buffer size increases without limit.

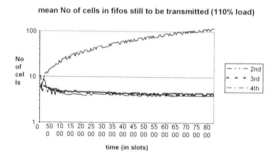

Fig. 4. Buffer size evolution with time.

Except these two scenarios, a set of simulations were performed where the occupied bandwidth in comparison to the total available is within the range from 60% to 110%, where the 10% of the upstream bandwidth was pre-allocated to the 1st priority (by means of programmed permits in the permit list as necessary for the polling of new bursts) and the rest equally distributed among the priorities. Figure 4 depicts the results from these simulations where the delay for the 2nd and 3rd priorities increases very smoothly with the load, leaving the 4th priority to suffer the congestion. The 4th priority delay increases asymptotically towards the 100% line, as expected in any queuing system.

At the same figure we can observe the behavior of the system when no priorities are available except the unsolicited permits of the 1st priority, i.e. the sources which were used for the three other classes are all feeding the same queue, marked "all" in figure 4. The total load is the same with 10% used for the 1st priority and the rest for "all". The benefits of prioritization are clear since no traffic class can enjoy bounded delay in this case. In contrast, when priorities are enforced the 2nd and the 3rd classes enjoy a seemingly lightly loaded medium. Only the last priority sees a performance reduction, which however is well equipped to handle using the TCP layer congestion

mean delay per priority

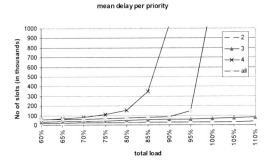

Fig. 5. Mean delay v. total load

Mean delay for 85% load

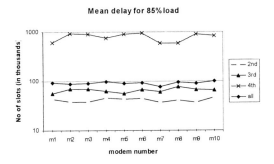

Fig. 6. Mean delay v. modem number

control tools. The delay increases linearly with the total load until the 90% point, where the congestion appears. The "all" line seems to increase asymptotically to the 100% limit.

Finally, as it can be witnessed from Figure 6, the system shares its resources fairly to all the modems involved. For the following instance, it is 25% for 2nd, 3rd and 4th and 75% for the "all", totaling a 85% load when the 10% for the 1st is added. There is an almost linear behaviour of the various modems, for each priority, as their traffic profiles are alike for occupying exactly the same amount of load percentage. The variance from the linear behaviour is justifiable from the random traffic profiles that each modem had.

4. Conclusions

Tree-shaped topologies present attractive cost advantages for broadband access networks by allowing many customers to share the expensive head-end equipment and the feeder section. In addition they offer reuse of the copper last drops to the customer at least during the crucial introductory phase and probably for many years to come. Tree-shaped shared medium access networks such as PONs and HFC effect a distributed multiplexing function which concentrates traffic from many users and many services with diverse requirements. To be able to guarantee that the QoS of sensitive traffic will not be disturbed by best-effort data traffic requires embedding differentiated support for flow aggregates with common requirements.

The under evaluation system thanks to prioritization can guarantee the QoS required by sensitive traffic while exploiting any unreserved bandwidth for the support of best-effort traffic. At higher layers, where the TCP resides, congestion avoidance would exercise flow control in response to losses, while re-transmitting lost packets, and reduces the rate, leading to a quite efficient overall system. The high priority traffic represents high revenue services, enjoying satisfactory performance, while the accommodation of 4th priority best-effort traffic exploiting the spare capacity, as it becomes available, guarantees a high system utilization. This is achieved with no harm to the TCP-based user applications, since in real-life instead of a tendency to infinite buffer overflows and unbounded delays, the rates of the sources would adapt to the possible bottleneck sharing the spare capacity equally among them.

A method to guarantee resource availability for the services with demanding quality, such as simple over-provisioning or fully dynamic SLAs (e.g. RSVP) would be of course necessary. Any such scheme should of course allow for some spare bandwidth to protect plain best-effort service users from the frustration of bandwidth starvation. However, as it would only involve longer term averages and not rule out brief intervals with the full capacity taken up by high priority traffic, (which can not use closed loop congestion control), the bandwidth devoted to the 4th and partly the 3rd priority would provide the necessary leeway for the statistical gain of the preventive control based services. In other words, if we did not have the lower class we would have either to accept less demanding- quality traffic or leave the excess bandwidth unutilized in an effort to avoid performance degradation during statistical extremes. It is the availability of closed loop congestion control of TCP that allows us to offer both guaranteed performance to demanding traffic and good system resource utilization, at the same time.

Acknowledgements

The work presented in this paper was partially funded by the EU ACTS project AC327 "AROMA". The opinions appearing here are those of the authors and not necessarily of the other members of the AROMA consortium. We thank also Prof. J. D. Angelopoulos for his fruitful guidance as and Th. Orfanoudakis and J. Sifnaios for their valuable help.

References

1. IETF, Differentiated Services Working Group, RFC 2475 "Architecture for Differentiated Services", December 1998
2. Van Jacobson, Kathleen Nichols, Kedarnath Poduri, Internet Draft, draft-ietf-diffserv-phb-ef-02.txt, "An Expedited Forwarding PHB", February, 1999

3. Juha Heinanen, Fred Baker, John Wroclawski, Internet Draft, draft-ietf-diffserv-af-06.txt "Assured Forwarding PHB Group", February, 1999

4. J.D.Angelopoulos, G.C. Boukis, I.S.Venieris, "Delay priorities enhance utilization of ATM PON Access Systems", *Computer Communications Journal,* Elsevier, Vol. 20, No , December 1997, pp. 937-949

5. J. D. Angelopoulos, Th. Orphanoudakis, "An ATM-friendly MAC for traffic concentration in HFC systems", *Computer Communications Journal,* Elsevier, Vol. 21, No. 6, 25 May 1998, pp. 516-529

6. J. D. Angelopoulos, N. I. Lepidas, E. K. Fragoulopoulos, I.S. Venieris, "TDMA multiplexing of ATM cells in a residential access SuperPON", *16 Journal on Selected Areas in Comm.,* Special issue on high capacity optical transport networks, Vol. 16, No. 7, September, 1998

7. ATM Forum Technical Committee, Traffic Management Group, "ATM Forum Traffic Management Specification Version 4.0" Doc. 95-013R11, March 1996

Performance Evaluation of RMTP Using NS
&
RMTP Integration in Current Networks

T. Asfourl and A. Serhrouchnil

ENST Ecole Nationale Supérieure des Telecommunications, Paris, France
{asfour, ahmed}@enst.fr

Abstract In this paper we focus on reliable multicast services. We present our work concerning the integration of RMTP in the network simulator NS. The resulting model allows us to simulate the performance of RMTP-based reliable multicast services. The simulation results of eight different scenarios presented in this paper show the poor performance of RMTP when used to multicast data to receivers in "heterogeneous" environments. Another question that we address in this paper is the scalability limitation of RMTP in spite of its hierarchical design. This limitation is due to several reasons like the use of the hierarchical acknowledgments (HACKS), the unlimited number of receivers that can be associated to a designated receiver, and the buffering space required in the sender and in the DRs when some or many receivers experience high loss rates. We discuss as well the challenges that face the integration of RMTP in current networks and its future coexistence with other multicast as well as unicast protocols.

1   Motivation

In nowadays networked, computerized world the word "reliability" is associated to almost every product and service. In protocol engineering terms the word "reliability" has more than one interpretation according to the requested service. We define the service in our context as a function of three main factors: time, data, and number of customers that request it. Reliability in time sensitive service is mainly a function of some or many temporal factors like delay and synchronization signals, while it is mainly a function of data factors like data size and loss rate in data sensitive services.

Services with two involved end points are called unicast services. In multicast more than two end points are involved in the communication. File transfer between a sender and a receiver is an example of a data sensitive unicast service. Videoconferencing and multi-party games are examples of time sensitive multicast services. In our paper we focus on reliable multicast services only.

Multicast services can be built directly using IP-multicast [6] at the network level. The main point of the architecture of this protocol is that senders and receivers don't need to know each other explicitly. It is sufficient to know one group address, which makes the distribution of data simple and scalable. In contrast, this "anonymity" prevents sources from necessary feedback information about sent data, which means that services built directly on IP-Multicast are not reliable.

Transport layer multicast protocols have been proposed to achieve reliability, like RMTP 10,161, LGMP [9], MFTP 151, SRMTP [3,4), MDP [12] SRM [8] OTERS [5], etc. In our paper we focus on RMTP as an example of a tree-based reliable multicast protocol. A multicast service based on RMTP is defined as a function of three factors: throughput, delay, and number of receivers. The quality of this service is evaluated. RMTP is simulated using the network simulator NS, and the source code will soon be available for public use. The integration of RMTP in current networks is discussed from both commercial and technical points of view. The rest of this paper is organized as follows: In the first section we introduce RMTP and in the second section we present a simplified RMTP model that we have integrated in the network simulator NS, and we describe as well the simulation environments and the five resulting agents and their conceptual protocol flow. The third section describes the simulation scenarios and the fourth section shows the results of the simulation. In the fifth section we discuss the integration of RMTP in current networks and its coexistence with other transport protocols in the network. Finally the last section will conclude this paper and give some perspectives.

2   A Tree-based Multicast Transport Protocol: RMTP

RMTP [10,16] is an example of a tree-based multicast transport protocol. Receivers are divided into subgroups depending on their geographical location and organized in a tree-like hierarchy. In each subgroup a designated receiver (DR) is responsible for processing receivers status messages and for local recovery of lost packets. The sender multicasts data to all the receivers, which share the same IP-Multicast address. Receivers send their status periodically to the associated DR, who in turn plays the role of the sender in its local region and multicasts lost packets, if possible, to the receivers in its local region. The DRs send their status periodically to the sender. The sender processes these status and performs global recovery. Figure 1 shows the topology of this protocol.

All receivers share one IP-Multicast address, and this is the address used by the sender for multicasting data. In each local subgroup, receivers share an IP-Multicast address that will be used by the subgroup associated DR for local recovery. Receivers send their status to the associated DR by unicast, and DRs in turn send their status also by unicast back to the sender. Figure 1 shows a one level hierarchy of DRs. It is quite possible to have multiple hierarchical levels, in other words a subgroup can be split into several subgroups, and in consequence we obtain a recursive hierarchical structure. In Figure 1 we see that the DRs send their status to the "Top Node" and not directly to the sender. The top node has a network control function and is mainly used to provide network managers with tools for monitoring and controlling the performance and network utilization of large applications.

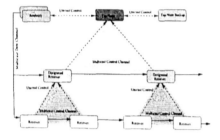 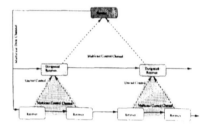

Fig. 1. RMTP Architecture

Fig. 2. RMTP Simulation Prototype Architecture

In previous version of RMTP this top node does not exist and status messages are directly sent to the sender. Any way the sender and the top node have a duplex unicast connection, and status messages will be aggregated or forwarded from the top node to the sender, and this is why we allow ourselves to say that DRs send status messages to the sender. RMTP receivers send HACKs (Hierarchical ACKS) to notify the reception and the loss of data packets. Additional options can be used like, NACKs and FEC for increasing scalability and average latency. In our paper we don't consider these options, and our simulation also does only consider the main RMTP features.

3 Simulation model

In this section we begin by presenting the main features of RMTP that will be simulated and integrated in NS, and then we show the simulation model entities, as well as the relation between these entities.

3.1 RMTP Model

In the previous section, Figure 1 presents the architecture of RMTP as described in [11]. Our model, shown in Figure 2, excludes the top node and therefore the DRs send their status directly to the sender. The following main features will be supported by our model:

- The sender multicasts data to all receivers and DRs using the multicast Data channel.
- Receivers in each subgroup send HACKs periodically by unicast to the associated DR.
- DRs send HACKS periodically by unicast to the sender.
- In each subgroup the DR multicasts loss packets to the subgroup using the multicast control channel.
- The Sender achieves global recovery.

– A congestion control and avoidance algorithms,that will be described later, is used by the sender to prevent overloading the network.

3.2 Simulation environment

The network simulator NS [7] is used to simulate RMTP. NS is an object oriented simulator from U.C. Berkeley/LBNL. It is written in C++, with an OTcl interpreter as a front end. The simulator is event driven and runs in a non real time fashion. A rich set of protocols in different layers have already been integrated

in the simulator. Some examples for available protocols are FTP, TELNET, HTTP, TCP, SRM, and UDP. NS is a free software for network simulations, and it is possible for users to integrate new protocols and new classes into. Some Reliable multicast protocols like SRM [8], MFTP [15], and OTERS [5] have already been integrated in the simulator.

3.3 RMTP Agents

To integrate RMTP into NS we defined a set of classes with a class hierarchy shown in Figure 3.

The *RMTP* class contains common functions and variables for both RMTP sender and receiver. The *Snd* class presents the sender, and the *Rev class* presents the receiver. The designated receiver is the combination of the two classes *SndDr* that inherits from the *Snd* class and *RcvDr* that inherits from the *Rev class*. This combination is due to the dual role of the DR as a sender, for local recovery in its local region, and as a receiver at the same time. The same class hierarchy in C++ is mirrored in OTcl.

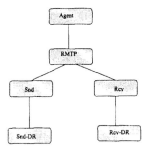

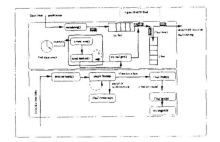

Fig. 3. RMTP: Class hierarchy within NS

Fig. 4. Sender Block Diagram

RMTP class contains common functions and variables for both RMTP sender and receiver. *Snd* class presents the sender, and *Rev* class presents the receiver. The designated receiver is the combination of two classes *SndDr* that inherit the *Snd* class and *RcvDr* that inherit the *Rev* class. This combination is due to the binary role of the DR *as* a sender, for local recovery in its local region, and as a receiver at the same time. The same class hierarchy in C++ is mirrored in OTcl.

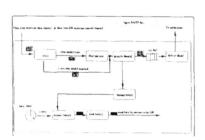

Fig. 5. Receiver Block Diagram

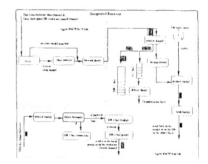

Fig. 6. Designated Receiver Block Diagram

3.4 RMTP sender functional block diagram

Figure 4 shows the conceptual protocol flow of the RMTP sender. The actual simulation of RMTP with NS is based on these diagrams. The sender is responsible for receiving data from the application layer and for splitting it into RMTP data packets. Packets are sent for buffering in what we call "transmission buffer", and there it waits for a send signal. The sender multicasts data packets at regular intervals defined by the parameter *interval-*. Regulating the send intervals is done by the use of a timer that expires after the delay *interval-*, and upon expiration the timer is rescheduled. The sender uses three variables to control the number of transmitted packets in each interval: Ws-, *cwnd_* and *use-win-* The maximum number of packets transmitted in *interval-* is Ws-. At the beginning of each interval the sender computes the number of lost packets *nb_lost_* in the previous interval by checking the bitmap of the last HACK processed. The following algorithm will be then applied for the calculation of *use-win-:*

```
if            nb_lost_  >  CongC_

    closecwnd();

else if nb_lost_ > CongA_

    mincwnd();

else

    opencwnd();

use _win_ = window();
```

The *window()* function returns the minimum of *cwnd-* and Ws-, while *closecwnd()* closes the congestion window i. *e. cwnd_* = *1* when *nb_lost_* exceeds a congestion control threshold *CongC_* , *and mincwnd()* minimizes the congestion window to Ws-/2 when *nb_lost_* exceeds a congestion avoidance threshold *CongA_*, and finally *openewnd()* open the congestion window exponentially, i. e. *cwnd_ = cwnd- + 1*. The above algorithm is a TCP-like exponential slow-start algorithm for congestion control and avoidance.

After the number *k = use* -win- of packets to be sent is determined, the sender calls *send-packet()* *k* times, at each call the *send-packet()* gets a packet from the transmission buffer and sends it to the underlying network layer to be multicast using the multicast data channel. A copy of this packet is sent as well for buffering, in what we will call "final buffer" for possible retransmission requests. In our current study we don't consider the buffer size limitation question for both transmission and final buffers, we suppose that both buffers are big enough and no loss due to their size can be occur.

The RMTP sender receives DRs status messages, which arrive periodically, and aggregates these messages. Based on the bitmap included in the aggregated message the sender checks for well received and lost packets. If a packet is declared lost from any of the DRs the sender will look for this packets in its final buffer and reinsert it in the head of the transmission buffer. The sender will remove all well received packets from its final buffer.

3.5   RMTP receiver functional block diagram

Figure 5 shows the conceptual protocol flow of the RMTP receiver. The RMTP receiver listens to both the multicast data channel and the associated DR control channel. If an error model is attached to the receiver, packets will be filtered using this error model, and packets that are not dropped will be forwarded to the *process-data()* function. If no error model is attached to the receiver, received packets will be forwarded directly to the *process-data()* function. This function sends the received packet for buffering before it can be delivered to the application. It is as well responsible for setting the bit corresponding to the received packet to 1 in the bitmap. Hacks will be sent in fixed intervals controlled by a timer. At the expiration of this timer a HACK packet is prepared and sent to the associated DR, and the timer is rescheduled.

In our simulation model we choose to attach error models to receivers and not to the sender, which seems more logical, since errors can occur anywhere: in the sender, in network or in receivers, but finally it is up to the receiver to detect these errors. In other words our simulation model considers an accumulated error model attached to receiver.

3.6   RMTP DR functional block diagram

Figure 6 shows the conceptual protocol flow of a RMTP designated receiver. The DR has a sender's and a receiver's role, and this is why we split it into two blocks. The first block represents the DR receiver block which has the main data reception function like end-point receivers. In contrast, the DR receiver block is not responsible for HACK reception and sending, this function is moved to the DR sender block. The DR sender block completes the reception data process and keeps a copy of received packets in a DR final buffer for possible retransmissions. The DR sender block is responsible for sending HACKS periodically to the sender or to upper DRs. It receives HACKS from end point receivers and aggregates these hacks. Then it checks the bitmap in the aggregated HACK and removes well received packets from its final buffer. The DR sender block looks for lost packets in its final buffer, and found packets are retransmitted by multicast to the local group using the DR multicast control channel. In contrast to the main sender, the DR sender block has a limited final buffer size, by consequence packets may not be saved for possible retransmission in the final buffer.

4   Simulation topologies

In order to create topologies for multicast scenarios we wrote a generic program that takes two parameters: the number of DRs and the number of receivers attached to each DR. Using NS facilities, bandwidth, delay and link characteristics are defined very easily which allows us to generate very different scenarios in order to simulate the behavior of RMTP in different network conditions. Eight different scenarios shown in Figure 7 will be presented in the paper. Each of the first six scenarios is a tree with the sender as the root, one internal node as the DR and 50 receivers. Different link characteristics are associated to each scenario, with error models attached to some or all receivers. Performance criteria are: Number of sender retransmissions *(Snd_ret)*, number of DRs retransmissions *(DR-rett))*, average throughput for each receiver and DR (TH), and average delay per packet for each receiver and DR *(D)*.

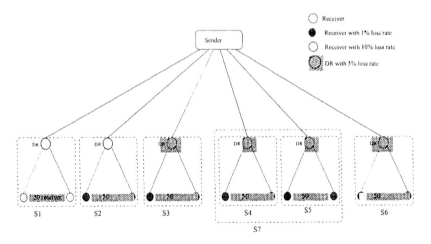

Figure 7. The topologies of all the nine scenarios

The first six topologies have the same multicast tree structure of one sender, one DR, and 50 receivers but with different link characteristics and attached error models. Table 1 give a summary of the different characteristics of the five scenarios.

Table 1. Bandwidth, loss rate and delay characteristics of the first six scenarios

Topo#	BW Snd-DR	D Snd-DR	BW DR-Rcv	D DR-Rcv	Error modell
1	512kb/s	40ms	64kb/s	10ms	10% first receiver
2	512kb/s	40ms	64kb/s	10ms	1% all receiver
3	512kb/s	40ms	64kb/s	10ms	1% all receivers, 5% DR
4	1Mb/s	40ms	512kb/s	10ms	1% all receivers, 5% DR
5	128kb/	40ms	28kb/s	10ms	1% all receivers, 5% DR
6	512 kb/s	200ms	64kb/s	30ms	11% all receivers, 5% DR

In the seventh scenario the sender will open a multicast session simultaneously with the fourth and fifth subgroups presented respectively in the fourth and fifth scenarios. While in the eighth scenario the sender opens a multicast session with two identical subgroups, each one has the topology of the subgroup in the fourth scenario. All scenarios have the same window and congestion control parameters as well as send data and HACK intervals. A summary of these parameters are given in Table 2. Finally, the simulation time of all simple topologies, i. e. which have only one subgroup, is 500s and for each combined topologies, i. e. which have more than one subgroup, is 250s. This is due to the heavy NS runtime load caused by memory and CPU consumption during simulation, which increases with the number of active agents.

Table 2. Send data and congestion control parameters

interval-	2 s	CongC_	30 packets
Rcvhack_int-	0.3 s	Ws-	10 packets
Dr_hack_int_	0.8 s	CongA_	20 packets
f-bu, f-7R	500 packets	f-bu f_snd_	not limited

5   Simulation results

The retransmission simulation results in Table 3 show that the local recovery role of DRs becomes limited when the network conditions become "worse", and in consequence in these conditions it is up to the sender to retransmit packets. In the first topology the DR is very efficient and it performs 327 retransmissions and protects the sender that has to retransmit only 9 packets. The 9 retransmissions does not mean that there were exactly 9 lost packets which the DR was enable to recover. Since HACKs take time to arrive at the sender, during this time the sender keeps sending data. When a HACK arrives to the sender, the sender considers that the packets sent and not acknowledged in the current HACK are lost, even if it is not the case. This is why it is important to send HACKS with a certain frequency to avoid this "misunderstanding". On the other hand if DRs send HACKS with a very high frequency, a lot of HACKs will be useless since they have the same information as older HACKs. This is why it is very important to choose the *DR_hack_in_* very carefully. This relation between sender-DR delay and the number of sender retransmissions illustrates the increase *of*

Table 3. Retransmission Results for all scenarios

Topo#	Snd_ret	DR1_ret	DR2_ret	simulation time
1	9	327		500s
2	141	1472		500s
3	727	3120		500s
4	116	541		500s
5	2284	370		500s
6	1021	2709		500s
7	816	135	374	250s
8	31	465	444	250s

the number *of* retransmissions from 727 in the third scenario to 1021 in the sixth one. The third and the sixth scenario have the same attached error models and the same bandwidth characteristics but differs in delay criteria. In consequence two very important questions have to be addressed: How often HACKS must be sent from DR to sender and respectively from receivers to DR, and what is the relation between HACKS frequency and the round trip time (RTT). In our previous illustration we referred to the 300ms delay between the sender and the DR in the sixth scenario as the responsible cause for the big number of sender retransmissions. It is rather the RTT difference that illustrates this result, but since the two topologies have the same link speeds and no other flows than the RMTP flow exist, we can base on the delay instead of the RTT.

The results of the third scenario show that despite the important role of the DR in local recovery (3120 retransmissions), the number of senders' retransmissions (727) is high in comparison to the second scenario (141). This result is due to the fact that the final buffer itself suffers from loss so more loss indications will arrive at the sender. To illustrate the impact of the error model attached to the DR we suppose that : r is the probability of packet loss associated to each receiver, and q *is* the probability of packet loss associated to the DR. If we consider that the underlying routing tree is the same as the virtual multicast tree then the probability P that a packet is received well by the n receivers is: $P = (1 - q)(1 - r)^n$. In consequence the probability P' that at least one loss indication per sent packet will arrive to the DR is given by the formula

$$P'=1-P=1-[(1-q)(1-r)'1] \tag{1}$$

For the second scenario where *(q = 0%, r = 1%, n = 50)* the value of P' is 39%, while for the third scenario *(q = 5%, r = 1%, n = 50)* P' is equal to 42%. Unfortunately the final buffer size in the DR is limited, and the DR itself suffers from loss which makes the DR unable to reply to all retransmission requests, and it sends in turn more *loss* indications to the sender. In Equation 1 we notice that when the number *of* receivers $n \rightarrow oc$, $P \rightarrow 1$, which means that every sent packet has to be retransmitted.

The retransmission results *of* the fourth scenario, which enjoys good bandwidth criteria, show that the sender "misunderstanding" of received HACKS is decreased. This is due to the small delay that a HACK takes to reach the

sender. In contrast the retransmission results of the fifth scenario where the sender "misunderstanding" of DR's hack is increased and the DR does not fulfill its role of local recovery due to bad adjustment of the *Rcv_hack_int_* parameter in function of the RTT. The impact of the bad adjustment of *Rcv_hack_int_* and *DR-hack_int_* illustrates as well the big number of senders' and DR' retransmission in the sixth scenario where propagation delay criteria are the "worst" among other scenarios.

In contrast to the first six scenarios, the simulation time of the seventh and the eighth scenario is *250s*. The retransmission results show that the DRS' role of protecting the sender from receivers' feedback become limited when the number of subgroups increases and especially when subgroups have different criteria.

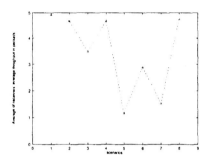
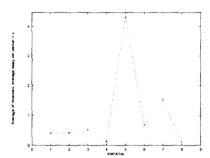

Fig. 8. Throughput results of all scenarios   **Fig. 9**. Delay results of all scenarios

Figure 8 shows the average throughput results for all scenarios. We notice that the best average throughput is achieved in the first scenario *(4.9packet/s*

Ws-/2) where only one receiver suffers from a 10% loss. The average throughput decreases to *4.6packet/s* due to the 1% loss rate in all receivers, and it decreases considerably to *3.5packet/s* with a 5% loss rate in the DR. Despite error models attached to the DR and to all receivers, good bandwidth criteria in the fourth scenario increase the throughput up to *4.67packet/s*. This is due to the quick recovery of lost packets. The impact of bandwidth limitation on the throughput average is very clear in the fifth scenario where the group achieves its worst throughput in comparison to other scenarios. In the six scenario where the multicast session is carried simultaneously to two bandwidth heterogeneous subgroups the average throughput is determined mainly by the worst subgroup. This result shows the limitation of RMTP performance when carrying one multicast session simultaneously to multiple subgroups with a big "intervariance" of link speed among subgroups. In the seventh scenario the sender carries one multicast conversation with two identical subgroups (identical to the subgroup of the fourth scenario). As we can see in Figure 8 the throughput of the multicast session of the seventh scenario is high and it is very close to the throughput of the fourth multicast session. In other words the average throughput of an RMTP multicast session carried to multiple subgroups with a small bandwidth "intervariance" does not decrease significantly with the increase of the number of subgroups.

Figure 9 shows the average of receivers' average delay per packet of the multicast session. The worst scenario in terms of delay results is the fifth scenario and this was expected due to the bandwidth limitation of the scenario's links. In contrast the average delay is minimal in the fourth and the eighth scenario due to good bandwidth criteria. We notice as well that the average delay increases when loss increases, and this is due to the increase of the number of necessary retransmission and in consequence lost packets experience more delay before arriving at receivers.

6   RMTP integration in current networks

In the previous sections we noticed that RMTP faces critical problems when used to multicast data to heterogeneous and high big groups. The use of periodic HACK messages increases the network load as well as the processing load on both sender and receivers. We noticed as well the importance of adjusting the intervals at which receivers and DRs send their HACKS. The use of RTT in order to adjust these intervals becomes very critical in heterogeneous environments where receivers, respectively DRs, have different RTT values. In RTP/RTCP [13,14], where the RTP data protocol is responsible for data transmission and RTCP is responsible for control functions, the receivers send their status periodically to the sender. In contrast to RMTP and in order to improve the scalability, the rate of RTCP periodic state messages is adjusted in function of the number of receivers in the multicast session.

Another obstacle facing RMTP is the coexistence with other protocols in the network. Until today most applications are unicast and use TCP. So to be able to survive, RMTP must not affect TCP sessions. In other words there must be a fair share of network resources between the two protocols. A lot of reliable multicast protocols are proposed in order to have a good adaptation between services and protocols. Any way RMTP and any other multicast protocol have to take into consideration the interoperability with other protocols.

RMTP uses the strategy of recovery after loss was experienced. This strategy is a direct consequence of the current best-effort network. A lot of work in the QoS domain is achieved and we think that it is a good idea to give this protocol some extensions based on QoS criteria in order to deal with the reasons for loss and not with loss recovery. This is one of our objectives for further works in the multicast domain.

7   Conclusion

In this paper we presented our work concerning the integration of RMTP in the network simulator NS. The resulting agents are used for the performance evaluation of RMTP. The results of eight different scenarios show the poor performance of RMTP when used to multicast data to a heterogeneous set of receivers. The mechanism used by RMTP in order to split the set of receivers into subgroups is based on the geographical proximity of receivers. In addition to the problem of determining the geographical location of hosts in actual IP networks, this mechanism does not take into consideration the receivers' QoS, therefore the resulting subgroups might be very heterogeneous. The intrasubgroup heterogeneity affects the DR role in local recovery, and in consequence more retransmissions have to be done by the sender. The intersubgroup heterogeneity leads to a very poor throughput performance, since the throughput of "good" subgroups is controlled by the throughput of the "worst subgroup". On the other hand the hierarchical structure of RMTP is a good strategy for protecting the sender from receivers' feedback and avoiding the implosion problem. This protecting role becomes limited when the DRs become unable to fulfill their local recovery mission due to the heterogeneity of the set of receivers in the associated local group. We discussed as well some challenges that face the integration of RMTP in current networks where TCP is still the dominant protocol. In a parallel work that we presented in [1, 2] we proposed a new mechanism for multicast group management that we called CGM. The main idea of this mechanism is to split the set of receivers in a multicast session into homogeneous subgroups based on their QoS criteria rather than their geographical distance like in RMTP. We used the CGM mechanism mainly with SRMTP [4] which stands for Scalable Reliable multicast Transport Protocols. The results that we have obtained encourage us to use the same mechanism with RMTP, and this is the subject of our current work in this discipline.

References

1. Taghrid Asfour, Stephan Block, Ahmed Serhrouchni, and Samir Tohme. Contractual group membership: a new mechanism for multicast group management. In *Proceedings of The Fifth IEEE Symposium on Computers and Communications* (ISCC'00), France, Antibes, July 2000.
2. Taghrid Asfour, Stephan Block, Ahmed Serhrouchni, and Samir Tohme. New *QoS* Based Mechanism for Heterogeneous Multicast Group Management. In *Proceedings of the 5th INFORMS Telecommunications conference, US,* Florida, March 2000.
3. Stephan Block. The Design of a Scalable Reliable Multicast Protocol. Technical report, Universität Stuttgart and Ecole Nationale Supérieure des Telecommunications (ENST), Paris, 1998
   Stephan Block, Ken Chen, Philippe Godlewski, and Ahmed Serhrouchni. Some Design Issues of SRMTP, a Scalable Reliable Multicast Transport Protocol. In Helmut Leopold and Narciso Garcia, editors, *Multimedia Applications, Services and Techniques: Proc. 4th European Conference - ECMAST'99,* volume *1629 of Lecture Notes* in *Computer Science,* pages 423-440, Madrid, Spain, May 1999. Springer Verlag.
5. D. R. Cheriton and D. Li Cheriton. OTERS (On-Tree Efficient Recovery using Subcasting): A Reliable Multicast Protocol. In *Proceedings of 6th IEEE International Conference on Network Protocols (ICNP'98),* pages 237-245, Austin, Texas, October 1998.
6. S. Deering. Host Extensions for IP Multicasting. RFC 1112, August 1989.
7. Kevin Fall and Kannan Varadhan. NS Notes and Documentation. Technical report, UC Berkeley, LBL, USC/ISI and Xerox PARC, January 2000.
8. Sally Floyd, Van Jacobson, Ching-Guang Liu, Steven McCanne, and LiXia Zhang. A reliable multicast framework for light-weight sessions and application level framing. *IEEE/ACM Transactions on Networking,* 5(6):784-803, December 1997.
9. Markus Hofmann. Impact of Virtual Group Structure on Multicast Performance. In C. Diot A. Danthine, editor, *From Multimedia Services to Network Services. Fourth International COST 237 Workshop,* volume 1356 of *Lecture Notes in Computer Science,* pages 165-180, Lisboa, Portugal, 1997. Springer Verlag.
10. Sanjoy Paul, John C. Lin, and Supratik Bhattacharyya. Reliable Multicast Transport Protocol (RMTP). *IEEE Journal ore Selected Areas in Communications, special issue on Network Support for Multipoint Communication,* 1996.
11. Sanjoy Paul, Brian Whetten, and Grusel Tascale. RMTP-11 Overview. Technical report, Talarian, Lucent Reuters, September 1999.
12. Luigi Rizzo and Lorenzo Vicisano. A Reliable Multicast data Distribution Protocol based on software FEC techniques. Technical report, Dipartimento di Ingegneria dell'Informazione, Università di Pisa and Department of Computer Science, University College, London, April 1997.
13. H. Schulzrinne. RTP Profile for Audio and Video Conferences with Minimal Control. RFC 1890, January 1996.

14. H. Schulzrinne, S. Casner, R. Frederick, and V. Jacobson. RTP: A Transport Protocol for Real-Time Applications. RFC 1889, January 1996.
15. StarBurst Software. StarBurst MFTP-An Efficient, Scalable Method for Distributing Information Using IP Multicast. http://www.starburstcom.com/white.htm, 1998.
16. B. Whetten, M. Basavaiah, S. Paul, T. Montgomery, N. Rastogi, J. Conlan, and T. Yeh. RMTP-11 Specification. Internet Draft, Internet Engineering Task Force, April 1998.

Active Virtual Private Network Services on Demand

Alex Gali[1], Stefan Covaci[2]

[1]University College London
United Kingdom
[2]PopNetAgentScape
Germany

Abstract This paper focuses on the role of the convergence of network and service management technologies in providing a support for the increased intelligence required in the provision of management solutions, and the mobility of such solutions. Such convergence benefits the interoperable services characterised by high distribution, a dynamic nature, and the complexity of used network resources. It is based on the research results achieved in the ACTS AC338 project MIAMI (Mobile Intelligent Agents for Managing the Information Infrastructure) [1], [14] and [9].

The MIAMI project focused on the use of mobility of software intelligence agents and the benefits of mobility in the context of *active service management*. MIAMI developed a unified Mobile Intelligent Agent framework able to meet the requirements of an emerging global open information infrastructure. Further, MIAMI created a new service management application: the Active Virtual Pipe (AVP): a dynamic Virtual Private Network Service on demand, to demonstrate the benefit of Mobile Intelligent Agent technology for network and service management.

This paper discusses the advent of mobile agent technologies and how they may enhance traditional connectivity management services in a form of Active Virtual Private Networks services, making them dynamically customisable by consumers

Keywords: Active Services, Virtual Private Networks, Virtual Enterprise, Mobile Agents, Miami Project

1. Introduction

A Virtual Enterprise (VE) is defined here as a temporary federation of autonomous, legally and economically independent companies that collaborate on a common business goal, taking place in electronic "virtual" space. This type of consortium is formed for a particular task, to share the knowledge, competence, resources, and business background of the participants, in sum, to contribute to the overall business goal of the VE. The federal character of the VE matches the diversity and complexity of the market offerings and products' structure. It also increases the flexibility for reacting to the rapidly shifting demands of today's markets. A VE usually consists of several partners situated at different geographical locations. The Active Virtual Pipe (AVP) provides location transparency, so that physical separation of VE members is not an obstacle to effective business activities. During the life cycle of the VE the various partners will be engaged in adding value to the other's product or service (build-up a value chain) according to the contract. A major characteristic of a VE is that all partners share the risk and the profit of the joint venture. Each member pursues the strategy of extensive "outsourcing" and concentrates on its core competencies.

Many VEs focus predominantly on information and communication based products, so there is a need for business data to be transported between the participants over an information transport network. This puts heavy demands on the underlying network infrastructure. The transportation of the business process data must cope with the different constraints of each individual business process. In addition, the stakeholders of a VE often belong to different network management domains as they are situated in different geographical locations. A successful VE requires a very close co-operation among all participants, as with a single company.

The above reasons highlight a VE's need for a highly dynamic and flexible information infrastructure able to provide connectivity with guaranteed quality on demand. The major drawback of conventional solutions such as the currently available virtual private networks (VPN), is their low-level adaptability to the changing requirements of the business processes, and, for the most cases they provide few QoS guarantees (e.g. bandwidth) or no guarantees at all (e.g. VPNs via the internet). The MIAMI project provides an extensive VE support environment, for the creation, operation and administration phases (including the necessary resources allocation) in an automated way. Figure 1 gives a simplified view of the proposed environment.

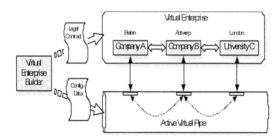

Fig. 1: A Virtual Enterprise Support Environment

2. Active Virtual Pipe service on demand

A VE usually consists of several partners situated at different geographical locations. The AVP will provide transparency between the members of the VE so as to create what is virtually a single company. The AVP is a programmable resource of the information infrastructure that supplies advanced communication and connectivity services with guaranteed QoS to the VE.

The Active Virtual Pipe (AVP) is defined as a programmable, dynamic Virtual Private Network with QoS guarantees that can be directly configured according to the demand of the VE. It as an abstract view of the dynamically (self) configurable global connectivity service that controls the transfer of telecommunication data streams. For instance, there is a possibility that the quality of service parameters of existing connections can be configured dynamically by the VE in accordance with the current business activities. Since the does not own the transport networks' resources, it makes use of the networks of several connectivity providers (CP), which can be dynamically selected and configured on demand. Due to this capability, the AVP is in the position to offer a cost efficient, dynamic, virtual private network, well suited to the needs of VEs.

The AVP is a programmable resource of the information infrastructure that supplies a Virtual Enterprise (VE) with advanced communication and connectivity services, and a guaranteed QoS. AVP is a novel example of an Active Service [34]. It provides an abstract view of the dynamically (self) configurable global connectivity service in charge of the transfer of telecommunication data streams. It is a programmable, dynamic, QoS guaranteed,

Virtual Private Network that can be directly configured according to the demand of the Virtual Enterprise, and it provides transparency between the members, regardless of their physical location.

Additionally, the AVP being the central "node" in the VE that represents the "glue" between all partners includes a generic workflow management facility [2], [3]. This facility can be programmed according to the needs of the business processes by the VEB. It enables an efficient, one-stop-shopping inter-organisational workflow management and provides a generic VE administration service to any VE. Furthermore the AVP allows for outsourcing of common VE management activities in order to reduce the expenditure of running a VE. Partners of the VE can focus on their own business logic and outsource the management overhead of the overall VE to the AVP. The AVP can establish a business-process-driven relationship between the service management layer and the network management layer because of its central management function and its role as an advanced connectivity provider. The AVP allows for a business process driven connectivity management.

The AVP monitors and controls the inter-organisational workflow and is able to derive appropriate the connectivity requirements used to configure the underlying network infrastructures. The workflow management facility provides the VE with a connectivity management interface at the level of workflow. The services of the AVP are provided through dynamically programmable interfaces that can be customised by the VE.

3. Miami Design Approach

Figure 2 shows an overview of the main components of the MIAMI system.

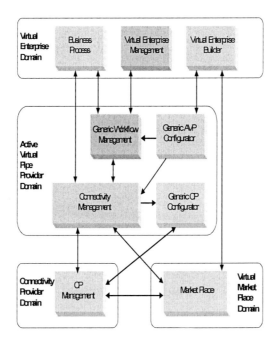

Fig. 2: MIAMI Components Overview

VE Domain: The VE domain contains three functional facilities, the VE Builder (VEB), the VE Management and the Business Process. The VEB provides support for the definition of business processes and connectivity requirements, the search for suitable business partners including AVP Providers, negotiation among these partners, creation of the VE contract, and configuration of the selected AVP according to the contract. The finding and selection of suitable business partners and AVP Providers is achieved by the use of mobile intelligent agent technology. The configuration of the selected AVP Provider is achieved via a programmable interface. The VEB generates configuration data, which includes the definition of the business process and connectivity requirements, and programs the AVP accordingly. As a result of the configuration, a number of interfaces are instantiated and provided to the VE. These interfaces are realised by specialised dynamically created static agents that give the VE the capability to adapt the interfaces exactly to its needs.

Active Virtual Pipe Provider Domain: The AVP Provider offers the VE a programmable information infrastructure. The Generic AVP Configurator receives the configuration data from the VEB and creates and configures all runtime facilities and a Connectivity Management facility, and interfaces accordingly. The AVP Provider negotiates & leases basic network service resources from different Connectivity Providers and provides high-level connectivity services to the VE. For the selection of the appropriate Connectivity Providers, the Virtual Market Place is searched for reasonable offers. The Connectivity Management facility and the Generic CP Configurator manage the Connectivity Providers. The VEB and Virtual Enterprise Management (VEM) together form a workflow management system.

Connectivity Provider Domain: The Connectivity Provider domain comprises one or more transport networks owned by different Connectivity Providers. Connectivity Providers give VEs the connectivity they need by supplying the Connectivity Provider Management facilities, which can be configured by the Connectivity Management facility and the Generic CP Configurator of the AVP. The Connectivity Providers (CP) manage the resources of the transport networks. The CP offers a service management interface to the AVP based on the management policies configured by the AVP Provider. The Connectivity Providers dynamically advertise on the Virtual Market Place service resources that are temporarily available, which can be bought by AVP Providers.

Virtual Market Place Domain: The Virtual Market Place domain contains a number of different market places, which can be accessed by the Subscriber, the VE Builder, the CP Management, and the Connectivity Management facility. These market places allow potential business partners, Virtual Enterprises, AVP Providers and Connectivity Providers to advertise their offers. The VE Builder searches for offers from product providers or sends its own product offers. The CPs contacts the market place to offer their services and free service resources. The AVP Provider searches the market place for reasonable CP resources, which can be dynamically allocated in order to fulfil VE's requests.

4. Network Scenario

The MIAMI project assumes a network scenario consisting of an end-to-end IP network which interconnects the participants of a Virtual Enterprise (VE). In addition to standard best-effort, Internet quality connectivity for general email and web browsing, the VE users also require access to higher quality connectivity facilities for real-time services such as high bandwidth video conferencing or for high speed access to large files. The users expect to pay a premium rate for guaranteed quality services, but they also wish to use lower cost, and correspondingly lower quality services for more general-purpose communications.

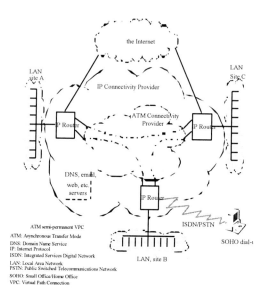

Fig. 3 Networking Scenario

In addition to providing access to the Internet, the IP Connectivity Provider (IP CP) also makes use of the services of an underlying CP - offering semi-permanent ATM connections in this case - which is able to provide guaranteed quality leased lines between the IP CP's routers. Figure 3 shows the network scenario we assume in the rest of this paper.

The service provided by the ATM CP is an end-to-end Virtual Path (VP) service offering PVPs (permanent VPs) between specified termination points. Associated with each VP are a number of parameters, which define the capacity of the connection and level of performance to be provided in terms of end-to-end delay, delay variation and tolerable cell loss ratio. VPs can be created, deleted and modified through client management actions. The clients may monitor VP usage and performance statistics, and initiate fault monitoring activities on their resources.

Issues associated with inter-administration connectivity and federation of management systems are outside the scope of this paper. The dynamic customisable approach presented for a single CP domain could also apply to a multiple CP, inter-domain environment. The remainder of this paper will concentrate on the ATM CP, although the issues discussed are also generally relevant to CPs offering managed services for any network technology.

5. Connectivity Providers

The responsibilities of the IP and ATM connectivity providers in terms of the network resources they manage are displayed in Figure 4. The figure shows the layered approach to AVP. Typically, responsibility for the ATM, and IP networks falls to separate organisations. Where the same organisation is responsible for both network layers, there would be a single CP.

Provider Interfaces: The ATM CP can be seen as a system with two main classes of interface:
- the interfaces to its clients (either the IP CP or the AVPP), through which it offers a set of services for the management of the connectivity; and

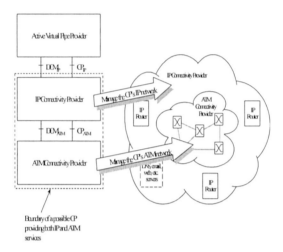

Fig. 4 Connectivity Providers Hierarchy

- the interfaces to the underlying Network Elements (NEs), which are ATM VP cross-connects in our scenario. The interfaces to the NEs will be based on the technology offered by the vendors of the switches - either SNMP or CMIP management interfaces; if CORBA-based interfaces [10], [11] are available then this is also an option. It is unlikely that switches will have mobile or intelligent agent enabled NEs in the near future.

There are two major interfaces between the ATM CP and its clients: the CP API (Connectivity Provider Application Programming Interface) and the DCM (Dynamic Connectivity Management) interface. Both of these interfaces are based on agent technology. A non-agent based approach to the design of the Connectivity Management is depicted in [17] and [15]. *Agencies* are made visible across each of these interfaces. These are locations where an agent may execute and may be addressed remotely. Within each agency a number of fixed, i.e. permanent, agents will be present together with a number of mobile, or visiting, agents that execute on a temporary basis. In general the fixed agents are created by the host environment-the ATM, and the clients of the ATM CP create the mobile agents.

Agent-Based Communication (ABC) refers to the mechanism by which agents communicate with one another. The term implies that specific protocols and interface definitions are used. These protocols could either be based on general distributed systems techniques for remote method invocation (CORBA or Java RMI, for example) or on higher level semantic/AI languages such as KQML [5], [12] or FIPA's Agent Communication Language (ACL) [4], [22] which support interactions with "semantic heterogeneity". The specific languages and communications protocols are dependent on the chosen agent platform and are not of direct concern in this paper. Mobile agents may communicate with fixed agents and mobile agents may communicate with other mobile agents using ABC. The communication between agents could take several forms: to raise asynchronous notifications, to query specific agents, to retrieve information and to invoke operations. A mechanism for publishing the facilities offered by an agent operating in the server role is assumed- i.e. a way of formally specifying an agent's interface. In the design of our system, the Unified Modelling Language (UML) [23] is used to specify formally an agent's interface, being subsequently mapped to the Java language. It is further assumed that an event/notification service is offered by the host environment for disseminating the events raised by agents based on filtering criteria. Today's agent platforms, including those based on OMG's MASIF specifications [8], do not currently offer event/notification services. In our environment it has been necessary to implement notifications in a non-generic, ad hoc basis. It should be noted that the MIAMI project is currently in the process of extending the Grasshopper agent platform [24] to allow communication between agents using FIPA's ACL.

ABC extends beyond the local agency to allow communication with agents in remote execution environments. This implies two methods for communications in agent systems:

either remote operations may be invoked through ABC (in a similar way to traditional distributed systems based on statically located objects); or

mobile agents may physically travel to the remote agency where they may run in the local environment and invoke the *same* operations through *local* (i.e. intra-node) rather than remote (i.e. inter-node) ABC mechanisms.

By relying on mobile agents, an active and dynamically adaptive management system can be built which is not fixed or limited by initial deployment decisions at system design or build time. The choice of using a remote or mobile agent-based communications method- is an issue, which may be decided dynamically, even at system *runtime*. For example, it is possible to create and deploy a mobile agent when the communications overhead between remote systems rises above a certain threshold (although this would be at the cost of physically transferring the agent to the remote execution environment).

Figure 5 shows the overall architecture of the CP's management system. The management system consists of three separate agencies for the main management activities of the CP: one for each activity: configuration management, performance management and fault management.

Configuration management: creates connection (ATM VPCs); modifies connection parameters (e.g. bandwidth); tears-down connection; maintains a database of network resources; schedules resource management and routes management.

Performance management: monitors QoS (utilisation of resources, performance of the network); logs monitored data; QoS and reports on performance.

Fault management: deals with Fault detection; Fault isolation and alarm filtering; Fault reporting; Testing

In addition there are two agencies, which represent the two classes of interface to the clients of the CP: the CP agency supports the CP interface, and the DCM agency supports the DCM interface. Within these latter two agencies a number of fixed and mobile agents may execute. The fixed agents are provided by the CP, at initialisation time, and form the agent-based interfaces to the basic management services of the CP. The mobile agents belong to the clients of the CP and are dynamically created by remote clients.

Initially, there are two main obstacles to a potential client invoking CP management services: legally: the client may not have a contract with the CP and physically: it may not have access to a management interface. The first step is to negotiate a contract. A fixed agent in the CP agency offers an interface to allow contract negotiation. This negotiation can be achieved in two ways: either the client creates a mobile agent to move to the CP agency and negotiate locally with the fixed agent; or the client may communicate remotely with the fixed agent.

Following successful contract negotiation, the CP creates an agency and DCM interface for the client. This involves the creation of one or more fixed agents in the DCM agency to offer specific interfaces to the management services, which feature in the contract. The fixed agents tailor (in a *static* sense) the management services of the CP to the requirements of the client and to limit access to the services according to the terms of the contract. For example, not all management services may be made available to all clients, or the geographical coverage of configuration management may be limited to specific locations. In other words, the fixed agents operate as *proxys* to the configuration, performance and fault management services of the CP.

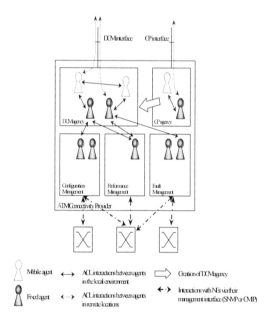

Fig. 5 Agent-based ATM Connectivity Provider

A management service could be invoked, either by a locally running mobile agent or by a remote operation from the client. When a management service is invoked, the fixed (proxy) agent triggers the corresponding operations on the agents in one of the configuration, performance or fault management agencies within the CP. It is within the latter agencies that the real management work - such as the creation of a VP - is achieved. Through the activity of the CP's configuration, performance or fault management systems, modifications are made to the network elements through their management interfaces (SNMP, CMIP) to reflect the original requests made by the clients at the DCM interface.

The MIAMI System provides a mobile intelligent agent platform which serves the needs of the enhanced European and Global Information Infrastructure defined in the ETSI Document 'Report of the Sixth Strategic Review Committee on European Information Infrastructure [35]. The MIAMI Agent platform enhances a MASIF-compliant platform by an agent communication transport service, high-level communication via ACL (agent communication language), wrappers, resource management, logging services, move transparency, and security services and, thus, builds a FIPA-compliant agent platform. These aspects are presented in more detail in [1]. This platform has been used in the development of the above scenario.

In the scenario above the interactions between the client and the CP were discussed for contract negotiation, tailoring of offered management services and dynamically invoking specific management services. This section discusses the way in which the configuration, performance and fault management systems *within* the CP are organised.

In general, network management systems are hierarchical with a network-wide view at the top of the hierarchy and an element-specific view at the bottom, with zero, one or more intermediate levels according to the needs of the system. This hierarchical approach can be seen in both TMN [6] and TINA [6] architectures. It is assumed that the configuration, fault and performance management systems in the agent-based CP will also follow a hierarchical architecture so as to maintain scalability and compatibility with existing management architectures and information models.

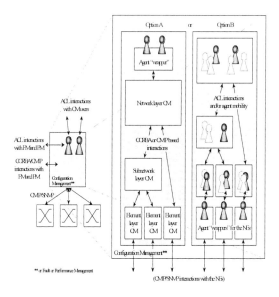

Fig. 6 Agent-based management functions

There are two ways in which the CP's management systems could be deployed:

- either through building agent wrappers on existing management software - to represent the highest level of the TMN or TINA compliant system in the agent environment; or
- through building the entire management system from scratch in an agent-based way and through building agent wrappers to represent the SNMP or CMIP interfaces of the network elements in the agent environment.

These two approaches are shown in Figure 6. Although the figure shows options for configuration management, the same holds for fault and performance management. Option A shows the approach of wrapping legacy systems with agents at the highest level; option B shows agent wrappers at the NE level and the entire system being built using agent technology.

It is also possible to take a hybrid approach. In this case, agent wrappers would be provided *at each level* of the hierarchy of a legacy management system. Mobile agents would then be able to visit the hierarchical level relevant to their operation and interact with fixed agents representing the legacy software at that level. Mobile agents in such a hybrid environment could relocate by traversing the hierarchy horizontally between subnetworks, or management functional areas, or they could migrate vertically to "zoom-in" or "zoom-out" to the level of detail with which they are concerned.

The hierarchical nature of management systems in TMN and TINA (option A, Fig. 6) is fixed at system design time and to a certain extent at standardisation time. For each management service to be deployed, the system designers make decisions on the placement of functionality at each hierarchical level and on whether to distribute or centralise functionality within a particular hierarchical layer. These decisions are based on many factors, including the degree of parallelism required; the quantity and complexity of information to be passed between components, the question of whether existing information models need to be modified to support the required information flows; the scalability of the solution; balancing of processing load between management workstations; the complexity of each component.

A promising application of mobile agents for network management is in deploying each management component as a set of co-operating agents (Option B in Figure 6). The way in which these agents are grouped and placed is initially determined by the system designers according to similar criteria as those for the design of static hierarchical TMN or TINA systems. However, now it is possible to revise the grouping and placement decisions during the operation of the management system through the mobility of agents. There are a number reasons for the migration of agents on-the-fly or for spawning new agent instances. These include: to reduce the processing load on an overloaded management workstation, to cater for an expanding set of managed resources or reduce the quantity of management traffic or information lag, between remote systems when it crosses an unacceptable threshold.

With the basic operation described in the previous section there is very little apparent advantage in adopting an agent based management system - very similar facilities are available in a traditional system based on distributed systems: TMN or TINA for example. However, there is one distinguishing advantage in the system as presented above: the way in which the DCM interface could be *customised* - with client-owned agents - to *tailor* the services offered to specific clients - this is one area where agents gain advantage over traditional software systems.

In the scenarios below mobile agents are able to autonomously interact with one or more management functional areas in the server to add value to the original management services. In traditional client-server distributed systems for network management (e.g. SNMP, TMN, TINA) the client is limited to working with the in-built facilities of the server. Any further manipulation of management information, beyond that which was generically provided by the relevant standards or by the developer of the server, must be performed in the client application code.

If customers wish to add value to the basic management services offered by the provider they must have the capability of running a management platform in their premises which supports the protocols and information models offered by the server. In addition, the customers must deploy suitable applications running on their local platform to house the required logic. The client applications must interact with the remote server to receive notifications and to initiate management operations. The quantity of information to be exchanged between client and server is a function of the management activities being undertaken and on the efficiency of the protocols and information models supported by the server for the task in hand. The delay and cost associated with each remote operation is a function of the network interconnecting the client and server. Given the particular protocols and information models supported by the server and the characteristics of the network interconnecting them, it may be not be cost effective or even possible to perform certain management tasks remotely in the client. For example, the cost of communication may outweigh the benefits of performing some management tasks (such as fine grain, real-time monitoring of performance parameters) or the information may be out of date by the time an appropriate course of action has been determined by the client application, if the network delays are too large.

With mobile agents it is possible for a client to program a mobile agent with specific functionality which may then be deployed *at run-time* in the server to add value above and beyond the server's basic facilities. Following deployment, the mobile agent may act autonomously to interact with the local environment and make *local* decisions which may then be implemented as *local* management actions without needing to interact with the remote client every time a decision is required. Through this approach it is possible for a remote client to deploy management behaviour and algorithms *inside* the remote server.

To illustrate the use of dynamically deployed agents the following three integrated scenarios we have identified in the MIAMI project:

Intelligent reporting: Mobile agents may respond to reports from both the performance and fault management systems. According to their programmed policies, rather than relaying *all* fault and performance reports back to the remote client, they will only report when certain conditions have been fulfilled. An example might be performance degradation on one connection following the failure of a connection in a remote part of the network, which forms an alternate route. Only the correlation of these two events might be relevant to the client. Alternatively, observed performance degradations might cause the agent to initiate tests to verify that unreported failures have not taken place. This scenario integrates the facilities of fault and performance management.

Fault repair: A mobile agent is programmed to listen to fault reports from the fault management system when connections have been interrupted by network failures. According to a pre-programmed policy it can initiate new connection requests between the same end points as the failed connection to restore connectivity. This scenario integrates fault and configuration management.

Bandwidth management: Assuming that performance monitoring agents have been deployed (either by the client or the bandwidth management agent itself) to monitor and report on the utilisation of connections, a bandwidth management agent will be programmed to listen to utilisation reports for certain connections. Depending on the policy for a specific connection the bandwidth management agent may decide to request increased bandwidth on highly utilised connections or to reduce the capacity of a lightly utilised connection. The decision may depend on the cost of changing the bandwidth and so a negotiation between the configuration management agents and the bandwidth management agent may take place. This scenario integrates performance and configuration management.

The integrated scenarios introduced above combine the basic facilities of the configuration, fault and performance management services offered by the server with additional customer specific logic. In other words the customer is able to program the offered management service to a certain degree. This concept has its parallel in traditional management systems through the use of the OSI management Systems Management Functions (SMFs) [25] for event forwarding and logging, resource monitoring [26] [27], and testing [28], albeit in a more limited way. Previous research work [29] [30] has demonstrated how clients can take advantage of these generic facilities to simplify the construction of intelligent clients.

We now consider how these basic facilities could be implemented through the use of mobile agents. Rather than being restricted to standardised capabilities such as the SMFs it is now possible to build entirely arbitrary and powerful behaviour into mobile agents which will be physically located in the managed system's environment. This embedded intelligence not only allows event reports tailored to the client's requirements to be emitted, but it enables the migration of the client's logic and decision-making algorithms to the server. This has an obvious impact of reducing the quantity of management traffic between remote systems and achieves more timely access to information generated by the remote server.

In OSI management, SMFs were standardised by international organisations and encapsulated in the compiled functions of OSI agents. In CORBA-based management systems the SMF-like facilities could be determined by the designers of the systems and embedded at design and system-build time. With mobile agents and intelligent reporting in agent-based management systems, the SMF-like facilities can be enhanced and extended almost infinitely and deployed *at run time*.

As seen in the examples above, it would be very difficult to capture such behaviour in traditional TMN or TINA systems without standardising such a bandwidth management service at the Xuser or ConS interface. If such a service was to be standardised it would be difficult to capture all possible potential behaviours that clients may request without making a comprehensive and therefore complex specification of the service in GDMO or IDL. However, this proposal allows the client to direct a dynamic and customisable network management through the use of programmable, intelligent agents based on mobile code.

6. Analysis and Conclusions

Traditional and emerging frameworks for network management such as TMN and TINA allow customers electronic access to management services. However, these services are fixed in the sense that new features can only be added after a lengthy research-standardisation-deployment cycle. In this paper we have discussed the advent of mobile agent technologies and how they may enhance traditional connectivity management services making them dynamically customisable by consumers. Consumers may introduce their own value-added logic during service operation to cater for the dynamics of their environment and to enforce their own policies.

This prompts a new paradigm for building network management services. Instead of providers building services that attempt to encapsulate the requirements of all clients, they build the necessary hooks and let the clients apply their logic. Customisation and programmability of management services was always possible in traditional systems based on client-server paradigms, through the development of client applications (in the customers' premises management platform), but at high cost and low efficiency compared to the proposed agent-based approach.

The initial approach for the configuration management domain in MIAMI was to base it on an existing TMN system for ATM PVP set-up [13], [31] and [15], to which an agent interface would be added (option A in Figure 6). For logistical reasons, the final implementation is based on *static* agents internally (Figure 6, option B), which communicate using remote method calls. This implementation could also be based on distributed object technology e.g. CORBA. Agents were chosen for two reasons: first for uniformity, since there is no need for an adaptation agent-based interface; and second, for evaluating mobile agent platforms in the same role as distributed object frameworks. It should be finally noted that we do not see any immediate benefits from applying mobile agent technology to configuration management.

On the other hand, the performance and fault management systems use agents internally (Figure 6, option B) in a way that mobility is exercised and exploited. In the performance management domain, customised agents replace, augment and allow customisation of the functionality of the TMN/OSI-SM metric monitoring and summarisation objects [26], [27] and [16] while in the fault management domain customised agents do the same for TMN/OSI-SM testing objects [28]. In both domains, mobile agents are instantiated at the "network management level" of a management hierarchy according to requests originating from the DCM domain, migrate to network elements and perform relevant tasks locally, enjoying minimal latency and reducing network traffic. Details of the performance and fault management approaches are described in [32] and [33] respectively.

In summary, agent mobility in the presented network management architecture is used in a fashion, which we would term "weak mobility". Mobile agents are instantiated at a control point by a master static agent and then move to another point (i.e. network node) where they stay until their task is accomplished. This can be considered as an intelligent software deployment activity. The key benefit of this approach is *programmability*, allowing clients to "push" functionality to a point offering elementary hooks, which can be accessed to provide derived, higher-level services. In a similar fashion, we could term "strong mobility" as a situation in which a mobile agent moves from point to point using its built-in logic, adapting to changing situations in the problem domain where it is involved.

We propose a dynamic framework for creating and realising business in the information infrastructure environment. The framework capitalises on the benefits of the Mobile Intelligent Agents technology to offer an easy-to-use, automated and efficient solution to the operation of a VE across all its lifecycle phases. The framework supports customisation and as such enabling innovation in the area of future infoware products. A key mediator role of the VE framework, the Active Virtual Pipe (AVP) between the business and the service and network management levels has been developed. This mediator role is a novel way to integrate legacy network and service management solutions into a business driven virtual enterprise environment using Mobile and Intelligent Agent technology. AVP is a novel and key example of an Active Service. In this paper we have discussed the advent of mobile agent technologies and how they may enhance traditional connectivity management services in a form of AVP services, making them dynamically customisable by consumers.

Acknowledgements

This paper describes work undertaken in the context of the ACTS MIAMI (AC338) project. The MIAMI consortium consists of GMD FOKUS – Germany, Alcatel Bell – Belgium, Alcatel CIT – France, Algosystems SA – Greece, CRIM– Canada, France Telecom – France, Hitachi Ltd. – Japan, Imperial College of Science Technology and Medicine – U.K., University College London – U.K., The University of Surrey – U.K. The ACTS programme is partially funded by the Commission of the European Union. We would like to thank Dr Volker Tschammer, MIAMI project manager in the last part of the project, for his support.

References

1. MIAMI Project Public Web Site, http://www.fokus.gmd.de/research/cc/ima/miami/
2. Workflow Management Coalition, The Workflow Reference Model (1994)
3. Workflow Management Coalition Web Site, http://www.wfmc.org/
4. http://drogo.cselt.stet.it/fipa/
5. Labrou, Y., and Finin, T., A semantics approach for KQML - a general-purpose communication language for software agents, Third International Conference on Information and Knowledge Management (CIKM'94), November 1994.
6. ITU-T M.3010, Principles for a Telecommunications Management Network.
7. TINA consortium, Overall Concepts and Principles of TINA, Document label TB_MDC.018_1.0_94, TINA-C, February 1995.
8. Mobile Agent System Interoperability Facilities Specification OMG TC Document orbos/97-10-05, November 10, 1997. ftp://ftp.omg.org/pub/docs/orbos/97-10-05.pdf
9. MIAMI project deliverables -, http://www.fokus.gmd.de/research/cc/ima/miami/
10. CORBAservices: Common Object Service Specification", Object Management Group, 1996
11. "Common Object Request Broker: Architecture and Specification", r2.0, OMG, 1995
12. Finin, T. et. al. (1994) KQLM as an Agent Communication Language, Proc CIKM'94, ACM Press
13. Galis, A., Brianza C, Leone C, Salvatori C, Gantenbein D, Covaci, Mykoniatis G, Karayannis F - Towards Integrated Network Management for ATM and SDH Networks supporting a Global Broadband Connectivity Management Service in "Intelligence in Services and Networks: Technology for Co-operative Competition", Springer- Verlag, Berlin, 1997, ISBN 3-540-63135-6
14. Galis, A., Griffin D., Eaves W., Pavlou G., Covaci S., Broos R -"Mobile Intelligent Agents In Active Virtual Pipes Support For Virtual Enterprises" In Intelligence In Services And Networks – Springer Verlag, June 2000, ISBN 1 58603 007 8
15. Galis, A., –"Multi-Domain Communication Management Systems"- CRC Press, July 2000, ISBN 0-8493-0587-X
16. Griffin, D., Pavlou, G., Tin, T., "Implementing TMN-like Management Services in a TINA Compliant Architecture - A Case Study on Resource Configuration Management," Intelligence in Services and Networks: Technology for Co-operative Competition, ed. A. Mullery, M. Besson, M. Campolargo, R. Gobbi, R. Reed, pp. 263-274, Springer Verlag, 1997.
17. ACTS MISA AC080- Consortium and Deliverables- Management of Integrated SDH and ATM Networks, http://misa.zurich.ibm.com/

18. "Network Resource Architecture," The TINA Consortium, Version 3.0, Feb. 1997.
19. "Network Resource Information Model," The TINA Consortium, Document No. TB_LRS.011_2.0_94, Dec. 1994.
20. Ranc, D., Pavlou, G., Griffin, D., Galis, A., "A Staged Approach for TMN to TINA Migration", proceedings of the TINA'97 Conference, Santiago, Chile, November 1997, Proceedings to be published by IEEE
21. Kurt Rothermel, Radu Popescu Zeletin (Eds.) - Mobile Agents, First International Workshop, MA'97, Berlin, April 1997, IBSN 3-540-62803-7 Springer Verlag
22. Agent Communication Language, FIPA 97 Specification, Version 2.0, October 1998, http://www.fipa.org/spec/FIPA97.html
23. Booch, G., Rumbaugh, J., Jacobson, I., The Unified Modelling Language User Guide, Addison-Wesley, 1999.
24. http://www.ikv.de/products/grasshopper/index.html
25. ITU-T Recommendations X.730-750, Information Technology - Open Systems Interconnection - Systems Management Functions.
26. ITU-T X.738, Information Technology - Open Systems Interconnection - Systems Management: Metric Objects and Attributes, 1994.
27. ITU-T X.739, Information Technology - Open Systems Interconnection - Systems Management: Summarisation Function, 1994.
28. ITU-T X.745, Information Technology - Open Systems Interconnection - Systems Management: Test management function, 1994.
29. Georgatsos, P., Griffin, D., Management Services for Performance Verification in Broadband Multi-Service Networks, in - Bringing Telecommunication Services to the People - IS&N'95, Clarke, Campolargo and Karatzas, eds., pp. 275-289, Springer, 1995.
30. Pavlou, G., Mykoniatis, G., Sanchez, J., Distributed Intelligent Monitoring and Reporting Facilities, IEE Distributed Systems Engineering Journal, Vol. 3, No. 2, pp. 124-135, IOP Publishing, 1996.
31. Karayannis, F., Berdekas, K., Diaz, R., Serrat, J., A Telecommunication Operators Inter-domain Interface Enabling Multi-domain, Multi-technology Network Management, Interoperable Communication Networks Journal, Vol. 2, No. 1, pp. 1-10, Baltzer Science Publishers, March 1999.
32. Bohoris, C., Pavlou, G., Cruickshank, H., Using Mobile Agents for Network Performance Management, to appear in the Proc. of the IFIP/IEEE Network Operations and Management Symposium (NOMS'00), Hawaii, USA, IEEE, April 2000.
33. Sugauchi, K., Miyazaki, S., Covaci, S., Zhang, T., Efficiency Evaluation of a Mobile Agent Based Network Management System, in Intelligence in Services and Networks - Paving the Way for an Open Service Market, Zuidweg, Campolargo, Delgado, Mullery, eds., pp. 527-535, Springer, 1999.
34. Covaci, S., (Editor) - "Active Networks"- Proceedings First International Working Conference, IWAN'99 - Berlin June 1999, Springer ISBN 3-540-66238-3
35. Report of the Sixth Strategic Review Committee on European Information Infrastructure, www.etsi.org/specrec/SRC6/SRC6.htm

Convergence of IP-based and Optical Transport Networks

Artur Lason[1], Antonio Manzalini[2], Giorgos Chatzilias[3], Lampros Raptis[3], Didier Colle[4], Piet Demeester[4], Mario Pickavet[4], Monika Jaeger[5]

[1] AGH – Department of Telecommunications,
Poland
[2] CSELT - Centro Studi e Laboratori Telecomunicazioni
Italy
[3] NTUA - National Technical University of Athens,
Greece
[4] Ghent University – IMEC
Belgium
[5] T-Nova Deutsche Telekom Innovationsgesellschaft mbH
Germany
lason@kt.agh.edu.pl
antonio.manzalini@cselt.it
lraptis@telecom.ntua.gr
didier.colle@intec.rug.ac.be
monika.jaeger@telekom.de

Abstract. Today Network and Service Providers are aware of the increasing data traffic volumes and as such they are strategically moving investigations toward a single integrated voice and data infrastructure. In this context IP is gaining the role of the integration layer for multiple services. Nevetheless incumbent NSPs that build a multi-service IP network are going to need connectivity to its preexisting legacy networks (e.g. ATM, SONET, SDH). This reason motivates the introduction of a client-independent Optical Transport Network (OTN) as a missing link to guarantee a smooth evolution from legacy networks to a data-centric OTN. The scope of this paper is to give some guidelines about the definition of functionality and architectures of a multi-layers infrastructure supporting the transport of data and circuit-based services. Particularly, the identification of the different service requirements, as well as the understanding of the allowed degradation, provide a picture of the needed survivability mechanism of IP over OTN scenarios.

1 Introduction

The emerging of new application services and the exponential increasing of data traffic (mainly Internet) pose the problem of the evolution of Network Service Providers' infrastructures to face such trends.

If on one side existing TDM-based infrastructure of incumbent NSPs should evolve to support efficiently the exponential growth of data traffic, on the other side New Comers should deploy cost-effective infrastructures to gain rapidly market shares. As a matter of fact many new NSPs are entering metropolitan areas with single integrated voice and data infrastructure where IP is gaining the role of integration layer for multiple services. This is a first strong driver.

Another main driver is the introduction of the DWDM technology. WDM techniques, which were originally introduced around '95 to increase transport capacity, have laid the foundations for implementing network functionality directly in the optical layer. This is what is known as Optical Transport Network [6].

Thanks to these two main drivers (increasing of IP traffic and introduction of DWDM), the convergence of IP-based and Optical Transport Networks is becoming one of most strategic areas for NSP.

2   Multi-layers optical Transport networks

Historically NSPs have used several layers to build their networks: adopting for example IP routers over ATM switches over PDH or SDH network elements. Figure 1 reports a typical layered architecture based on legacy systems.

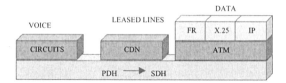

Fig.1   Example of legacy current scenario

Nevertheless even when deploying a new IP-based infrastructure, there is the need to guarantee the connectivity to preexisting legacy networks belonging to the same or to another NSP. SONET\SDH framings and rates seem to be the most common standard to support connectivity to preexisting infrastructures.

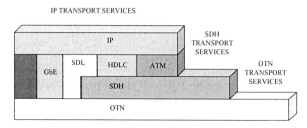

Fig.2   Introduction of the OTN as a server layer network

Furthermore DWDM point-to-point systems have been already introduced in backbone networks to increase transport capacity. Recently DWDM is a technology mature to deploy also networking functionality by means of Optical Network Elements (ONE) such as OADM and OXC. Again the Optical Transport Network (OTN) represents another layer added to the stack (Figure 2). Thanks to DWDM technology, the OTN can provide a large amount of raw bandwidth supporting the delivery of big volumes of IP traffic. Furthermore it can deliver other data and circuit-based services in a more cost-effective way.

In this context the optical internetwork is defined as a multi-layers transport network composed by two or more connected OTN or client networks connected to one or more OTN. In Figure 3 and Figure 4 respectively show examples of reference scenario and architecture of an optical internetwork.

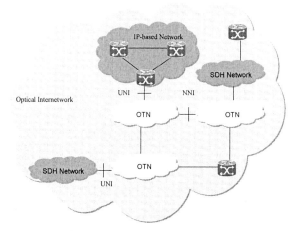

Fig. 3   Reference scenario of an optical internetwork

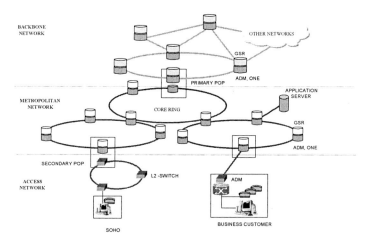

Fig. 4   Architecture of an optical internetwork

3   Application services

The design of the architecture for the different network segments (access, metropolitan and backbone) of the infrastructure is strictly related to the requirements of the services to be provided to the customers. The technical requirements of the envisaged application services will impact the choice of the enabling technologies and viceversa the infrastructure capability of supporting certain transport services will impact the providing of new application services.

Applications could be classified into two general categories based on the relative technical requirements, namely Streaming and Elastic Services.

The first category of applications is based on information sent to the customer in a streaming way: the source continuously produces data stream during the end-user session. The source is bursty, mostly due to the adaptive nature of video coding scheme. The application classified into the Streaming Service is sensitive or highly sensitive to the delay introduced by the transport network (Table 1). The Elastic group of applications are characterized as a much less delay sensitive. Data bit rate variation of this group is much higher in comparison to the applications of Streaming Service.

A proposal of assignment of sample applications to the Streaming and Elastic Services is presented in Table 1. The application delay sensitivity has been indicated with the grade ranging from 0 up to 5. The highest grade means that the application is highly delay sensitive, whereas 0 means that the delay in the transport network can be neglected. The same rating has been used for the need for protection.

The need for protection for given applications may vary with the type of end-user. For example the transport of digital video signal between movies server and end-user and between TV studios is characterized by different protection requirements. The most important conclusion from applications analysis is that the Streaming Service generally requires to be supported with protection mechanism of the future transport network. The data bit rate variation of Elastic Service is very high. This effect is observed because Elastic Service applications are in most cases based on information retrieval. Such characteristics of Elastic Service are reflected also in different requirements for protection. Implementation of restoration algorithms in an integrated IP over OTN network would likely be suitable to assure the integrity of Elastic Service.

Each of the applications has also been analyzed with regard to its expected importance in the short and mid-term perspective. It was assumed that the short-term prognosis means the three years perspective and mid-term forecast refers to the application significance in next 10 years. The importance of the application is given in the scale from –5 to +5. The rate of 0 means no changes in the applications importance. The negative ratings is adequate to the forecasted degradation of the application importance. The aggregate rating of the prognosis is higher than zero due to the confidence in continued development of network services. The permanent development of new services should be taken into account in studies on integrated transport services. Data presented in Table 1 suggest that increasing impact of Streaming Services on the integrated transport networks will still be observed in the future.

Table 1 Overview of application and their relationship to the application services

	Application	Bit rate	Bit rate variation	Delay sensitivity	Need for protection	Importance	
						Short-term	Mid-term
Streaming Service	POTS	64 Kbps–32 Kbps	const	5	5	-1	-3
	VoIP	8 Kbps–32 Kbps	const	5	5	3	5
	Video-telephony	256Kbps–1920Kbps	high	5	5	2	3
	Video- conf.	256 Kbps at least	high	5	5	3	4
	Tele-working	64 Kbps–2 Mbps	very high	5	4	3	5
	TV Broadcast	2 Mbps–8 Mbps	high	4	4	1	0
	Distance Learning	64 Kbps–2 Mbps	very high	5	5	2	4
	MoD	750Kbps–4 Mbps	high	4	3	1	3
Elastic Service	News on Demand	64 Kbps	very high	2	2	2	4
	Internet Access	64 Kbps–2 Mbps	very high	1	2	5	5
	TV Listing	64 Kbps	very high	2	2	1	2
	Tele-shopping	64 Kbps–2 Mbps	very high	2	2	5	5

4   Transport Network Services

Network transport services are those functions and utilities supporting connectivity, communications and control required by applications operating across the network. The definition of these network tranport services will determine the flexibility of an enabling infrastructure to support current and unforeseen new network applications.

4.1   SDH-based transport services

In principle, an optical internetwork could be able to provide SDH-Section transport services and SDH-VC transport services.

For the SDH-Section transport services the OTN acts as a virtual fiber connection.

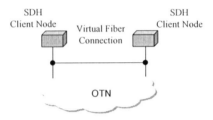

Fig. 5   SDH-Section fiber connection

For SDH-VC transport services, the optical internetwork acts as a virtual container link. For example in order to achieve payload bandwidths larger than the basic VC-4 container, the principle of concatenation has been defined. Concatenation is a procedure whereby a multiplicity of Virtual Containers is associated one with another with the result that their combined capacity can be used as a single container across which bit sequence integrity is maintained [1].

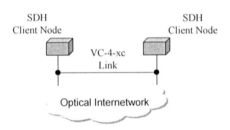

Fig. 6   Example of SDH VC transport service

In [2] SDH-based transport services include PLR, STE and LTE circuits. Particularly they are SONET/SDH rate and framed point-to-point circuit between two UNIs.

A PLR-Circuit is a circuit that preserves all SONET overhead bytes between clients. The SONET/SDH signal may be concatenated or channelized but cannot be SONET/SDH TDM demultiplexed or multiplexed within the optical internetwork.

An STE-Circuit preserves all SONET/SDH line overhead bytes between clients but is not required to preserve the section overhead bytes. The SONET/SDH signal may be concatenated or channelized but cannot be SONET/SDH TDM demultiplexed or multiplexed within the optical internetwork.

An LTE-Circuit preserves the SONET/SDH payload, but is not required to preserve the section or line overhead bytes. The SONET/SDH signal may be concatenated or channelized and may be SONET/SDH TDM demultiplexed or multiplexed within the optical internetwork to allow the subrate circuits to be individually routed, or to allow multiple LTE-Circuits to be multiplexed within the network to better utilize a network link. Thus, an LTE-Circuit implies timing and synchronization requirements not required in PLR-Circuits or STE-Circuits.

180

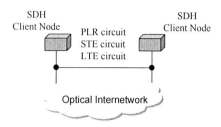

Fig. 7   Example SDH-based transport services

4.2   IP-based transport services

The Internet protocols consist of a suite of communication protocols of which the two best known are the Transmission Control Protocol (TCP) and Internet Protocol (IP).

The IP is a network-layer (Layer 3) protocol that contains addressing information and some control information that enables packets to be routed. IP has two primary responsibilities: providing connectionless, best-effort delivery of datagrams; providing fragmentation and reassembly of datagrams to support data links with different maximum-transmission unit size.

In general, an IP Transport Service could be defined as the capability of an IP-based network to deliver datagram payloads from a Service Access Point to anyone of the interfaces with the IP-address for that SAP. In this sense, an IP Transport Service could be intended as an added value service to the simple data transmission. If the end-user always acts via an application protocol, the Service Access Point of a NSP could be either at the network-layer, or at the transport-layer, via intermediate protocols (e.g. TCP, UDP) or at the application layer via application protocols (FTP, HTTP, SNMP, SMTP, etc).

An IP Transport Service could be qualified by a set of parameters (terms of the Service Level Agreement) such as: access mode, rate of service availability in hours per year, multiple service quality classes, routing configuration indicating the beloging to one or more IP-based VPN, multicasting, throughput management and traffic shaping, security issues, provisioning time is the time required to set-up the service.

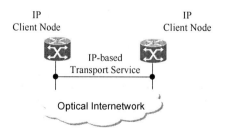

Fig. 8   Example of IP-based transport service

4.3 Optical transport services

Three main kinds of Optical Transport Services can be identified:
- Leased OCh service
- Leased Wavelegth
- Leased Dark Fiber

Bandwidth management of leased OCh is a further service option considered as to the bandwidth management of leased OChs. This special service might be offered when the inverse multiplexing of OTN, e.g., OCh virtual concatenation is available.

Different potential ways to set-up OCh services are envisaged:
- a permanent set up from the network management system by means of network management protocols;
- a soft permanent set up from the management system, which uses network generated signalling and routing protocols to establish connections;
- a dynamic set up by the customer on demand using signalling and routing protocols (switched optical channel).

Regarding leased wavelegth, this transport service is offered to clients equipped with coloured line terminals and in case of Leased Dark Fiber who is providing this service doesn't control the degree of use of the fiber itself.

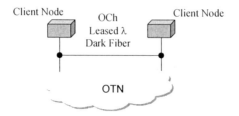

Fig. 9   Example of Optical transport service

5 Resilience of IP over OTN

IP is very dynamic, adaptable and self-healing, whereas OTN has superior link-level protection and restoration capabilities, but is static and slow to provision on a network scale. This prevents the optical layer from effectively addressing the rapid changes in the IP topology.

It is important to provide survivability at multiple layers, in order to recover any service in all expected failure conditions and to combine positive features of the different recovery schemes. In such a case, a proper coordination of the recovery schemes at multiple levels is necessary. A sequential strategy could be implemented with a hold-off timer or a recovery token signal from one layer to the other.

In order to define an efficient, flexible and fast resilience strategy, a clear understanding of the current protection mechanisms available in the different network technologies is needed.

5.1 Restoration in IP networks

Internet, considered by many as the integrated service layer, was not designed with fast restoration/protection mechanism in mind. IP can recover from any defect in the network, but the time needed is in the order of seconds, making IP unsuitable for streaming applications. It takes IP some time to detect a failure, propagate the information to other routers and then each router compute a new path.

However, there are some techniques that can be used to make IP more efficient. For example equal cost multi-path forwarding (ECMF) enables routers in case of a failure to utilise more than one path for packets that have the same destination. The selection can be done in a per-packet round-robin fashion or on a per-flow basis. Moreover, in order to minimise convergence time of the routing algorithms, the IP network can be segmented into multiple domains. Hierarchical link-state routing protocols like OSPF and IS-IS allow such partitioning, restricting the flooding of the topologic changes in smaller areas of the network.

5.2 Recovery in the OTN

On the other hand, protection/restoration mechanisms for the OTN have already been defined. Many ideas of the SDH protection has been shifted and adopted by the OTN, thus OTN has capabilities similar to the SDH achieving fast protection/restoration within 50 msec. The flexibility of these mechanisms is of vital importance since the disruption of a fiber carrying Terabits of traffic affects thousand of connection. Emerging technologies like DPT, MPLS and MPlambdaS can play a major role in-between the IP and the OTN, allowing a smooth interworking.

5.3 Survivability in MPLS networks

Multiprotocol Label Switching (MPLS) binds packets to a route based on labels. Protection of traffic at the MPLS layer is desirable because MPLS:

- has faster responses in failures compared to IP(layer 3).
- allows the direct integration of IP and WDM layers. The IP packets can be transferred directly over a WDM channel without the need of intermediate layers like SONET/SDH.
- allows finer granularity of protection.
- allows the differentiation of the protection based on the traffic type.

MPLS fast protection switching is similar to the ATM, occupying resources along a back-up path only when needed, combining the advantages of protection (fast recovery) and restoration (capacity efficiency). In addition, label stacking makes it possible to provide a single back-up Label Switched Path (LSP) to protect multiple LSPs, although there are some problems concerning the consistency of the labels inside the back-up LSP. Different protection mechanisms can be identified [4].

Residing at the higher layer, recovery at MPLS level may incorporate more service specific details. For example, along the alternate path low priority traffic (both working and rerouted) will be preempted first (instead of preempting all existing working traffic along the alternate path or not allowing traffic to be rerouted over the alternate path). Also a finer granularity of protected objects can be obtained at this higher layer.

5.4   Deploying DPT rings for fast network recovery

Dynamic Packet Transport (DPT) is a L2 ring technology, proposed by CISCO Systems. DPT resides between the IP and the OTN and this is why it is mentioned here. WaRP is a protocol that enables the provisioning, routing, protection, and restoration of virtual wavelength paths (VWP). DPT delivers 50 ms restoration timeframes in a mesh network topology, improved fiber utilization via 30% spare capacity reduction, provides rapid end-to-end provisioning on the order of seconds

5.5   Recovery in MPLambdaS networks

The Multiprotocol Lambda Switching (MPLambdaS) can be considered as the glue between the IP layer and the optical layer. The control plane of OXC can be the same as that of MPLS-capable routers. The recovery in MPLambdaS networks is performed by dynamically setting up back-up optical connections The issue of dynamic connection set-up is an area of intense research in different standardisation bodies (OIF, IETF, etc).   When LSPs are passed to the optical layer, no explicit encapsulation is needed because each wavelength channel represents a LSP label and the encapsulation is done implicitly. Moreover, new fiber-restoration schemes which aggregates (stacks) all the wavelengths of one link into a larger optical LSP.

5.6   Multilayer Network Recovery

The previous paragraphs discussed recovery techniques for each technology independent from each other. However, an IP over OTN network is a typical example of a multi-layer network. Indeed, there is a need to deploy recovery techniques in multiple layers.

Only from a failure-coverage point of view, we can raise following arguments. Only providing recovery in the OTN, would make it impossible to recover from a router failure in an acceptable short timeframe. On the other hand, only relying on the IP restoration capabilities, would make it impossible to restore an affected optical transport service (e.g., a leased OCh service), which have been discussed in section 4.3.

Thus, it is necessary to provide recovery at both IP and OTN level. However, the IP restoration capabilities are not sufficient for all types of traffic, and therefore more advanced techniques as in MPLS or DPT may be required.

5.7   Coordination of multiple single layer recovery schemes

As discussed in the previous paragraph, it is recommended to deploy a recovery mechanism in multiple layers. Of course, one has to assure that they work properly together.

A first solution is the most simple and straightforward one. The recovery schemes in the different layers are running in parallel and independent from each other. This may work in some cases, but at least it is inefficient. Although it may be sufficient that the traffic is restored by a single mechanism, both the IP (or MPLS/DPT) and

OTN mechanism are doing their job and thus requiring twice spare resources. Even more, it may be possibly that in some cases (e.g., in severe failure conditions) the OTN has simply to restore the connectivity of its client layer, before this client layer would be able to restore any traffic. For instance, Figure 10 shows a double failure, which requires that MPLS switches over to the backup LSP and that OTN restores the link along the backup LSP. In such a case it is necessary that MPLS does not give up before the OTN restored its backup LSP.

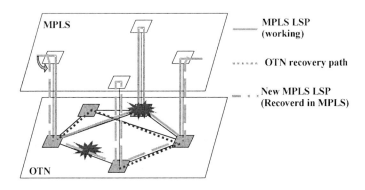

Fig. 10   link + node failure

To solve these problems, one can deploy a sequential strategy: one recovery scheme starts trying to perform the recovery, and only if it fails, the control is handed over to the next recovery scheme. The most intuitive approach, is a bottom-up approach, where the lowest layer (in our case the OTN) tries to restore the traffic. However, the interesting features of e.g. MPLS or DPT may promote a top-down approach, where the recovery starts at the highest layer.

A first way to implement such a sequential approach is to apply a hold-off timer: when a recovery mechanism starts its attempts to restore the traffic, a hold-off timer is launched and only when this timer goes off, the next recovery scheme starts to recover the remaining affected traffic. Although this solution is rather simple to implement, such a timer may introduce a significant delay.

To avoid such an additional delay, one could hand-over the control to the next layer by sending a recovery token signal to that layer. Although it does not introduce a delay, this solution requires a modification of the interface (and the corresponding standards) between those layers.

5.8   Integrating IP and OTN

The fact that the control plane of a MPLambdaS network is similar to the one of a MPLS network, opens up the opportunity to integrate both control planes into a single one. Such an integrated control plane makes it possible to deploy a recovery scheme, covering both the optical and electrical MPLS network, which avoids interworking issues, being discussed in the previous paragraph. Even more, since the control plane now has a full-view of the network (covering both electrical and optical level), new opportunities are raised to take more intelligent recovery actions. However, one has to make a trade-off between algorithmic complexity (required for more intelligent recovery actions) and recovery speed.

6   Conclusions

The increasing data traffic and the evolution of the WDM technology are driving the convergence of IP-based and Optical Transport Networks. Particularly, the introduction of the OTN will guarantee a smooth evolution from legacy (e.g. SDH, ATM) to a data-oriented architectures, that initially could even coexist in the same infrastructure.

As such, if the first evolutionary step for NSPs is the introduction of the OTN [6], supporting the transport of legacy clients, the next one seems to be the development of interworking and interlayer communications between optical network elements (OADM and OXC) with IP Routers in order to deploy a data-centric OTN.

In a multi-layers scenario it is important to provide survivability at multiple layers, in order to recover any service in all expected failure conditions and to combine positive features of the different recovery schemes. In such a case, a proper coordination of the recovery schemes at multiple levels is necessary. A sequential strategy could be implemented with a hold-off timer or a recovery token signal from one layer to the other. The identification of the different service requirements, as well as the understanding of the allowed degradation of services will give a clear picture of the needed survivability mechanism.

References

1. ITU-T Rec. G.707;
2. OIF-99-161;
3. ITU-T Rec. G.709;
4. Makam et al, work in progress, internet-draft October 1999: Protection/Restoration of MPLS networks. http://search.ietf.org/internet-drafts/draft-makam-mpls-protection-00.txt
5. C. Metz in IEEE Internet Computing, March-April 2000: IP restoration and protection"
   ITU-T Rec. G.872.

Acknowledgments

This paper has been prepared in the framework of the IST project LION – Layers Interworking in Optical Networks. The Authors wish to thank all project Participants.

QoS ad hoc Internetworking: Dynamic Adaptation of Differentiated Services Boundaries[1]

Michael Smirnov,

GMD FOKUS, Kaiserin-Augusta-Allee, 31
tel. +49 30 34637113
fax.: +49 30 34638000
E-mail: smirnow@fokus.gmd.de

Abstract. The Internet needs QoS internetworking to become QoS aware in the end-to-end sense. We address only one issue of this - adaptation of the differentiated services architecture, in particular, dynamic creation of DiffServ virtual boundaries. Our solution is fully distributed and data driven, therefore it could be considered as QoS ad hoc internetworking on contrary to statically configured DiffServ. The proposal is two fold. First, to support invariance under aggregation we suggest to maintain Per Domain Behaviours (PDB) based on Per Path Behaviours, and, second, to use group communication based on native IP multicast for needed QoS signalling and resource control. The paper shows that due to flexible grouping policies the approach has high scalability and good deployment potential.

1 Introduction

The motivation to add dynamic adaptation to the DiffServ is many fold. First, DiffServ as a QoS enforcing technology is only needed in the Internet when and where there is a congestion. Then, statically configured DiffServ routers will loose buffering capacity in the absence of priority traffic. This buffering capacity could be used to accommodate more packets when unavoidable bursts occur. Last but not least: in the absence of any congestion, packet classification, metering, marking, shaping and policing performed by DiffServ edge router will waste router resources. The latter was shown by multiple measurements by TF-TANT: *"... the minimum nodal delay added by the DiffServ router ..., is equal to the transmission time of 2 BE packets"* [1]. We interpret this result in favour of best effort service for transmission of flows even with various priorities being set if there is no congestion. Following these motivations the ideal case would be to have DiffServ in a router up but not running until a congestion occur.

The paper is structured as follows. First, we shortly summarise known DiffServ issues, then present our requirements and framework. One of the main requirements is

---

[1] Research outlined in this paper is partly funded by the IST project CADENUS "Creation and Deployment of End-User Services in Premium IP Networks" http:// www.cadenus.org/

to support dynamic service creation over DiffServ networks, i.e. to be able to maintain for particular microflows of IP datagrams the so called invariance under aggregation. Practically that means, that terms and conditions defined in a service level agreement (SLA) for this microflow are to be met Internet wide while the micro flow itself becomes anonymous within a DiffServ bahaviour aggregate. This major requirement is addressed then in the main part of the paper presenting a new paradigm for resource control in a distributed environment allowing dynamic creation of services. We argue that only when based completely on a group communication paradigm a resource control framework will meet this requirement. A short introduction of a particular mediated group communication service applicable in this case is followed by a number of brief examples. The paper is concluded by a summary of QoS ad hoc internetworking.

2. DiffServ issues

An early explanation of the DiffServ stated: *"Code-points should be looked at as an index into a table of packet forwarding treatments at each router"* [2]. In the naive understanding this might mean that differently marked packets can actually take different routes within the network. Not excluding this possibility, the paper focuses mainly on a traditional differentiation of packet forwarding paths inside a router.

It is well understood currently that DiffServ as such is not a service but a building block. Services will be built by adding rules to behaviours, such as:
- Rules for initial packet marking,
- Rules for how particular aggregates are treated at boundaries,
- Rules for temporal behaviour of aggregates at boundaries.

The latter aspect can be addressed e.g. with regard to treatment of out-of-profile portion of the traffic from a particular class. Below we concentrate on the common denominator for all the behaviour rules – resource control, and show how to achieve efficient and scalable resource control in a DiffServ friendly manner, as well as being also conformant to SLAs.

3   Requirements and Framework

Services built on top of DiffServ should have desired end to end QoS properties. There is a need to have standard common understanding of how these services should be constructed, e.g. which rules should be applied at which points. While this work probably will be started soon by the IETF, this paper proposes another framework, outlining how a set of rules defining services should be created, configured and controlled in a DiffServ domain, and between domains, while meeting three main requirements oultined below.

The Internet mainly needs QoS enforcement when and where a congestion occurs. While congestion is always dynamic the first requirement is dynamic reaction to congestion situations. This reaction should take part of dynamic QoS enforcement protecting high priority flows in the congested area.

The enforcement needs to be scalable, or/and *evolveable* [3], which is our second requirement. The latter assumes systemic nature of the framework facilitating and supporting many services. We argue that operating with groups rather than with individual flows or flow aggregates provides both scalability and evolveability.

Lastly, QoS Enforcement should conform to end-to-end requirements, i.e. preserve invariance under aggregation [4].

We distinguish a general framework which is actually a service creation framework – a structured set of user and network service components and associated operations at the boundary of the DiffServ domain - and QoS specific framework outlining the QoS inter-working between network entities participating in the service creation.

These entities, defined through the business model, are: user (customer), wholesale (reflected e.g. by TCS at DiffServ meter) and retail (reflected e.g. by SLA mapped to TCS and SLS) network providers and subsequent interfaces at the service creation level[2].

4. The new paradigm for network resource control

Resource control problems in IP networks are traditionally solved either in a Network provider domain by means of network management and/or over-provisioning, or by means of dedicated protocols using the same IP infrastructure as data flows carrying network payload. This paradigm does not scale when, e.g. dynamic service creation is required. Flow differentiation or simply network QoS is regarded here as an example of network services which is to be created dynamically, triggered either by a flow itself or by a combination of a flow and network conditions, e.g. congestion.

In a very generic form network resource control takes form of decison making supported by two operations: *GET(information on resource usage)* and *SET(parameters of resource usage)*. We may think of this two generic operations as of distinct interfaces from a control entity to a network.

Further, the decision making can be represented as a set of rules which usually take form of pairs (conditions, actions). If all of the required conditions are true, then the pre-defined set of actions is enforced. After any action is taken, however, a new condition, or a set of conditions are taking place. If resource control rules are to be combined in a sequence, then some care should be taken to avoid unwanted dependencies between a given set of conditions after an action (post-conditions) and pre-conditions of a next action in a Resource Control Sequence (RCS).

Fig. 1 represents a case when a resource control decision procedure is given as an RCS of three rules. Each rule is defined as a set of pre-conditions and an action to be taken. Some of the pre-conditions are those obtained from the *GET* interface and reflect particular aspect of the network state, other conditions can be internal to the

---

[2] The general architecture of the Cadenus framework is to appear as "Cadenus Framework" white paper at URL http://www.cadenus.org

decision making itself. These internal conditions may be timers' state and other conditions configuring the complete resource control sequence in such a way that there are no unwanted dependencies; such an RCS is referred to as *safe*. Similarly, actions defined in an RCS can be those invoked at the *SET* interface, or internal actions setting internal pre-conditions to guarantee safeness of the resource control.

Resource control sequence:
{(IF c1, c2, c3 THEN a1), (IF c4, c5 THEN a2), (IF c6 THEN a3)}

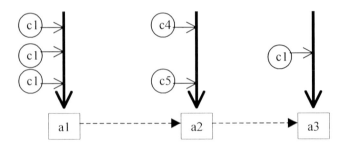

Fig. 1. Example of a simple resource Control Sequence

Safe configuration of conditions in a sequence of rules can be done for a *limited* set of rules and, more important, for a *static* set of rules. Usually safeness is guaranteed by some additional modelling of rules interaction, which is done off-line when an RCS is designed. In case of IP networks deploying dedicated resource control protocols it leads to an isolation of these protocols. When isolated, each protocol is safe by itself, however it is to be configured separately. Hence, the rule interaction problem is pushed towards the off-line configuration. Clearly, the complexity of multiple configurations combined with the complexity of safeness makes it close to impossible to achieve any level of dynamicity in service creation.

On the other hand, there are multiple attempts to centralise the resource control problem. These are motivated by an obvious benefit of having all resources state information at the disposal of a decision making. Theoretically, it facilitates dynamic creation of services, provided that the resource control latency of a centralised system is bounded acceptably. Centralisation, after all, imposes even more complexity induced by the need to maintain itself, usually by a hierarchy. The hierarchy of *GET* and *SET* interfaces tends to waste network resources on self-maintenance, to introduce additional heterogeneity of network resource control protocols at different levels of the hierarchy. In particular, additional heterogeneity may be wished due to different aggregation levels of data passing through *GET* and *SET* interfaces.

The new paradigm, discussed below, aims at dynamic service creation based on a uniform mechanism, thus providing better scaling properties. Instead of distributing network <u>state</u> information (pre-conditions) via dedicated distributed but isolated protocols, or via centralised hierarchcical architectures, the new paradigm distributes <u>event notification</u> information via a single mechanism of group communication.

The sequence of actions in Fig. 1 is safe only because it was designed to be safe. Any attempt to change the sequence

$$RCS = \{R1, R2, R3\} \tag{1}$$

to a sequence, say, $RCS1 = \{R1, R2, R4, R3\}$ should be checked with regard to all possible interactions of rules. Thus, assuming that

$R1 = (IF\ c1, c2, c3\ THEN\ a1)$;

$R2 = (IF\ c4, c5\ THEN\ a2)$;

$R3 = (IF\ c6\ THEN\ a3)$, and

$R4 = (IF\ c7, c8\ THEN\ a4)$,

we will need to check any potential unsafeness caused by interaction of at least post-conditions of R1, R2 and pre-conditions of R4, that is between a1, a2 and c7, c8, and to check any potential unsafeness caused by interaction of post-conditions of R4 and pre-conditions of R3, that is between a4 and c1. Given that pre-conditions and post-conditions may be complex objects, and safety checks are to be performed on whole ranges of values of state data items, this task is far from being trivial.

What does it mean actually to check safeness between an action and a set of conditions, like between a1 and c7 above? An action ai is unsafe with regard to a condition cj if post-conditions of ai are conflicting with cj. While it is not simple to define the nature of a conflict in a generic way, one distinction - between *global* and *action-local* safeness - helps to imagine possible taxonomy of safenesses. Let us define that ai is *action* ak *locally unsafe* to condition cj if cj is a pre-condition of ak. Then, action ai is globally safe to condition cj is there is no such ak that
1. cj is a pre-condition of ak, and
2. ai is action ak locally unsafe to cj.

Let us further define that *event* is the combination of an action and of all its post-conditions. Each action's post-conditions are at the same time pre-conditions of other actions, possibly subject to relevant aggregation. Defining the notion of event as above, we "freeze" a temporal continuity of actions. Fig. 2 illustrates this concept. Events e1, e2, e3 are defined as

$$\{ai: bi1, bi2, \dots, bin\}, \tag{2}$$

that is any of the post conditions bi is associated with the parent action, i.e. it is not *anonymous*. An event, as the association of an action and its post-conditions makes sense only
i.   until all possible conflicts are checked, and
ii. in the environment which allows dynamic creation of RCSs.

Imagine, that RCS in Fig.2 was generated dynamically, for example by a hot plug-in of R2 between R1 and R3. Then, event e1 is to be checked for safeness to c4 and c5. While, c4 and c5 are conditions which just appeared in the RCS it may happen e.g., that c4 is just a negation of b1, etc. If e1 is safe to R2, then R2 can be fired[3]. Real RCS is more properly modelled by a branching tree, rather than by a linear sequence.

In this case e1 is safe to all children in RCS tree (e.g. {R.2.1, R2.2, R2.3}) if there are no conflicts between post-conditions of e1 and any of pre-conditions of {R.2.1, R2.2, R2.3}. If a conflict occurs with, say, pre-conditios of R2.2, then the truncated RCS tree, i.e. {R.2.1, R2.3} still can be fired.

Resource control sequence:
(IF c1, c2, c3 THEN a1), (IF c4, c5 THEN a2), (IF c6 THEN a3)

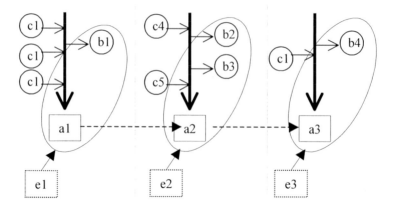

Fig. 2. To the definition of an event

Why not dynamically store all pre- and post-conditions in a single database and run a global safety check over this database before firing any rule? Such an approach is explicitely a non-target for this work. While there are some obvious advantages of a conditions data base approach, it happens to be practically feasible (scalable and efficient) for centralised service creation environments only. If rules are distributed across multiple network nodes, and the latency of the safety check is to be minimal, not to introduce additional delays for time-critical flows in the network, then there is the need for a "hop-by-hop[4]" safety check mechanism.

After all safety checks (against a single set of pre-codnitions in the linear case, and against multiple sets in the case of RCS tree) are performed, the particular event is obsolete. That means, we no longer need to keep an association (2); and post-conditions bi1, bi2, ... , bin may become *anonymous*.

---

[3]   The notion of firing is borrowed from Petri Nets.
[4]   Here a hop is a single rule in an RCS which, of course can coincide with a physical hop of an IP network.

By so introduced event based safety check, we achieve desired automicity of resource control in a distributed environment allowing dynamic service creation. There is still another problem in such environments. Namely, in order to perform a safety check as described above, each rule needs to know the set of child rules. These child rules are to be notified on the event by a mechanism which will be equally good for a one-to-one or for one-to-many parent-child relations.

5. Group Communication

What is the best event notification mechanism for such an environment? Bearing in mind, that RCS can be a branching tree, we inevitably should conclude that a multicast technology should be very advantageous for event notification. Unfortunately, it is not possible to use native IP multicast as the basis for event notification because it largely lacks controls, mainly for group membership. Service model for IP multicast allows any host to become a group member

There have been many discussions in recent years about practical applicability of IP multicast service model [5]. The lack of providers' control over multicast membership and content has been identified as the main drawback of IP multicast [6]. In order to make multicast more deployable many argued that the service model itself should be changed. An excellent background material– motivation and requirements - for this view could be found in [7].

Recently we have presented another approach [8]. We argue that there is no need to modify IP multicast service model while it should be considered as a building block for higher level services (we call these *Group Communication Services*) which should feature needed controls. Further, to facilitate controls over membership and content we propose to make service components programmable to an extent which is currently feasible. Rather than following active networks paradigm in tailoring network elements to meet user and provider requirements we follow slightly modified client server approach to achieve the same goal.

We introduce a new service model dubbed Mediated Group Communication. In this model we have a new logical entity above IP multicast – Group Communication Mediator (GCM). Main semantics of IP multicast are retained (it's our building block), i.e. we retain group membership as being under the supervision of receivers, we do not consider a sender to a group to be necessarily a member of a group. We build a GCM to be back compatible with native IP multicast [5].

Consider a set of hosts H and a set of groups G in which these hosts participate. We define then a meta group G_H as a union of all the groups $\{g_1, g_2, ... \}$ which belong to G and one special group g_0 which consists of one or more GCMs. We show in [8] that all other members of G are sending their control messages to g_0, and g_0 in turn controls group R – e.g. a group of all multicast border routers in a GCM domain.

The criteria for group membership can be derived from the conditions of safety checks, thus each network node, based on its state information and services requested will know which pre-conditions for its required actions it has to check continuously. Based on this, the node joins dynamically a number of *safety groups* which advertise event notifications with relevant conditions. It is interesting to investigate what is the relation of physical network topology and safety adjacency between safety group members, especially for such services as QoS on demand, e.g. triggered by network congestion.

6.   Groups: Engineering view

DiffServ architecture scales because it groups microflows into flow aggregates. Similarly, we propose PDBs to be constructed out of per path behaviours (PPB), which in turn, are path dependant groups of DiffServ per hop behaviours. Basically, the whole framework for SLA based internetworking can be built upon principles of multicast.

There are two obvious motivations for considering multicast as a primary technology to build a framework. The first motivation comes from, so to say, topological prospective – we see that QoS enforcement in the network should include a number of network nodes to collaborate in order to provide a preferential treatment for a flow traversing these nodes. The situation is more complicated when resource reservations are used. In this case, to be reserved resources are to be kept in mind foreseeing future reservations with stringent time and probability guarantees – basically it is an over-provisioning.

The second motivation is the intention to design a QoS provisioning framework with the requirement of flexibility and support of heterogeneous QoS provisioning paradigms and IP network technologies. Such a framework being designed over a group communication paradigm will have an essential feature of hot plug-in. That is a new framework component can be added by just subscribing to a particular multicast group. If the native IP multicast service model is taken as the basis for group communication of the framework components then it is not possible to control the process of joins and leaves. Therefore we plan to investigate into the area similar to mediated group communication applied for a QoS provisioning framework. To support required signalling we propose group communication over native IP multicast using private addressing space. We plan to link this to the DiffRes [9] architecture being currently under simulation.

7.   Examples

As one example let us consider a new authentication module plug-in into the framework. In the native IP multicast it will be quite simple for an intruder to inject a misbehaving component by simply joining even a secret IP multicast group address. We need membership controls. This seems to complicate the framework, however in reality it brings an advantage of flexibility at the level of services supported by dynamic creation and configuration.

Another example can be per path behaviours introduced earlier. PPB - a concept allowing inter-domain QoS transaction set-up and end-to-end tariffed transactions. In statically configured DiffServ network groups of micro-flows and flow aggregates

(PPBs and PDBs groups) automatically define groups of edge routers, which control membership for these groups and recover their end-to-end properties. When statically configured edge QoS enforcement (at the boundary of DiffServ domain) can't meet end-to-end requirements for a PPB or PDB group, then we propose dynamic creation of virtual DS boundary (by, e.g. activating DS mechanisms in interior routers following thresholds derived from committed SLAs).

As an example of economic perspective of this approach let us consider the use of airline industry analogy.

An airline company presents a wholesale entity with quasi-static allocation of quality classes, travel agents present a set of retailers with ability to create service dynamically based on service components provided by the wholesale.

When congestion happens at wholesale, then two things are possible : resource reallocation (if Business class is congested the boundary between Business and Economy classes is dynamically adjusted), or/and selective upgrade (if Economy is congested, and there are still available seats in the Business class) based on a relative "value" of an economy passenger(e.g. frequent traveller account state). Both cases are examples of dynamic creation, or re-configuration of a service conformant to an SLA but also utilising resources of an aircraft in an autonomous manner.

8.   Conclusions and future work

The summary of QoS ad hoc interworking:
*   Resources are partitioned dynamically between quality classes in a fully distributed manner,
*   Dynamic QoS and charging differentiation are enforced only when a network is experiencing congestion,
*   QoS/service creation is facilitated by [re-]enforcing of [virtual] boundaries and distributed resource control for PPB and PDB groups,
*   Resource control is facilitated by the event notification mechanism which allows dynamic creation of a service in a fully distributed manner,
*   Event notification is based on the notion of safety groups where group membership is dynamic, however controlled.

We plan to evaluate our architecture for a number of end-user services (especially for VPN and IP Telephony based on SIP), to prototype it and to measure actual bene-fits.

References

1. *Testing of EF in the wide in presence of stream aggregation and congestion*, TF TANT, Quantum   project,   Feb,   2000,   URL: http://www.cnaf.infn.it/~ferrari/tfng/qosmon/ef/wan/wan.html
2. K. Nichols, *Update on the IETF Diffserv Working Group* , NANOG 13 , Detroit, MI, 1998, URL http://www.nanog.org/mtg-9806/ppt/nichols/ tsld006.htm

3. D. Hutchison, *Systemic Quality of Service*, Dagstuhl seminar Quality of Service in Networks and Distributed Systems, May, 2000, URL http:// www.dagstuhl.de/DATA/Seminars/00/

4. K. Nichols, B. Carpenter *Definition of Differentiated Services Behaviour Aggregates and Rules for their Specification*, IETF work in progress, Feb., 2000 URL http://www.ietf.org/internet-drafts/draft-ietf-diffserv-ba-def-01.txt

5. Deering, S., *Host Extensions for IP Multicasting*, IETF, RFC 1112, August, 1989

6. IP Multicast Backgrounder. How IP Multicast alleviates network congestion and paves the way for next-generation network applications, An IP Multicast Initiative White Paper, URL: http://www.ipmulticast.com/community/ whitepapers/backgrounder.html

7. Diot., C., Levine, B.N., Lyles, B., Kassem, H., Balensiefen, D., *Deployment Issues for the IP Multicast Service and Architecture*, - IEEE Network Magazine, Special Issue on Multicasting, Jan./Feb. 2000

8. Smirnov, M., Sanneck, H., Witaszek, D. "*Programmable Group Communication Services over IP Multicast*", Proceedings of the IEEE Conference on High Performance Switching and Routing, June 2000, Heidelberg, IEEE, pp. 281-290.

9. D. Sisalem, S. Krishnamurthy, and S. Dao, "*DiffRes: A light weight reservation protocol for the differentiated services environment,*" tech. rep., HRL Laboratories, Malibu, USA, Dec. 1999. Under submission, URL:
   ftp://ftp.fokus.gmd.de/pub/step/papers/Sisa9912:DiffRes.ps.gz

An Integrated and Federative Approach to QoS Management in IP Networks

Daniel Ranc, Jacques Landru, Anas Kabbaj

Institut National des Télécommunications, Rue Charles Fourier, F-91011 Evry Cedex
Daniel.Ranc@int-evry.fr
ENIC, cité scientifique, rue G. Marconi, F-59658 Villeneuve d'Ascq
Jacques.landru@enic.fr

Abstract. This paper focuses on challenges related to the management of Quality of Services in IP networks and proposes a solution relying on a CORBA-based Service Architecture tightly and dynamically linked to the managed network. This architecture relies on Management Layers: EML, NML and SML inherited from the TMN architecture in order to deliver a more deterministic behaviour of the IP network. It makes also use of concepts defined by the TINA architecture as well as the TMF regarding key SML aspects. EML management is taken in charge by in-depth use of JDMK agents hiding network heterogeneity.

1. Introduction

The large increase of IP networks penetration makes it progressively replace legacy technologies such as X.25 and proprietary technologies such as IPX, DSA or SNA. On the other hand the deployment of ADSL and radio at the local loop will introduce new requirements for Services Management, particularly for new types of high bandwidth-related services such as Video on Demand, Multiconferencing or Teleteaching. This shift however brings new requirements at the network level regarding isochronous signal transmission in order to realistically carry e.g. voice data. Other services related to Network Management formerly using deterministic X.25 technology are replaced with IP services exhibiting a much more probabilistic behaviour, even if the overall mean efficiency of the latter is high.

This situation may be taken in charge by two approaches: a first network-oriented approach where the equipments themselves would be in charge of Quality Management – the DIFF-SERV, MPLS or tag switching techniques would represent this approach; a second approach where a Management System would rule the use of network resources and organise the dispatching of services to users – the TINA architecture has been an attempt to do so.

The opinion of the authors is that both approaches are unavoidable: a resilient, QoS-based IP network matching the requirements of a Service Management Layer that delivers high quality on-demand services in near-real time to users.

The proposed system which is developed by the GET[1] within the GESTICA [GEST] project is an attempt to demonstrate the usefulness of this combined integrated approach. This paper will therefore first briefly recall fundamentals of current service management architectures. The core part will develop the proposed architecture which combines a layered CORBA-based TMN architecture with Service Management components interpreted from the TINA architecture. Finally a last section will provide the author's conclusions.

2. Current Network Management architectures

2.1 The Telecommunication Management Network framework

The TMN (Telecommunication Management Network) [8.] architecture defined by ITU-T sets the framework for large-scale, hierarchically layered network management systems.

Four layers are defined apart from the equipments: the Element Management layer, where one single equipment is managed at a time; the Network Management Layer, where the whole network is considered; the Service Management Layer which takes in charge network services; the Business Management Layer, realising network-related strategical tasks. The figure 1 summarizes this classical layered scheme.

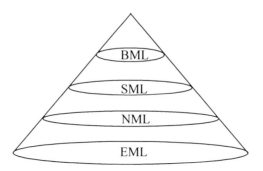

Fig. 1. TMN layered architecture.

Another key feature of the TMN is its dual buildingblock: the manager-agent couple. In this scheme a managing entity, the manager, issues requests to a managed entity: the agent, which replies asynchronously by issuing notifications. In the TMN, Operations Systems (OSs) are dual manager-agent while Network Elements (NEs) are agent entities which use an OSI 7 layer stack and CMIS/P [12.] as the communication service and protocol. In this scheme, the manager issues CMIS/P requests via the OSI stack which are processed by the agent, with the results being passed back later in an

---

[1] Groupement des Ecoles de Télécommunication

asynchronous fashion. The agent and manager are *roles*, e.g. this notion is dynamic in time, with OSs taking either of the two roles in different instances of communication.

The information presented to the manager by the agent is standardized and consists of publicly available Information Models. The latter are specified using the Guidelines for the Definition of Managed Objects (GDMO) [13.] formalism which is object-oriented and consists of classes, attributes and methods (actions and notifications). Values of GDMO attributes, action and notification parameters are represented using the Abstract Syntax Notation One (ASN.1) [X208] structuring language and particular coding rules in order to be transmitted by the OSI stack. An agent entity is composed of three fundamental elements: the *Managed Information Base* (MIB), the collection of *implementation objects*, and the *resources*. The MIB consists schematically of the *naming tree* and the *Managed Objects (MOs)*. The naming tree is the access structure for the Managed Objects via a standardised hierarchical naming scheme. The managed objects, specified in GDMO, are the entities subject to manipulation by the manager through CMIS/P operations.

The scenario depicted above is, however, not cast in stone. New, powerful distribution infrastructures and telecommunication architectures are seriously challenging the TMN. The following two sections summarise the two main frameworks in this area.

2.2. CORBA From a TMN Point of View

The Object Management Group (OMG) has specified the Object Management Architecture (OMA) and is now defining higher level business-oriented services and entities, making the overall architecture more and more complete [2.].

The distribution architecture of CORBA is built upon a software bus, the Object Request Broker (ORB), which provides the infrastructure allowing distributed software components to communicate. Key aspects are: ubiquitous use of the client-server paradigm, true object orientation, implementation language independence, and implicitly distant objects i.e. no difference between local and remote objects and access transparency.

The latter property is particularly powerful for TMN systems designers. Within the OSI communication framework, the programmer has (even with sophisticated software layers between the local representation of the Managed Object, and the distant one) to deal explicitly with communication aspects; whereas in a CORBA-based environment, the programmer does not even distinguish (after some simple initialisation) between local and remote objects: the remote object appears and is manipulated in the same manner as a local one in the programming language. The remote object is in an implicit distance situation. Additionally, data representation burdens which are taken care by the ASN.1 standard in the ITU-T world, are dealt with in a completely transparent way by the ORB.

An additional benefit of CORBA compared to OSI Systems Management is its ease of use. An OSI programmer is faced with relatively expensive tools and complex software Application Programming Interfaces (APIs) such as the low-level

XOM/XMP2 or the slightly more sophisticated TMN/C++ API defined by the former NMF, whereas CORBA systems are comparatively cheap and easy to learn. Typically, competent C++ programmers are able to build simple client-server systems in half a day under CORBA; this is, unfortunately, not the case with OSI-SM technology. These shifts in distribution technology are the main motivations that make CORBA an attractive environment when projecting new TMN systems.

This decision-making point however hides some key questions regarding the TMN architecture. CORBA does not provide powerful access aspects of CMIS/P such as scoping/filtering, neither does it provide a suitable architecture for credible telecommunication management e.g. thinking of building blocks like the Event Forwarding Discriminator, an essential entity for scalable, fine-grain event dissemination in telecommunications systems, and the fundamental Systems Management Functions (SMFs) which support generic Fault, Configuration, Accounting, Performance and Security (FCAPS) functionality, and which are not covered suitably (in a TMN requirement perspective) by COSS [3.] services.

In summary, the TMN system designer is very much tempted by the comfort of the CORBA architecture, but has no real valid support regarding the management of telecommunications systems. The temptation may be so strong however that some had-hoc implementations may arise and solve particular local network management problems but at the cost of ITU-T standards respect and, as a consequence, at the cost of the hope for general interoperability between all stakeholders of the telecommunications service architecture, which is a strategic issue.

2.3. The TINA Architecture

The Telecommunication Information Networking Architecture Consortium (TINA-C) [1] aimed at providing an advanced object-oriented software architecture for integrated telecommunication network and service management and control. In summary, the TINA architecture consists of two major building blocks:

- the *Service Architecture*, which represents a step forward beyond the specifications delivered by the ITU-T, which are limited to the network and element management, and which addresses service control. The Service Architecture introduces the concepts of access and service sessions and integrates service control with service management.
- The *Network Resource Architecture*, which is the TINA view of TMN network and element management. This uses the Network Resource Information Model (NRIM) to model connection-oriented networks in a technology-independent fashion.

The distribution infrastructure proposed by TINA is the Distributed Processing Environment (DPE) which is based on CORBA but is enhanced with telecommunication-oriented features. The DPE runs on an overlay network that is used for control and management, known as the Kernel Transport Network (KTN). TINA makes an extensive use of the Open Distributed Processing (ODP) [14.] methodology in its specifications.

---

2 The XOM/XMP API has been specified by the X/Open consortium, now known as the Open Group.

A discussion on the architectural relationship and potential migration of the TMN to the TINA architecture can be found in [9.], which was written with the idea that TINA will eventually replace today's Intelligent Network (IN) and TMN functionality. However, it is no longer clear whether TINA will be fully applied in a unifying telecommunications software architecture. More realistically, a selected set of TINA concepts and components are expected to be found in future telecommunications systems.

2.4 The JDMK framework

In the belief that new management architectures should be more decentralized, more adaptive and more dynamic, system designers created a particular flavour of JavaBean for network management purposes. Such objects can be used to develop modular open network management agents [7.]. Sun Microsystems has created specific management components called *managed beans* (MBean) [4.] which are Java objects featuring management capacities (get, set internal values or event notification) in addition to the properties of normal JavaBeans (modularity, dynamic loading, ...).

An MBean instance is manageable as soon as it is registered within a framework..Associated with this framework several protocol adaptors can be used. MBeans can then be accessed in an open manner through several protocols (SNMP, CORBA, RMI, HTTP, ...). The MBean concept is the core technology to build modular open multi protocol manager/agent components.

Furthermore, the Java Dynamic Management Kit [4.] is a set of tools enabling the development of Java-based managers or agents as well as a number of useful classes based on the MBean concepts and delivering a high-level API for communication.

Although property of a specific vendor, Sun Microsystems Inc., this framework will probably gain a rising attention and its status is likely to evolve in the future. Sun Microsystems has started the standardization process to produce a universal open management extension of the Java programming language : the Java Management eXtensions JMX [6.]. These extensions can represent the transition from current management technologies (static agents and protocols) to an open model where:

- management agents are plugged in dynamically and immediately available
- management applications are freed from dependancies on a fixed information model or specific communication protocols,
- new management services can be fetched through the network and plugged dynamically into the manager.

The JMX architecture is divided into three levels:

- the instrumentation level: implements the Mbean, the Java object that represents a manageable resource;
- the agent level: provides management agents, the containers that provide core management services which can be dynamically extended by adding new resources. An agent is composed of a Mbean server, a set of Mbeans representing managed resources, and at least one protocol adaptor. An agent may also contain management services, also implemented as a Mbean.
- The manager level: implements management components that can operate as a manager or an agent for distribution and consolidation of management services. It

provides an interface for management application to interact with the agent, distribute or consolidate management information. Additional management protocol APIs provide a standard way to interact with legacy management technologies (SNMP, WBEM, TMN, ...).

The latter level is used to implement the EML layer in the GESTICA architecture.

3. The GESTICA architecture

3.1 Project objectives

The GESTICA project aims at providing an integrated quality control over high bandwidth services such as Video on Demand, Multiconferencing or Hi-Fi Audio. The main concern in this context is that available protocols (e.g. IP) are not oriented towards isochronous transmission, overlaying applications having the burden of smoothing the unpredictable behaviour of underlying resources.

The challenge of the proposed structure is to combine the probabilistic nature of IP and ethernet with a management environment controlling the feasibility and quality of service criteria in a session-based manner. A TINA-inspired session management layer is then be in charge of the control of the availability of network resources in order to insure the required functionality.

The SML components are tightly linked to realistic network information through a network data repository which is constantly updated.

This section presents the GESTICA architecture starting with a global overview, after which it will detail the different Management Layers SML, NML and EML.

3.2 Architecture overview

The GESTICA architecture is mapped on the layering of TMN :

As can be seen in figure 2, GESTICA combines the three classical TMN layers:

- The Service Management layer which is in charge of the user session access and its parameters, as well as service maintenance; this layer is based on a series of components distributed over CORBA.
- The Network Management Layer which delivers on demand connectivity to SML, as well as playing the role of network information repository regarding the network topology, the status of network resources, performance statistics etc. This layer makes use of a particular CORBA-based Management Agent: the TRAC agent.
- The Element Management Layer which manages equipments regarding their configuration on behalf of the NML, and track equipment status and network information in order to be delivered to NML. This layer makes use of Java Mbeans which hide the heterogeousness of equipments and protocols.

In other terms, the EML layer uses specific protocols (telnet, snmp etc.) to access equipments, while all the rest of the system runs over CORBA.

202

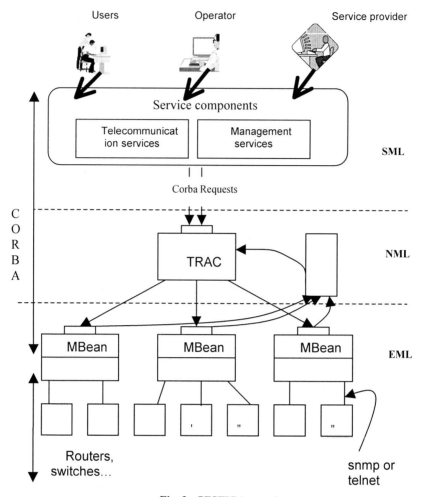

Fig. 2. GESTICA overview.

From a functional point of view the GESTICA system runs simultaneously in two modes:

- A planning mode where the users – in a session-oriented way mimicking the TINA approach - connect to the system at the SML level which then retrieves their user profile, connection classes, tariff selection etc. and allows them to choose the desired service with its parameters, notably the desired Quality of Service. Furthermore the session verifies if the requested service can be delivered in the required quality, given the status of the corresponding resources.
- A supervision mode where network information and statistics are constantly gathered and updated, keeping track in near realtime of e.g. ethernet collision statistics etc., in order to keep the NML layer informed of the status of the network. It is the SML layer which, regularly, interrogates the NML layer to verify the soundness of the network, and if it is not the case, makes the appropriate decisions.

3.3 EML layer

The EML layer is composed of JDMK-based modules insuring a uniform presentation of equipments to the NML layer, disregarding the heterogeneity of routers, switches etc. of different vendors. They take in charge the conversion of high level requests issued by NML, to the specific protocols (telnet, snmp or even sockets) of the particular equipment. Reciprocally, they poll equipment status information regularly (again in the particular protocol of the attached equipment) in order to transmit it to NML.

The available implementation makes use of SNMP mainly for configuration purposes, as well of the `telnet` protocol. It allows to build on demand end-to-end VLANs in the available platform (two routers, 6 switches) as well as to accept requests to upload relevant network statistics such as collision percentages or packet flow measures.

Figure 2 summarizes the situation of the Mbeans in the Gestica architecture.

3.4 NML layer

3.4.1. Architectural motivations

Given the different constraints and requirements to build a realistically deployable TMN agent based on CORBA, a number of considerations and options come to the mind of the system designer. In this particular project, several options were taken as main architectural options:

- Referring to the scalability requirement, it has been chosen to design the agent such as to hide the managed objects inside the agent as the default option. This way, it is possible to accumulate managed objects by the millions without dealing with ORB limitations. An option to publish, on request, the CORBA reference of a particular sub-tree of the MIB is kept open (e.g. if repetitive access to the physical view of an equipment is foreseen, the manager could be interested in such a shortcut through the naming tree).
- The external access to the agent conforms to the JIDM pseudo-CMIS specification. This has the double advantage of a standardised publicly available IDL specification, and of a total independence from the information model used by the agent. Full implementation of all CMIS features, such as scoping and filtering, were put at a top priority.
- A heavy decision has been made: to discard the existing ASN.1 definitions and to replace them by native IDL types by hand. This has the drawback of deriving from strict standard handling, at the benefit of a considerable simplification of the designer's life. A close examination of ASN.1 types delivered by standards indeed leaves an impression of pointless complexity (as an example, all ASN.1 `graphicString`, `printableString` etc. types, many in the standards, are thus replaced with the unique IDL `string` type). The cost of hand-typing the types has been tested on the existing implementation, and has been felt as quite acceptable.

3.4.2 Realisation

The architecture of the agent has been designed as a GDMO template handling engine at the heart of which is the naming tree. The latter is implemented using a scheme holding the managed objet skeletons, which in turn refer to their packages, and finally to attributes and even later to values. The complexity of the system relies on list and memory management, thus quite classical algorithms could be used.

A special note has to be made on multiple result handling. In the CMIS/P context, the agent answers to requests by replies, as many as the result requires (quite many, for a wide scoped request for example). This scheme had to be emulated in the CORBA infrastructure, because the team did not project to depend on a notification service. The actual workaround has been to encapsulate multiple results in a IDL sequence, where each element represents one result.

3.4.3 Implementation notes

The implementation of the agent is based on a quite standard environment: a Sun workstation running Solaris 2.5.1, the Orbix CORBA system and the Sun C++ compiler.

The large use of dynamic memory allocation mechanisms and pointer management has to be emphasized. As an example, the instantiation of a managed object through a M_CREATE request determines the allocation of dynamic memory for each and every entity of the instance (the instance itself, its attributes, their values). Reciprocally, the deletion of an instance requires to free all the memory allocated previously. All the MIB management is implemented using pointers which model the relations between all instance entities. An efficient tree management structure using a son-brother machine representation has been used. This scheme links only one descendant node to its parent node, all other nodes of the same level building a linear list with the former one. The following figure shows partially how these entities link together.

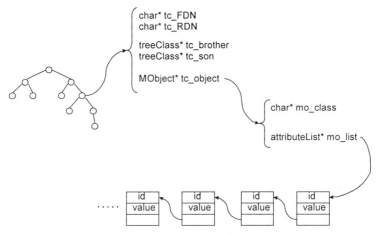

Fig. 3. Object relationships within the agent.

The agent's IDL interface contains all the information and structures needed to run pseudo-CMIS requests under CORBA. This interface is unique e.g. independent of the Information Model. All functions are confirmed e.g. returning results. The IDL definition followed the JIDM standard with the notable exception of the ASN.1 structures, which have been considerably simplified.

The agent has been run with a minimal manager reading requests from ASCII files. This scheme allowed to test the agent with very large request sets. A hierarchical scheme were the cities contain districts, the latter containing zones, and zones containing equipment, has been used. This network is variously meshed depending on the importance of nodes and their geographical situation.

The information model used for this experiment has been the TINA NRIM. The main class used to model networks is the subNetwork class which is subject to a recursive containment relationship e.g. subNetwork instances of layer (n) contain subNetwork instances of layer (n-1) and so forth. Other classes defined in the NRIM information model that were used are link, topologicalLink, linkTerminationPoint and topologicalLinkTerminationPoint (see figure below).

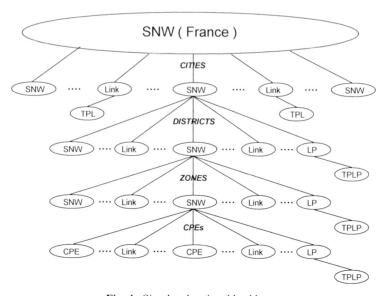

Fig. 4. Simulated national backbone.

Scalability tests included a large scale simulation of a french backbone including instantiation of 430,000 objects, demonstrating the scalability of the approach. Performance results were rewarding. One M_CREATE request including manager to agent and return paths through the available laboratory IP network takes less than 30 milliseconds. The agent has been extensively tested and its stability has been remarkable, both in terms of constant response time and of resilience.

3.5 SML layer

The Service Management Layer implements the User Session Manager (USM) and the Network Quality of Service Supervisor (NQSS).

3.5.1 The User Session Manager

The USM takes in charge the management of the user session in two steps. First, the USM realises the function of session access portal : to invite the user to identify himself and to choose a particular service including its parameters if necessary. Second, it verifies that the required service can be insured properly given the actual status (their load, in particular) of the corresponding resources as viewed at the NML layer. A service that would endanger the quality of already running services would be refused : this way, existing services are guaranteed to remain stable all over their session (in the hypothethis that the network carries only GESTICA-managed bandwidth).

3.5.2 The Network Quality of Service Supervisor

The NQSS implements the function of quality maintenance over the network. It constantly polls the NML layer in order to gather load and status information. In the case of node criticality (e.g. more than a certain percentage of collisions at EML, an information which is mapped into NML information in the form of load statistics), the NQSS makes appropriate decisions, for instance rerouting of certain services into nodes that are less critical. In certain extreme cases, the NQSS would even interrupt certain (less important) services in order to relieve the network. The priority of a service is a parameter which is a session characteristic and related to tariff classes.

The figure 5 shows how these components run over NML.

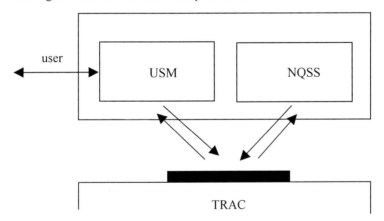

Fig. 5. The Service Management Layer.

3.6 Overall considerations

Two remarks can be made on the GESTICA system :
- The proposed approach is largely technology-independant because 1) the EML components hide the underlying equipments and 2) the Information Model that is

used at the NML layer is the NRIM which is technology transparent. In principle, GESTICA could insure the same services over ATM networks for example, or even mixed networks.

- GESTICA is largely independant of the actual Service as well. The proposed Service, for economical reasons, is VPN configuration but the architecture would not have changed if it would have been vIP or VoD.

4. Conclusions

The work presented in this paper is an attempt to integrate in a single system all components involved in Services Management as well as to demonstrate the feasibility of this approach. At the same time, the system shows how the TMN architecture may be used in CORBA-based systems, making profit of the features of both frameworks. This aspect could be the base for new, enhanced management systems both easier to develop and to deploy than CMIP-based TMN.

On the other hand this system demonstrates also how an IP network can be managed to exhibit a more deterministic behaviour, and this is a key feature for many users which appreciate this technology, but are reluctant to accept its low determinism.

Last, the proposed approach which implements LAN-based VPN management, can be easily extended to other services such as VoD or VoIP because the actual architecture remains independant of the managed service.

References

1. M. Chapman, F. Dupuy, G. Nilsson, *An Overview of the Telecommunications Information Networking Architecture*, in Proc. of the Telecommunications Information Network Architecture *(TINA'95)* International Workshop, Melbourne, Australia, 1995
2. Object Management Group, The Common Object Request Broker: Architecture and Specification (CORBA), Version 2.0, 1995
3. Object Management Group, CORBA Services: Common Object Services Specification (COSS), Revised Edition, 1995
4. Java Dynamic Management [TM] Kit (JDMK) SUN Microsystem Inc., Dynamically Extensible Management Solutions for today and tomorrow, http://www.sun.com/software/java-dynamic/
5. NMF - X/Open, Joint Inter-Domain Management (JIDM) Specifications, Specification Translation of SNMP SMI to CORBA IDL, GDMO/ASN.1 to CORBA IDL and IDL to GDMO/ASN.1, 1995
6. Java Management Extensions (JMX) SUN Microsystems, Inc. http://www.sun.com/products/JavaManagement/index.html
7. J.Landru, H.Mordka, P.Vincent *Modular Open Network Agent for Control Operations* IEEE Noms'98 New Orleans 15-20 february 1998 http://www.enic.fr/people/landru/publications/ieee-noms98/index.htm
8. ITU-T Rec. M.3010, Principles for a Telecommunications Management Network (TMN), Study Group IV,1996

9. D. Ranc, G. Pavlou, D. Griffin, *A Staged Approach for TMN to TINA Migration*, in Proc. of the Telecommunications Information Network Architecture (TINA'97) International Conference on Global Conference of Telecommunications and Distributed Object Computing, pp. 221-228, IEEE Computer Society, Santiago, Chile, 1997

10. T. Rutt, ed., *Comparison of the OSI Systems Management, OMG and Internet Management Object Models*, Report of the NMF - X/Open Joint Inter-Domain Management task force, 1994

11. ITU-T Rec. X.701, Information Technology - Open Systems Interconnection, *Systems Management Overview*, 1992

12. ITU-T Rec. X.710, Information Technology - Open Systems Interconnection, Common Management Information Service Definition and Protocol Specification (CMIS/P) - Version 2, 1991

13. ITU-T Rec. X.722, Information Technology - Open Systems Interconnection, Structure of Management Information - Guidelines for the Definition of Managed Objects, 1992

14. ITU-T Draft Rec. X.901-904, Information Technology - Open Distributed Processing, *Basic Reference Model of Open Distributed Processing*, 1995

An Integrated Fixed/Mobile Network for 3G Services

J. Charles Francis [1]

[1] Swisscom, Corporate Technology, CH-3000 Bern 29, Switzerland
JohnCharles.Francis@Swisscom.com

Abstract. Today, there are separate networks infrastructures for fixed and mobile communications, and for Internet access support. In future, however, a layered architecture with diverse service provision based on common network technologies is foreseen. The network evolution towards this goal is considered in the context of the standardisation work for UMTS. The scope of this paper includes 3GPP work for Release 2000 and wireline access to 3rd Generation mobile services.

1 Introduction

Traditionally, there have been separate networks infrastructures for fixed and mobile communications, and for Internet access support. In the EURESCOM P919 project, however, an alternative approach has been investigated using a layered architecture with diverse service provision based on common network technologies (Figure 1). As part of this study, Multi Protocol Label Switching (MPLS) technology has been investigated as a possible means to support seamless service provision across diverse access media, with potential cost savings in network operation and infrastructure.

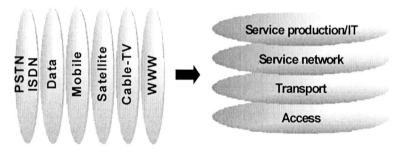

Fig. 1. Shift from dedicated service networks towards a layered architecture with common network technologies

2   Evolution of GSM towards 3rd Generation Capabilities

The cellular system deployed by European mobile operators is based on GSM technology and operates in the 900 MHz and 1800 MHz frequency bands. Work within the ETSI Special Mobile Group (SMG) has specified an evolved radio technology, the Enhanced Data rates for GSM Evolution (EDGE), which uses the existing GSM frequency bands to achieve bitrates of up to 384 kBit/s. The radio access for UMTS, the so-called UMTS Terrestrial Radio Access Network (UTRAN) operates in a higher frequency range (2 GHz) and achieves bitrates of up to 2 MBit/s. The same core network standards are defined for both EDGE and UTRAN, which will ensure compatible services. In this context, specification work for the future mobile core network is carried out by the *3rd Generation Partnership Project (3GPP)*.

When UMTS systems based on the so-called 3GPP *Release '99* specifications are deployed by GSM operators in 2002, it is highly probable that the UTRAN will be interconnected with the existing GSM and General Packet Radio Service (GPRS) core networks (Figure 2). The initial deployment of UMTS may well be limited to isolated islands (e.g., city centres, business areas, industrial plants) and in such a case, the GSM infrastructure will provide limited service support between UMTS islands. To allow the UTRAN to communicate with the GSM and GPRS networks, so-called Interworking Units are needed. The interface between the UTRAN and the core network is the *Iu*-interface and is based on ATM.

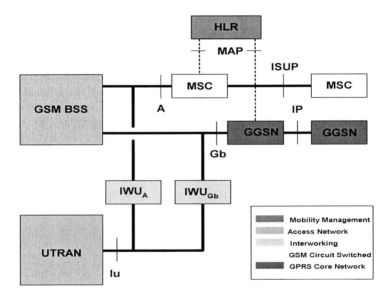

Fig. 2. Interconnection of a UMTS Terrestrial Radio Access Network (UTRAN) with the existing (legacy) GSM/GPRS Core Network

3   The 3GPP Release 2000

At the beginning of the year, work started within 3GPP on *Release 2000*. This specification will include functional enhancements to Release '99 and transport-independent support of UMTS with an "all-IP" option. Below are some of the 3GPP architectural principles for Release 2000.

Transport Independence (to control heterogeneous bearer mechanisms): The GSM/UMTS Core Network architecture will be independent of the underlying transport mechanism (e.g. STM, ATM or IP). Furthermore, operators will have the freedom to utilise a single or any combination of transport technologies.

Standardised alternatives for transport mechanisms: Alternatives for the signalling transport (e.g. SS7, SIGTRAN) for the service control, call control and bearer control protocols as well as the alternatives for the user plane transport will be standardised for relevant transport mechanisms.

Decomposition of network functions: The GSM/UMTS reference architecture all-IP option will be defined in terms of separate functions and clear interfaces such that it is possible to separate transport from signalling. This has the objective of separation of call/session, mobility and service control. This is intended to give operators the freedom to provision, dimension and upgrade these network functions in a modular fashion, providing flexibility and scalability of network implementations.

Flexible traffic processing function placement: The GSM/UMTS reference architecture will allow operators to place the traffic processing functions in the most practical, cost-effective part of the network.

Use of Internet protocols: The GSM/UMTS reference architecture will use, as appropriate, existing/evolving Internet protocols, e.g., to support multimedia services, interoperability with other next generation fixed or mobile networks, and media gateway controllers.

Support for a variety of mobile equipment: The GSM/UMTS reference architecture will support a range of different terminal types (simple speech only terminals, multimedia terminals, PDAs, Laptops, etc.). One particular aspect is that not all terminals may be able to support end-to-end IP capabilities.

Independence of access technology: The GSM/UMTS reference architecture will be designed to ensure that a common core network can be used with multiple wireless and wireline access technologies (e.g., xDSL, Cable, Wireless LAN, Digital Broadcast, all IMT2000 radio access technologies).

Support for roaming onto other 2G and 3G mobile networks: The GSM/UMTS reference architecture will be designed to facilitate roaming between different network types.

Support of Service Requirements: The GSM/UMTS reference architecture will include mechanisms for operators and third-parties to rapidly develop and provide services and for users to customise their service profile

Support of regulatory requirements: The GSM/UMTS reference architecture will include features to support regulatory requirements such as legal intercept, number portability and other regional requirements.

Separation between Bearer level, Call control level and Service level:

— *Use of different access technology to connect the "IP multimedia core network subsystem":* The IP multimedia domain will be connected to the bearer network at a fixed reference point (anchor point), thus hiding the micro mobility of the user equipment (terminal), but it will not hide roaming. This reference point will be independent of the access technology which can be GPRS, UMTS packet-switched or any relevant wireless, wired-line access technology as long as they provide transport of user packets up to this reference point and hide the micro-mobility of the User Equipment. As a consequence, the behaviour of the multimedia call control server can be the same whatever the access technology (radio or wired-line).

— *The access to the IP multimedia core network subsystem will be supported by the packet-switched domain at the, so-called, Gi interface:* The packet-switched domain provides bearers that will be used by the user equipment for its signalling and to provide user plane exchanges with multimedia (H323/SIP) call control servers and gateways. These servers / gateways will be located behind the GGSN, acting as an anchor point for the mobility, implying that when the terminal is moving, the call control server is not changed as long as the user equipment (terminal) is registered on this server.

The specifications need to support both circuit-mode and packet-mode domains: Considering the traffic mix resulting from the set of 3G services and the need for flexible evolution paths, it will be necessary to have separate circuit-switched and packet-switched domains. Each domain will handle its own signalling traffic, switching and routing.

Keep network functions separate from radio access functions: The same network should support a variety of access choices, and access technologies may evolve further. Therefore, network functions such as call control, service control, etc. should remain separate from access functions and ideally should be independent of the choice of access. This implies that the same core network should be able to interface to a variety of radio access networks.

4 Wireline and Cordless Access to 3rd Generation Mobile Services

From the perspective of an integrated fixed/mobile core network, wired and wireless access are just different ways of accessing the same services. xDLS modems can be introduced to upgrade the capability of the copper local loop allowing a bandwidth of several Mbit/s. A UMTS, or other *Iu*-compliant, basestation can be added at the customer site to provide cordless access to 3G mobile services. An Interworking Unit (IWU) can be introduced at the *Iu*-interface to separate traffic into the legacy circuit-switched and packet networks (Figure 3). This is analogous to the separation of UTRAN traffic into GSM circuit-switched and GPRS types, as shown in Figure 2. Direct connection to a UMTS core network can integrate connection-oriented and packet-based traffic (Figure 4).

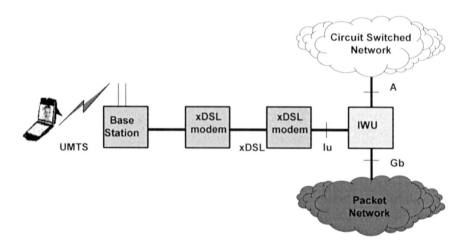

Fig. 3. Fixed network evolution towards UMTS. The copper local loop has been upgraded to xDSL and a UMTS basestation providing cordless access has been added at the customer premises. At the operator side, an Interworking Unit (IWU) separates circuit-switched traffic into respective (legacy) networks in a similar manner to Figure 2

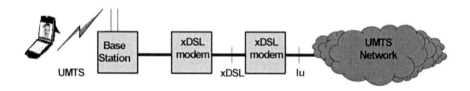

Fig. 4. In a future integrated scenario, xDSL can be connected to an integrated UMTS core network which handles both circuit-switched and packet traffic

5   Protocol Considerations

Over the next few years, telecommunication backbone networks will implement IP-based connectionless packet switching techniques supporting voice and multimedia traffic. Based on new QoS mechanisms and supported by centralised architectures, real time services will be enabled.

The evolution from the present circuit-switched networks to connectionless packet-switched networks will be made possible by the introduction of gateways. Gateways handle all the interworking functions, translating information between packet-based VoIP networks and circuit-switched networks. The entry gateway encodes the speech into a set of compressed voice frames, packetizes them into Real-Time Protocol packets and forwards them over IP or ATM core networks via the network interface. The gateways can be divided into three groups:

1. Media Gateways (MG) that handle the conversion of media streams from circuit format to packet format.
2. Media Gateway Controllers (MGC, Call Agents) that manage the packet network connections. In some configurations the MGC has integrated the Signalling Gateway function.
3. Signalling Gateways which constitute the interface to the circuit-switched networks with SS7 signalling.

For the communication between the MGCs (Call Agents) and the MGs, a media gateway call control protocol is used. Based on precursors MGCP and SGCP, the IETF MEGACO/ITU-T H.248 protocol is currently in the process of being standardised jointly by IETF and ITU-T. The evolution of the IP-network may be as follows:

1. In a first step, the CCS7 signalling trunks from the national/international telephony networks can be connected to the Media Gateway Controllers. The voice trunks are terminated at the Media Gateways. For the transport of the BICC protocol (between the Media Gateway Controllers) and the media streams (RTP between the Media Gateways), an ATM / MPLS Core Network can be used.
2. In a second step, the Media Gateways can be used for the connection of existing local exchanges. The LANs are directly connected to the MGs which are encoding the media into a set of compressed voice frames, packetizing them into RTP packets over IP. The CCS7 signalling from other National/International telephony networks can be connected to the Signalling Gateway Function (which may be collocated with the Media Gateway Controllers) using SCTP over IP.
3. In a third step, the IP (or another packet-switched network) can be connected over Multi Protocol Lambda Switching Network to DWDM. Other algorithms or Multi Protocols are under consideration.

Indicative solutions are shown in Figures 5 and 6.

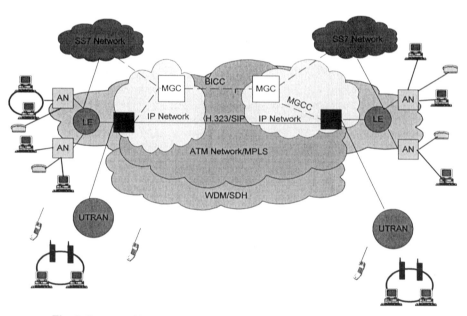

Fig. 5. Support of 3G mobile services by Multi Protocol Label Switching (MPLS)

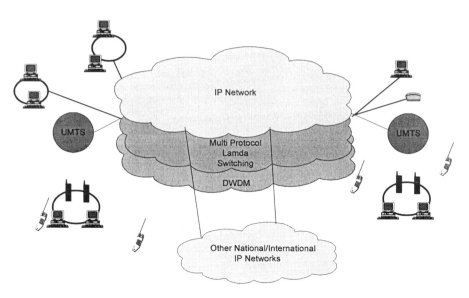

Fig. 6. An integrated fixed/mobile network for 3G services

6 Conclusions

The technical work for the evolution of GSM towards UMTS is carried out by the 3rd Generation Partnership Project (3GPP) and the currently approved specifications are referred to as Release '99. The specifications have been developed from a GSM-operator perspective and are aligned with the ETSI Special Mobile Group (SMG) work on GSM evolution based on EDGE radio technology. ETSI core network standards for 3rd Generation Mobile are based 3GPP specifications and the 3GPP is work is therefore critical for both future GSM and UMTS. Mobile telecommunications in Europe is highly constrained by ETSI standards as appropriate international signalling for roaming users and common terminal standards must be supported. The specifications for Release '99 have been based on interworking to the legacy GSM/GPRS network. Current 3GPP work for Release 2000 is looking to transport-independent solutions including an all-IP option.

To achieve an integrated fixed/mobile network for 3G services, a primary issue is to bring wireline/cordless access into the 3rd Generation mobile perspective. Options exist to upgrade the fixed network to provide a 3G cordless service and a variety of radio access technologies are available including UMTS, BLUETOOTH, BRAN, DECT and Wireless LAN. The same mobile services should be available independently of access media.

The use of wireline/cordless access to 3G mobile services will offer several benefits. For the operator, it reduces the demand on the cellular infrastructure and utilises existing investment in the local loop. For mobile subscribers, wired/cordless access to 3G mobile services offers a dedicated high-speed link which can provide access at higher bitrates than the UTRAN.

The future mobile services, whether based on UMTS or EDGE, will be multimedia. In the context of the new investment for these services, the opportunity will exist to add wired access / cordless access and achieve the goal of diverse service provision based on common network technologies.

Abbreviations

AN	Access Network
BICC	Bearer Independent Call Control
BSS	Base Station Subsystem
EDGE	Enhanced Data Rates for GSM Evolution
GGSN	Gateway GPRS Support Node
GPRS	General Packet Radio Service
HLR	Home Location Register
IWU	Inter Working Unit
LE	Local Exchange
MAP	Mobile Application Part
MG	Media Gateway
MGC	Media Gateway Controller
MGCP	Media Gateway Control Protocol

MPLS	Multi-Protocol Label Switching
MSC	Mobile Switching Centre
SGSN	Serving GPRS Support Node
SIP	Session Initiation Protocol
UTRAN	UMTS Terrestrial Radio Access Network
WDM	Wave Division Multiplexing
xDSL	DSL (Digital Subscriber Line) technology

Acknowledgement

This paper reports work carried out within the EURESCOM P919 project in which the author has worked. The views expressed are those of the author and do not necessarily reflect the views of participating companies.

Smart Card-Based Infrastructure for Customer Date Collection and Handling in the IST E-Tailor Project (IST-1999-10549)

James Clarke[1], Stephen Butler[1], George Kartsounis[2] , Simela Topouzidou[2]

[1]LAKE Communications, Business Innovation Centre, Ballinode, Sligo, Ireland
Jim.Clarke@lakecommunications.com
Stephen.Butler@lakecommunications.com
[2]Athens Technology Centre, 6 Egialias Str., Paradisos, Maroussi, 151 25, Athens, Greece
Gkart@atc.gr, Simela@atc.gr

Abstract. The E-Tailor project (Integration of 3D Body Measurement, Advanced CAD, and E-Commerce Technologies in the European Fashion Industry *(Virtual Retailing of Made-to-Measure Garments, European Sizing Information Infrastructure)* is the largest project in the European Fashion Industry and one of the largest projects in the Information Society Technologies (IST) programme under the fifth framework programme of European research. E-TAILOR aims to develop advanced infrastructures, which will establish a **new paradigm for Virtual retailing services of customised clothing** *(under Action Line (AL) II.3.3- personalisation of goods and services and AL IV.4.1 real-time clothing simulation and visualisation).* In addition to presenting the E-Tailor project, this paper will concentrate on work being carried out in relation to innovative and secure Smart card applications being developed in E-TAILOR.

1 Introduction

A number of initiatives have arisen recently in many European countries, evolving around the concepts of Made to Measure Manufacturing and Retail shopping via the Internet. The combination of these new services is now possible by the emergence of technologies, systems and practices, such as 3D whole body scanners, automatic body measurement S/W, 3D CAD systems for the customisation of existing styles, Virtual - try -on visualisation techniques and new smartcard technologies (multi-application Javacards)

The main objective of E-TAILOR is to develop a comprehensive innovative platform enabling the integration of new specialised forms of clothing retailing (Virtual-home shopping) with new high value-added services to the customer, namely the production of personalised garments at reasonable prices, in short time and with a close to perfect fit.

These new technologies promise a significant potential for the stimulation of the European Clothing Manufacturing and Retailing sectors; however, significant impediments must be removed in order to achieve a successful launching of the related applications and services.

E-TAILOR is founded on a global approach, global in terms of tackling all problems, global in terms of establishing a critical mass of all major players, global in terms of it's methodological approach combining both R&D and real state-of-the-art applications (demonstrators). Figure 1 contains a schematic of the entire E-Tailor system.

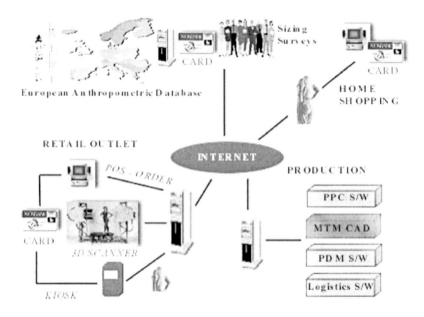

Fig. 1. E-Tailor Schematic Diagram

E-Tailor has the following technical/scientific objectives:
- To develop tools that will contribute to the solution of the sizing problem (non-uniformity of size designations, aggravated by the lack of updated body measurement data on European populations), in the form of a European Sizing Information Infrastructure (ESII). ESII will serve as a unification platform ensuring conformity in the storage and processing of sizing data, acquired in future full-scale sizing surveys, organised either by the member states or by the EU. The specifications for the content and data representation formats of a European Anthropometric Database (EAD) will be defined in common by the E-TAILOR User Group, prominent Clothing Institutes and the relevant standardisation committee (CEN/TC 248/WG 10, for the development of a standard size designation system of clothing).

  ESII will consist of:

- A European Anthropometric Database (EAD), linked to advanced tools for 3D body shape analysis, leading to the generalisation of body shapes and the production of a Library for Generic Body models of European populations.
- ESII will be embedded in a web-based Sizing Information Service for the European fashion industry.

- To develop a Customised Clothing Infrastructure (CCI), consisting of:
  - intelligent design systems, based on rules and practices of human tailors (computer version of the tailor's brain)
  - Order clustering and specialised ERP elements for the optimisation of the entire supply and production chain for overall quick response and cost effectiveness.
  - CAD interoperability standards

- To develop a realistic and efficient clothing simulation and animation platform, accessible by kiosks and via Internet, and enabling the customers to visualise realistic models of themselves (obtained from 3D scanners), wearing "real" clothes with selected colours and accessories from the retailers collection (Virtual Try-on). This will tackle the need for specific interfaces required for Virtual shopping of clothing (the mirror or dressing room problem). The development of these tools will be based on existing state-of-the-art Virtual Reality systems used to dress and animate virtual mannequins. Further R&D will enable:
  - the dressing of customers 3D models with garment descriptions imported from CAD systems;
  - the displaying of clothes with texture rendering different aspects of garments (such as fabric texture, seaming, buttons, zips, etc.);
  - simple animation of the customers models, moving them to different positions, e.g. front, rear view, raising their arms, etc.

- To develop a generic 3D framework tackling the problem of inconsistent body measurements obtained from different scanners and related measurement software. Combined efforts of 3D scanner manufacturers and R&D centres will lead to the development of:
  - Generic body representations at various levels of symbolic abstraction for exportation to other applications, such as shape analysis, CAD systems and Virtual Try-on S/W;
  - Generic representations of measurement data, as well as suggestions for standard EDI messages for the communication of measurement data;
  - System independent measurement extraction S/W.

- To develop a comprehensive Customer Data Infrastructure (CDI) for the collection, storage and legitimate exploitation of customer data (including body measurements, personal preferences, etc.), thus tackling the problem of customer acceptance for the whole service and the problem of issuing multiple non-interoperable cards for the same customer.

- To integrate current state-of-the-art systems (3D scanners, MTM CAD, Virtual Shopping applications) in the form of across the value-added chain demonstrators. Demonstrators will, however, serve both as reference for the development of advanced systems, as well as testing and evaluation sites for new developments.

The combination of R&D and Demonstrators will thus result in a new paradigm for comprehensive Virtual retailing services of customised clothing.

Since the project is so large and covers a variety of areas, The E-Tailor project is split into three sub-projects, the European Sizing Information Infrastructure (ESII), The Customised Clothing Infrastructure (CCI) and the Virtual Shopping Infrastructure (VSI). The following three sections will explain the sub-projects in detail.

2. Sub-project 1: The European Sizing Information Infrastructure (ESII) and related automatic body measurement technologies (3D scanners)

2.1 European Sizing Information Infrastructure (ESII)

The actual situation of the different size systems utilised by European clothing industry and retailers is confusing and unsatisfactory. The size labeling is different from country to country. Up to now there is no common size system available in the European Community. As the current sizes are not comparable between the countries, industry and retailers face enormous problems when they want to export or import garments within the European Market. This is known as a huge impediment to the further development of the EC Market.

A system of standard sizes to be used in the European Community can only be established if there are qualified data available concerning the measures of the population. Some Member States carried out more or less reliable surveys. But mostly these surveys are very old and the reliability is not too good. Most of these data can not be used for the new common approach. New coordinated measurements have to be carried through.

A clothing technical committee has been set up in the CEN with the task to establish the future basis for a European standard size system. The target is to create a system which can be used European-wide with standardised size designations and after that to develop common size charts so that clothing fits to as many people as possible. This can only be defined if the CEN group has data available about the distribution of body proportions in the Member States.

In 1990, the European Commission stated in a paper that "it would be useful to carry out a measurement survey of the European population in order to build up a system of size normalisation available for the clothing industry and valid for all Member States." In the face of the upcoming new MTM production the necessity has increased enormously. The objectives of establishing an ESII are: -

• To develop a standardised framework and platform (infrastructure) for the deployment of a large scale European Sizing Survey (funded by external sources) for setting up a system of standard sizes to be used in the Member States

- To set up a Web enabled European Anthropometric Database (EAD) in order to offer a new service to the European clothing industry and other interested industries
- To develop tools for standardised interpretation of anthropometric data (advanced shape analysis software, generic body models)

The ultimate target will be to organise and carry out a measurement survey in the European Community in order to get clear and extensive information about the distribution of the body measures in the population. A pilot technology proofing survey (approximately 1,000 persons) will test, populate and validate the European Anthropometric Database. The European Anthropometric Database (EAD) and the related innovative tools will offer a new service instrument to the European clothing industry, retailers, mail-order companies and other related industries.

2.2   3D Body Measurements and Systems

Existing whole-body scanners permit fully automated sizing surveys and rapid individual measurements for custom clothing applications. However, the different 3D data acquisition techniques, the different scanned data formats and the varying measurement application software, result in inconsistent measurements across various systems (hardware and application software dependencies). The main objectives are: -

- To assess current systems (3D whole-body scanners, automatic measurement techniques) with respect to measurement consistencies;
- To develop standard representations for 3D raw data and measurement data, generic 3D body models, enabling the development of configurable interfaces to CAD systems, shape analysis and Virtual-Try-on S/W;
- To develop body measurement software, which is system independent, with the aim of achieving, as far as possible, unified and reproducible results obtained from different scanning systems.

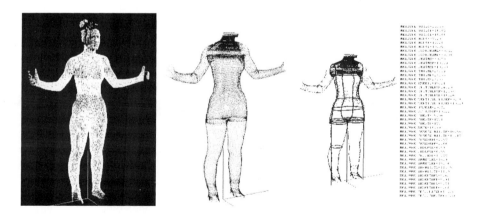

Fig. 2. From scanned data to measurements – Different body segmentation and body landmark detection techniques determine measurement results.

3. Sub-Project 2: Customised Clothing Infrastructure

The successful launching of Made-To-Measure services in the clothing industry is critically dependent on the following criteria:

- The *fitting* of the garments should be much better than the fitting of mass-produced clothing;
- The process for adapting existing styles to personalised measurements (alteration of existing patterns) is mainly based on geometrical transformations, based on the body measurements. However, in a true made to measure environment, only a handful of people have the skills and knowledge necessary to custom tailor a garment.
- The *additional costs* and *delivery times* involved in customisation. The whole ordering, design, production and logistics chain involves significant additional costs, compared to mass production. The client would not in general be prepared to pay more than an excess of 20%-30% to buy a custom-made garment. Furthermore, he would expect his order to be fulfilled in less than a week (3-4 days).

The processes that affect cost and delivery times are:

- Order handling processes. MTM orders are essentially unit orders;
- Communication processes;
- Design and production processes. Pattern alteration, marker making, cutting, sewing, finishing, packaging and delivery have to be done to the unit and, therefore have to be optimised.
- Raw material management is also critical. Fabric management must also be adapted to unit manufacturing.

The combination of these requirements lead to the need for the design and implementation of intelligent pattern alteration systems and an IT system (MTM - ERP) to optimise the whole MTM supply chain so that:

- Fitting is close to what is expected from an experienced tailor;
- A fast response time (less than 4 working days) is achieved;
- A minimum Inventory of fabrics and other material is kept maintaining a good service level; and
- Machine set-up times and material handling are minimised.

E-TAILOR's Tasks to meet the above requirements are:

- The design and development of an intelligent pattern alteration system and a morphological editor, which will enhance the design and customisation capabilities of current MTM CAD systems;
- The development of critical components for a highly specialised (MTM focussed) Internet enabled ERP system. The following figure (fig.3) illustrates the main elements of an integrated MTM MRP.
- Suggestions for interoperability standards, which will enable the exchange of data between CAD systems of different suppliers.

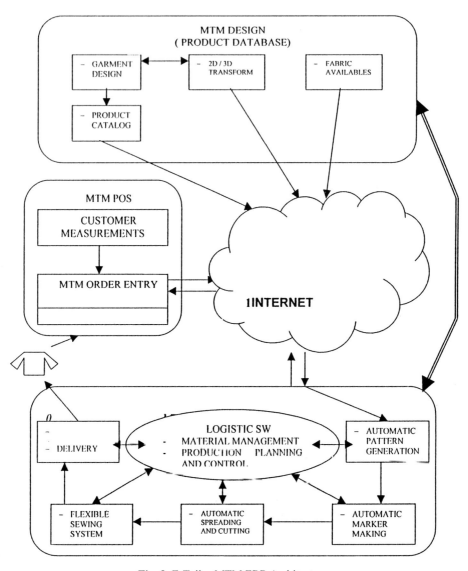

Fig. 3. E-Tailor MTM ERP Architecture

This approach requires the developments of interactive manipulation tools for 3D pieces of fabrics. The 2D garments are then imported into a 3D environment in two steps.

4. Sub-Project 3: Virtual Shopping Infrastructure

Most of the existing clothing simulation and animation systems are either non-realistic, or too slow, they all require extensive user interaction and none of them permits clothing simulations on bodies of real humans, with real garments from the collection of a manufacturer, or retailer. Furthermore, data volumes and processing times do not permit simulations on the web.

4.1 Objectives. The Objectives of sub-Project 3 are:

- To create a realistic and functional virtual shopping environment in the form of a *comprehensive interactive service* to the customers at *retail kiosks and on the web.* The core innovative tasks relate to the development of advanced clothing simulation techniques and software, enabling the customers to visualise *realistic models of themselves* (obtained from 3D scanners) *wearing "real" clothes* with selected colours and accessories from the retailers collection (*Virtual Try-On software*) and order with their customised smart cards;
- To integrate the Virtual Shopping Service with 3D scanners, CAD-CAM scanner-Web interfaces, Multi-application Smartcards and electronic catalogues.

The approach is to take the following as inputs: -
- 3D human body modelled with 3D scanners.
- 2D garments from clothes CAD/CAM software.

To process the following: -
- Assemble and seam the 2D garments to create a 3D garment;
- Dress the virtual mannequin with the 3D garment;
- Visualisation of the dressed mannequin for the web.

Step 1. From geometric 2D patterns to 3D physics-based pieces of fabric
This first step implies to import 2D CAD/CAM patterns, which only contain geometric and topological information (shape of the patterns, and seaming associations between patterns) and then convert them into 3D pieces of fabrics, which requires to convert the 2D patterns into a data structure that can be supported by the fabric mechanical model. The basic aspect of that process is constrained by the discretisation needed to perform the mechanical model simulation:

1	2D CAD/CAM importation
2	2D patterns discretisation
3	2D patterns physical, texture, colour properties assignments

Step 2. Assembling fabrics and pieces on the virtual mannequin - automated mannequin dressing.

The approach is similar to the one used by tailors when they assemble 2D paper patterns onto a mannequin. They use pins to assemble the paper patterns directly onto the mannequin. For that purpose, there are some specific features (lines) on the mannequin that allow the tailor to know where to pin which pattern of paper. The result is a "dressed" mannequin, although the garment was directly assembled onto the mannequin. Similar features will be added onto the 3D scanned body in order to allow a direct placement of the 3D patterns on the mannequin. Once the garment is assembled and the mannequin is dressed, the physical simulation will give its precise shape to the garment, so that is fits to the mannequin it is attached to. Different postures of the mannequin can be chosen.

4.2 Smart card-based Infrastructure for Customer Data Collection and Handling (CDI).

4.2.1 Introduction.

The smart card market in market in Europe has matured and grown significantly in the last few years. A number of proprietary electronic purses, telecommunications payment cards and loyalty schemes are available. There are a number of significant manufacturers of smart cards in Europe and hardware and software innovation continues rapidly. In Sub project 3 of the E-Tailor project, participants will develop and trial software/hardware components for the delivery of smart card mediated point of sale applications. It is foreseen with the widening of e-Commerce, the point of sale will move out of the retail unit and into the home, work place and public spaces through a variety of telecommunications enabled devices. It was decided to store on a smart card as the Secure Personal Data Storage (SPDS) because it will give a citizen full ownership and control of personal data, thus addressing privacy issues. An alternative would be to store the data in encrypted form and allow access by supplying a decryption protocol using a smart card.

The industry is now at the stage where it needs to converge on standards to gain widespread acceptance of the smart card by the citizen and the consumer. This section outlines some of the current options and standards available and makes some recommendations about an appropriate platform for E-Tailor personal data storage.

4.2.1 Broad requirements of the Secure Personal Data Storage (SPDS).

The requirements are broken into two categories – Customers and Retailers.
    Customer's requirements of E-Tailor-enabled businesses: -

> A technology that protects their ownership of personal ESII data;
- The ability to allow limited secure access to their data;
- Portability of their data from one point-of-sale context to another - retail kiosk to home PC connected to the web;
- Convenient inter-working with payment mechanisms and loyalty schemes;
- The data set will include 3-D body data, Front and side face images, Personal data, Private keys and Certificates;
- Customers are unlikely to accept anything greater than a 5 second delay for transfer of all of their data from the SPDS into the client application;
- It is likely that SPDS will be asked to return only certain parts of the data set so that some kind of indexed or tag-based file system will be necessary;
- Data stored on the SPDS should be protected from theft or misuse by someone other than the customer in two ways:

a) The customer should own the SPDS and carry it on her person.

b) Data and applications on the SPDS should be protected by means of a pin or some form of biometrics protection (fingerprint, voiceprint, etc.).

Retailer's requirements when providing E-Tailor body scanners: -

- A means of storing data extracted from a scanner;
- A way of associating this data with a customer;
- A means of assuring the customer of the privacy of her data;
- Standards-based software and hardware components to add to their scanner;
- In order to support structured and secured data access and hosting of applications, the SPDS should support multiple applications;
- It is of concern to all operators of an application distributed on smart cards that one application should not be able to corrupt of access data belonging to another application. The SPDS must support a security architecture that ensures the integrity and verifiable downloading of applications.

4.2.2 Hardware/Software components of the Secure Personal Data Storage (SPDS).

The CDI prototype hardware components will be cost-effective, secure and robust card readers that conform to chosen standards, which will utilise cost-effective multi-application or secure data storage smartcards that conform to chosen standards.

The CDI prototype software components will include a generalized card-based application for the definition, access and modification of an XML-based ESII data structure on a multi-application smart card. We have selected XML because it is the obvious choice for the representation and manipulation (via a myriad of tools) of complex data structures with a high semantic content. The CDI components will provide enhanced access to this data structure to allow retailers limited access and customers full access since retailers and manufacturers will only need to know those aspects of body data essential for the manufacturing process. The rest of the data will be private to the citizen. The CDI will include standard software component proxies for a number of common smart card readers.

4.2.2 Standards Compliance of the Secure Personal Data Storage (SPDS).

The following standards are relevant to the work being carried out in the development of the E-Tailor SPDS: -

- Public key infrastructure – PKSC#15

PKSC#15 is a new proposed standard from RSA Inc. to deal with the problem of incompatible formats of smart cards for use in a Public Key Infrastructure.

- OCF – Open card Forum

OCF is intended to overcome the proprietary and complex nature of smart card reader programming.

- OpenCard architecture

To address the requirements and objectives described in OCF, the core architecture of the OpenCard Framework features two main parts: theCardTerminal layer and the CardService layer. Further, the problem of card issuer independence is addressed separately by OCF's ApplicationManagementcomponent.

- PC/SC – PC Smart Card

The PC/SC Specifications 1.0 are completely platform independent, and can be implemented on any operating system.

- VOP – Visa Open Platform

Visa's open platform, which is now migrating to a multi-vendor standard as Global Platform, is an integrated environment for the development, issuance, and operation of multi-application smart cards. The standard consists of two parts – that covering the card terminal and that covering the smart card. The card specification defines an initial personalisation process and standard ways for customising the card after it has been issued. The goal is to provide tight control of the card and of the process of loading new applications onto the card.

- ISO 7816

The ISO 7816 standard "Identification cards – Integrated circuit cards with contacts" is the most significant standard for cards that establish electrical contacts with card readers when the card is inserted into the card slot.

- EMV ICC

EMV '96: ICC standards were agreed in 1996 and are commonly known as EMV. These are derived from ISO7816 and are intended by the sponsors (EuroPay, MasterCard & Visa) to ensure interoperability in the payment business. They state a minimum functionality for smart cards and card terminals.

- Java Card

Java Card allows on-card applications to be written in Java, and at the same time provides a good base for multi-application smart cards. All the major smart card manufacturers have implemented a Java Card. The latest specification is version 2.1.

- Multos

The industry consortium Maosco, which owns the specification, owns Multos. Multos is a multi-application smart card OS which provides an ISO compliant file system interface and, in addition, and execution environment for custom applications. Application developers develop these applications in a new application language called MEL (Multos Executable Language).

- SmartCards for Windows

This card is a combination of the traditional ISO 7816-4 compliant operating system and a programmable platform. Windows for Smart Cards is an 8-bit, multi-application operating system for smart cards with 8K of ROM.

- CEPS and electronic purses

The 'killer application' that will drive the widespread introduction of multi-application smart cards is the electronic purse. The electronic purse is a particular instance of a 'stored value' card, where monetary value, or something that can be equated to monetary value, such as loyalty points, electronic tickets etc., are stored on the card in a secure manner.

5.   Conclusions

The research and development undertaken in E-Tailor will feed into Smart card-based products with the following competitive features: -

- Facilitate multi-application smart card applications for payment, transport, individual authentication etc.

- Be portable across a wide range of hardware / software architectures including embedded devices and main-stream e-commerce applications
- Be portable across a wide range of point of sales venues – phones, PDA's, kiosks, public communications terminals and PCs
- Abide by emerging standards for smart cards, embedded systems and e-commerce applications to ensure maximum portability of our intellectual property.

Acknowledgements

The results described in this paper are an outcome of the work being carried out within the context of Information Society Technologies Programme (IST 1999-10549) E-TAILOR project, which is partially funded by the European Commission.

References

1. IST-1999-10549 E-Tailor. Deliverable Document 0201F01_D4Part1_5: T1.5 Personal Data Protection, July 2000. Authors – Hohenstein Institutes (DE), Somatometric Institute (EL), LAKE Communications (IRL), Nottingham Trent University (UK).

Provision of QoS
for Legacy IP Applications
in an ATM-over-HFC Access Network

Jürgen Jähnert [1], Stefan Wahl [2], H. C. Leligou [3]

[1] Supercomputing Center University of Stuttgart
Stuttgart, Germany

[2] Alcatel Corporate Research Center
Stuttgart, Germany.

[3] National Technical University of Athens
Athens, Greece
Jaehnert@rus.uni-stuttgart.de
swahl@rcs.sel.de
Nelly@telecom.ntua.gr

Abstract. With the increasing needs for delivering higher bandwidths to residential areas, new technologies (e.g. xDSL) are emerging that are enabling broadband access. Network technologies, which are based upon upgrades of existing network infrastructures, are, economically, one of the most promising approaches. One of these, competing with several other approaches, is hybrid fibre coax networks (HFC) currently being used for CATV broadcasting. The various access network technologies will be interconnected via broadband core networks resulting in heterogeneous networks which, have to provide end-to-end applications based on the Internet Protocol. Since the CATV network is of a shared nature, the provisioning of IP-based Quality of Service (QoS) requires suitable inter-layer signalling in order to provide QoS-related information as soon as possible to the relevant instances. This paper presents a system architecture using ATM-over-HFC for high-speed Internet access providing IP-based QoS. The architecture's seamless interoperability with other network technologies supports both IP-based and native ATM applications.

1. Introduction

Much progress towards high-speed networks has been made in recent years with respect to the bandwidth capacity of the backbone infrastructure. Also, in face of higher available resources in this section of the network, an effective exploitation of available resources is highly recommended since, from the application point of view, an increasing demand of this scarce resource can be identified. Solutions are required that enable customers to take advantage of the possibilities that future core networks could provide.

One promising approach is the Hybrid Fibre Coax (HFC) network. It consists of a fibre part for large-scale data distribution in city areas and a coaxial part into customer households. HFC networks form the basis of current cable TV infrastructures for the distribution of broadband TV signals. Thus, HFC systems combine two main advantages; firstly, they are widely installed and secondly, they are specifically designed for broadband communications.

HFC networks are characterised by a tree and branch topology. At the root of the tree, a head-end entity controls the traffic. The HFC network is a point-to-multipoint bus access network in the downstream direction, and a multipoint-to-point bus in the upstream direction. However, in order to support bi-directional broadband communication, the network has to be upgraded. For the physical layer, this requires the extension of the unidirectional amplifiers to realise an upstream transmission capability, introduction of medium access control (MAC) mechanisms and additional effort to guarantee privacy in a shared medium. On top of this physical layer, a transport layer is required to provide data traffic with a distinct QoS and a dynamic response to fluctuating demand. Futures applications will be based upon IP and will be supported by several QoS mechanisms. Since interoperability between the QoS mechanism and existing applications is highly recommended, the chosen QoS protocol should be modular and transparent to these kind of applications. For this reason, within the ACTS project AROMA, such a system was developed, implemented and verified.

Section 2 describes briefly the overall architecture of the AROMA system, its individual network elements, concepts and issues concerning the physical and transport layer.

Section 3 follows with an introduction of the QoS service-provisioning concept of the AROMA access providing seamless interoperability with both, other access technologies and the core network.

Section 4 provides a summary of initial measurements made in the real AROMA network focussing on IP-based QoS for legacy IP applications.

The paper closes with a conclusion.

2. The Aroma HFC Architecture

Within the European funded ACTS project AROMA 13, an HFC access network has been designed and implemented that provides both a multiple services environment and Quality of Service support 1.

ATM is chosen as the mechanism to provide connection oriented traffic with a distinct QoS. Since the most important protocol used at the application level is the Internet protocol which is deployed in a classical IP-over-ATM environment, a suitable inter-layer signalling is required in order to provide QoS-related information as soon as possible to the relevant instances. This is to effectively exploit the scarce resources of the shared medium and to offer a core-network compatible QoS concept.

Fig. 1 depicts the deployed system architecture. The relevant access network elements are the Access Network Adaptation (ANA), the Access Network

Termination (ANT) and the Access Control Unit (ACU). The ANA is connected to the ATM core network and adapts and routes the ATM data stream to the HFC cell.

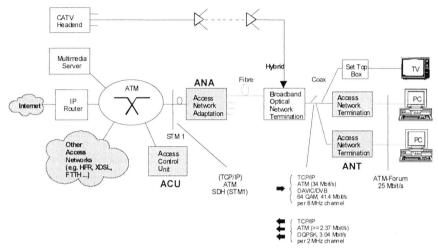

Fig. 1. The AROMA system architecture

A MAC controller in the ANA is defining, on a slot basis, which ANT is allowed to transmit data, controls access to the shared upstream channels. At the subscriber premises, the Access Network Termination (ANT) – also known as cable modem – provides an ATM Forum 25 Mbits/s interface to the AROMA end-system. Within the ANT, multiple upstream queues with different serving priorities are implemented. Furthermore, a hardware based piggyback feedback allows an individual and fast upstream bandwidth allocation on a per service-class basis. The Access Control Unit (ACU) performs the overall HFC access network management comprising the network element initialisation, user authorisation and the HFC resource management.

2. 1 AROMA dynamic MAC

The allocation of upstream slots in a tree-topology access system is based upon a reservation method which allows a dynamic adaptation of the bandwidth distribution to traffic fluctuations. The service policy of the MAC governs the distributed multiplexing from a central point situated at the head-end. Because of the much larger reservation delay and statistical behaviour of the aggregations from many customers, special care must be taken to safeguard QoS to sensitive traffic.

In the AROMA system, a mix of native ATM, IP-based quality services, better than best-effort and plain best-effort services are concurrently supported. The main tool is a QoS aware MAC protocol 7 with a suitable mix of bandwidth allocation mechanisms. This protocol relies on the following three innovative features: a DiffServ-aligned prioritisation architecture, the simultaneous collection of multiple requests from all class queues and polling of requests to guarantee worst case performance. The speed of the upstream is 7.2Mbps. The MAC protocol that was designed and implemented within the AROMA project supports four different priority connections, which are sufficient to offer QoS as justified in 12. Logically separate queues for each priority connection are necessary on the cable modem side for the proper operation of the prioritisation scheme. Their service is based upon reservations.

Services with a similar set of requirements are grouped into 4 behaviour aggregates (classes) supported by four different priorities connections:

All delay sensitive periodic Constant Bit Rate (CBR) traffic is serviced through the highest priority connection (1st priority) which is supported by unsolicited permits scheduled during the connection establishment stage. Traffic generated from voice and video applications should be injected into the system through this class. In the DS context, this would be the expedited forwarding PHB. This topic is discussed in greater detail in section 3.

Real-time Variable Bit Rate traffic, such as video applications, should be injected through a 2nd priority connection, which employs the reservation scheme and is guaranteed to be serviced up to the contracted peak rate. This class is targeting the higher assured forwarding classes.

The third priority traffic class is suitable for services requiring a better than best-effort treatment. A minimum cell rate service is guaranteed, while traffic exceeding this rate is relegated to the 4th priority. This mechanism is suited to the support of all four or the lower three AF classes.

Plain best-effort services are grouped into the 4th priority, which is serviced on a request basis without any guarantee but with a fair distribution of the bandwidth not consumed, by the three higher priorities amongst all users in a round-robin fashion. This priority can accommodate the lower assured forwarding classes.

This flexible and robust scheme offers different QoS levels, bounded worst case performance, efficient bandwidth utilisation and a flexible pricing strategy. Under light load conditions, all the connections are serviced leaving the prioritisation scheme to provide the highest priority service with a better performance. Under heavy load conditions, the performance of each priority connection cannot be worse than that guaranteed leaving the 4th priority to suffer more system congestion. This behaviour will be verified in Section 4.

2.2 The role of ATM in AROMA

The AROMA HFC access network is a pure ATM-based system. It provides the user with an ATM Forum UNI 3.1 interface and the capability to establish Switched or Permanent Virtual Connections (SVC, PVC) with accommodated QoS and traffic parameters. To support IP in the ATM environment, the concept of CLIP 16 has been applied. CLIP defines the address resolution for IP-over-ATM networks and the data formats of the data packets.

PVCs with dedicated traffic and source parameters are established between the connected AROMA end-systems and the IP gateway router. If an end-system has to send an IP packet to another IP instance, it has to request the corresponding ATM VC using the ATM Address Resolution Protocol Server. Knowing this ATM VC, it can encapsulate the IP packet according to 11 and send it via this ATM VC. The traffic shaping functionality on the ATM Network Interface Card (NIC) has to take care of the negotiated/agreed traffic parameters for this VC. Leaving the ATM NIC, the ATM cells are transported via the ATM Forum 25 Mbps interface to the Access Network Termination (ANT). Within the ANT several ATM layer functions are performed:

- ATM cell filtering
- VCI/VPI translation for the HFC segment
- Payload encryption using a connection specific encryption key
- Routing the ATM cell to the connection-defined MAC service class

Reaching the Access Network Adaptation (ANA) at the head-end site, the VCI/VPI values of each ATM cell are extracted and are used to address the ATM connection context table. On the basis of the connection context, it is first verified whether the received ATM cell belongs to a known and active ATM connection. If it this is not the case, the ATM cell is immediately dropped and counted. Otherwise, the connection context is applied to decrypt the ATM cell payload and the VCI/VPI values are translated for the ATM core section. The ATM connection context table defines also whether the ATM cell is routed towards the ATM core network or whether the ATM cells are looped back into the HFC network. In the downstream direction, the ATM cells experience the same processing with the exception that there exists no multiple downstream service priorities.

PVCs can be installed and released via the management interface of the AROMA system involving the HFC Resource Manager (HRM). The HRM is responsible for the management of all HFC resources. These include, for example, the traffic classes, service priorities, upstream/downstream bandwidth, VPI and VCI ranges. If HRM accepts the set-up of a new PVC, it ensures that required ATM connection context values are installed within the related network elements (ANA, ANT). Furthermore, the HRM calculates the MAC control values from the service class and traffic parameters and transmits these values to the MAC controller in the ANA. In the case of a PVC, it frees the allocated HFC bandwidth and the ATM connection is set to the unknown state in the ANA and the ANT.

Ranging and keep-alive checks running on the ANA identify quickly whether an ANT is switched on or off. Each change of an ANT activity state is forwarded to the HRM. Beside the above-mentioned HFC resources, the HRM also manages all the network elements in the HFC network and performs the user authentication and authorisation. These comprise the user-related functionality, which is required for supporting Differentiated Services with user individual Service Level Agreements. Thus, the HRM is capable of installing all the required PVCs for a user as soon as he switches on his ANT. This includes multiple PVCs between two identical endpoints, each with different service and traffic parameters. The ATM ARP server has been modified as a multi-service ATM ARP server to support the service differentiation, as explained in the following section 2.

For the introduction of Switched Virtual Connections (SVC), all the above mentioned functionality is also required. One difference in comparison with the installation of PVCs, is that the SVCs are set-up and released by the ATM connection end-points according to the procedures and messaging defined in the ATM Forum UNI specifications. An HFC Signalling Manager (HSM) processes all the ATM signalling messages and has also an interface to the HRM for requesting the set-up of a new connection within the HFC section and for informing the HRM to release an

ATM connection. Adding appropriate rules to the multi-service ATM ARP set-up of SVC with application dependent traffic parameters can be initiated.

3. The Service Provisioning Concept of AROMA

The objective of the AROMA project was the development of an intelligent resource management system to be applied to an existing multilayered ATM-based HFC access network connected to ATM core networks which, in turn, are connected to the global Internet.

Providing new innovative HFC-resource management components targeted at the support of ATM and harmonising them with resource management which, in turn, requires inter-layer communications between relevant network layers providing guaranteed QoS both to native ATM- and legacy IP-based applications.

The nature of the MAC layer as already presented and its interaction with the ATM layer managed by the resource manager on one side and the customers with their demand to run applications over the access network on the other side, requires a mapping or signalling of customer requirements as described in Service Level Agreements (SLAs) for the ATM network layer. Applications could be based on several technologies. On the one hand, there are QoS-aware applications based on native ATM or supporting IP based signalling such as RSVP, and on the other hand, there are various applications currently used by the whole Internet community which could be regarded as QoS unaware.

The Service-provisioning concept of AROMA addresses both groups. Since the Internet is based upon a quite simple technology which is widely accepted, from the customer point of view it is not relevant what kind of technology is provided by the network, but it is required that existing applications which are partly long-term investments will work over it. Various, mostly asynchronous services such as e-mail completely comply to the user requirements. Other, mostly synchronous applications do not yet fulfil user needs because of the missing QoS support from the network in general.

Targeting these kind of synchronous applications, e.g. videoconferencing, appropriate mechanisms interconnecting the application layer with the AROMA-specific ATM layer must be provided.

The MAC developed within AROMA supports a differentiation into 4 traffic classes with each class different serving strategies. This is priority based and can guarantee QoS only for aggregated data streams corresponding to the 4 MAC classes. Reaching this point, it became obvious not to introduce RSVP as proposed by relevant standard organisations like DOCSIS 3 for QoS provisioning in this specific section of the network, but to use the more appropriate DiffServ approach because of its queuing mechanisms were similar to the AROMA specific MAC protocol. A further reason is the goal of AROMA to provide QoS also for legacy IP applications (QoS-unaware).

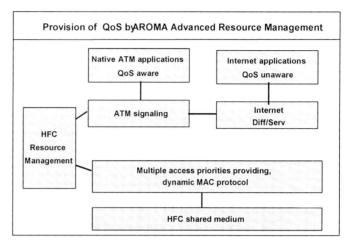

Fig. 2. AROMA Service provisioning concept

Fig. 2 shows the service-provisioning concept of the AROMA system. Since this paper is focusing on the provision of QoS for QoS-unaware applications, such as legacy IP applications, only the right vertical branch is dealt with within the following sections of the paper. Here QoS unaware applications are enriched with QoS in the AROMA system due to the deployment of the differentiated services (DiffServ) strategy. A DiffServ module groups the different services into behaviour aggregates (QoS classes) with a similar set of requirements. These service classes are attuned to the MAC control functions.

3.1 IP based QoS provisioning in AROMA

The differentiated services (DS) 9, 6 strategy recently adopted by IETF as a scalable and relatively simple methodology towards enriching IP services with a QoS, is applicable and quite appropriate for the case of tree-shaped access systems where IP services are dominant. To align such an access system to the differentiated services concept on the AROMA specific network infrastructure, requires the incorporation of provisions in the MAC function for the appropriate handling of each flow aggregation in respect to its requirements. Unlike the router case, it is not possible to offer such functionality as a software upgrade since the MAC requires a fast H/W implementation.

Each QoS class must encounter the specified Per-Hop-Behaviour (PHB) across the multiplexing points against the competing flows or at least not be delayed in a way that can not be retrieved in the next fully DS compliant node. This can not be realised in an HFC system without embedding suitable differentiation mechanisms into the MAC control function.

The basis of the approach is the use of access priorities in the reservation system, which can be programmed to fit in with required PHBs by means of the mapping of flows to priorities.

Within the ATM Forum, several proposals to implement DiffServ on top of ATM in general were discussed. Within AROMA, the so-called VC-bundle approach proposed by Cisco 8 was deployed.

A VC-bundle is a set of independent VCs terminating one IP address at each leaf end. Packets classified by the end-system are sent out according to their classification respectively and according to the Service Level Agreement (SLA) on different PVCs supported by the AROMA-specific MAC protocol over the HFC network to a DiffServ capable router terminating the VC bundle. Here, packets are leaving the AROMA specific Differentiated Services Domain (DS Domain) and are handed over to the Internet – to the next DS Domain. Exploiting features of the ATM NIC such as shaping the ATM layer could be regarded as access point for the DiffServ described in 8. On the other hand, the mapping from ATM connections to the different MAC queues combined with a piggy backing mechanism which in turn controls via MAC controller the resource allocation in the system up to the IP layer, a kind of interworking between DiffServ and the physical resources available in the system is established using ATM. Between the IP/DiffServ and ATM layer, the Cisco proposed VC-bundle concept comprising the second phase of the IP to ATM Class of Service (IP to ATM CoS) seems to be a promising solution for that specific network topology. ATM VC-bundle management allows configuring multiple virtual circuits that have different quality of service (QoS) characteristics between any pair of ATM-connected routers terminated each with one IP address. These VCs are grouped in a bundle and are referred to as bundle members. Using VC bundles, differentiated services can be created by flexibly distributing IP precedence levels over the different VC bundle members. It is possible to map a single precedence level or a range of levels to each discrete VC in the bundle, thereby enabling individual VCs in the bundle to carry packets marked with different precedence levels. Fig. 3 shows the AROMA adopted VC bundle concept.

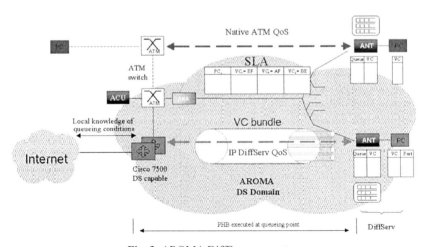

Fig. 3. AROMA DiffServ concept

To determine which VC in the bundle to use to forward a packet to its destination, an ATM VC bundle management software matches precedence levels between packets and VCs. IP traffic is sent to the next hop address for the bundle because all VCs in a bundle share the same destination. However, the VC used to carry a packet depends on the value set for that packet header in the IP precedence bits of the ToS byte.

3.2 MAC Parameter Mapping

The ATM cell sized slots are the quantum of MAC assigned bandwidth allocation in AROMA. Several slots, enough to accommodate an IP frame, can be successively assigned, and this can be used to advantage in the context of the DS architecture over a shared link such as HFC. Namely, it allows suspending the transmission of a lower priority packet on the boundary of a slot (cell), and transmit delay-sensitive packets before resuming the lower priority transmission. In addition, fixed slots are easier to handle in a H/W based MAC. Such ATM slotting is, anyway, the approach followed by main standards bodies, such as DAVIC 17, DVB/ETSI 4, and IEEE 802.14 5. As in the centralised multiplexer case, flows with demanding QoS (e.g. Expedited Forwarding or Assured Forwarding) must be identified and receive properly differentiated treatment from the plain best effort traffic. This is accomplished in the ANT by placing the corresponding cells in the high priority queues which will subsequently place higher priority requests and activate the higher priority permit allocation MAC algorithms.

To reduce complexity in the cost-sensitive residential access system, services are grouped into behaviour aggregates (classes) with a similar set of requirements. This is in line with the DS philosophy of flow aggregation for better scalability and flexibility. The characteristics of the four aggregation levels/priorities are as follows:

The high priority is devoted to delay sensitive periodic CBR traffic, which is supported by pre-allocated pre-arbitrated unsolicited permits issued on a periodic basis by the MAC controller. This class is suitable for services with very strict delay requirements, which undergo strict traffic profile control (traffic conditioning) such as the EF (Expedited Forwarding) service 14.

The second priority level is devoted to real-time variable rate flows, such as video services or VoIP and it is provided with peak rate policing for guaranteed QoS. MAC exercises a policing function by rate checking before issuing the permits. This check is based on credit allocation at the time of subscription or connection set-up. In the DS context it could be used for the top AF (Assured Forwarding) class 15.

The third priority is devoted to data services with higher requirements than best-effort. The traffic profile control assumed for this class aims at minimising the loss of packets and the disturbance to other traffic. The credit scheme is used to guarantee a minimum rate (while credits last) while traffic exceeding this limit is relegated to the 4th priority permit generation. The 3rd priority mechanism is suited to support all four or the lower three AF classes 15.

The fourth priority is reserved for plain best-effort services which employ loss based flow control at the TCP level and can be very disruptive to the other classes when sharing the same queue.

It can be stated that within the AROMA system an inter-layer signalling is realised. This inter-layer signalling spans from the IP transport layer down to the MAC layer.

IP source and destination ports and addresses are used by the multi-service ATM ARP or VC bundle management software to select the appropriate ATM VC, which has been set-up before with the appropriate parameters. This ATM VC is assigned to an appropriate MAC service priority whereby the according MAC parameters are derived from the ATM VC traffic parameters.

4. Concept verification

To verify the concept, initial tests in a heterogeneous ATM environment were undertaken. Here an ATM25 – ATM155 infrastructure as part of an underlying HFC infrastructure testbed was set up. Fig. 4 shows a conceptual testbed set-up to demonstrate the interworking of all three levels involved in QoS provision – IP – ATM – HFC/MAC.

Hidden from the terminal's CLIP driver, a VC-bundle was established between the AROMA end-system with the DiffServ module and the commercial edge router. Served by different MAC queues according to the MAC's QoS framework, the VCs of the bundle inherit the MAC QoS. Knowing the mapping VC-to-MAC which reflect the Service Level Agreement (SLA), IP packets eventually are fed into VCs according to the targeted QoS, the latter possibly covered by a SLA. According to the AROMA constraints, the VC-bundle consisted of 3 VCs and according to the IP application identified by the port number/IP destination, different VC's are selected to transport the data across the ATM network.

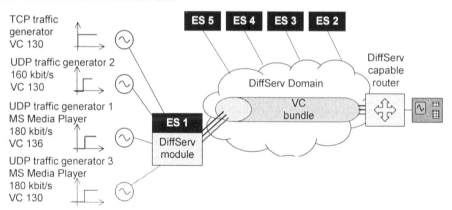

Fig. 4. :Measurement configuration

Next to this VC–bundle based leaf-end node, four further end-systems were connected to the shared access network receiving representative data streams. The tests were carried out with an ATM tester connected directly to the DiffServ capable router at the head-end side. Only non-corrupted cells were measured.

As mentioned, QoS for Legacy IP applications are targeted, whilst off-the-shelf applications were used as follows:

The TCP traffic used in the test was a standard WIN2000 shell application.

The two real-time applications used were standard MediaPlayer data-streams directed to different clients differentiated by the IP destination address parameters. Furthermore, a UDP generator was used to disturb the low priority data traffic.

In this test scenario, the total upstream capacity is about 6.9 Mbit distributed to three upstream channels. Of the five end-systems (ES) plus modems 1 – 5, configured in the test-bed, systems 2 – 5 were silent but given an "SLA" of 1.6 Mbit/s for Queue 1, the pre-arbitrated queue. This artificially confines ES/Modem 1 to the total upstream capacity of about 550 kbit/s, which is allocated according to Table 1. Because the ATM NIC drivers do not support VBR, all VCs are using ATM CBR exploiting only the Peak Cell Rate parameter of the latter. In the given set-up, these pre-configured VCs are mapped to MAC Queues 1 and 4 respectively. With respect to the bandwidth figures given in Table 1, we can identify the following:

Firstly, the TCP rate and the sum of all queue values (in kbit/s) could have been increased to the theoretical maximum remaining upstream capacity for System 1 which is 517 kBit/s, but the ATM NIC does restrict it to 500 kBit/s per each VC.

Table 1. : SLA-MAC parametrisation

ES/ Modem	Destination/ Modem	IP	ATM				HFC MAC		
		IP-Applic	VC	TC	Max Kbit/s	Mapped to HFC queue	Queue 1 Kbit/s	Queue 2 Kbit/s fix credit	Queue 4 Kbit/s
1	2	TCP	133	CBR	500	4			317
1	2	UDP	133			4			
1	3	UDP	136	CBR	500	4			
1	4	UDP	130	CBR	500	1	200		
2	-	-	-	-	-	1	1600		
3	-	-	-	-	-	1	1600		
4	-	-	-	-	-	1	1600		
5	-	-	-	-	-	1	1600		

For this setting the 200 kBit/s in Queue 1, allocated to the 1st priority queue of System 1, was definitely more than adequate. As can be seen from the graphs below (Fig. 6), the constant traffic shape reflects this fact. Initially, a TCP traffic stream using the FTP protocol was generated. This data flow was allocated the freely available capacity of the system.

The maximum transfer rate is limited by the ATM NIC parameter. After five seconds, three different UDP data streams (two real-time streaming applications – Windows MediaPlayer and one UDP generator) with different SLAs were generated (see Fig. 4).

According to Table 1, one real-time streaming application had a different SLA than the other three applications. The three data sources competing with each other mapped to the 4th quality group, having the lowest priority, transmitted at this point of time with a total bandwidth of about 800 kbit/s which was more that the system can transport since only about 350 kbit/s were available to this traffic class. Streaming data with the highest priority had an allocated bandwidth of 200 kbit/s, which results in an undisturbed traffic profile. It can be seen, that the running TCP application had immediately suffered from the UDP data transfers and rapidly decreased its sending rate. Cell loss occurs in the system but is restricted to the traffic mapped to the 4th queue – the traffic with the lower SLA. The measurements further show, that the TCP

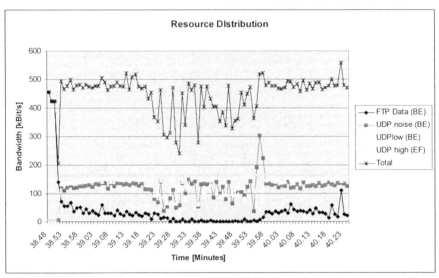

Fig. 5. Measurement result

flow does not really disturb the UDP streams in case of limited capacity because of its flow control mechanism. Further, the measurements verified that the MAC, which executed the QoS is highly dynamic and free capacity not allocated to a high priority traffic, but not used, can dynamically distributed to low priority traffic. This initial measurements verified the resource distribution and interworking between relevant layers in the simpler case of bandwidth distribution allocated to one end-system. Further measurements are required to verify the bandwidth distribution between different end-systems competing with available resources. The lack of early packet discard mechanisms in the system lead to the observed variance of the total traffic since only non-corrupted IP packets are measured.

Conclusion

The introduction of multimedia applications raises the need for adequate access networks capable of satisfying residential users with their requirements concerning bandwidth, delay or latency. This work has described an overall concept for a new network and service architecture, as implemented in AROMA, enabling high speed Internet access to residential users and providing a QoS concept to effectively distribute resources, especially on the scarce upstream branch of this specific network architecture.

The article focussed on the provisioning of QoS for legacy IP applications, but the architecture is flexible enough to support other technologies.

The architecture is partly based upon existing standards and extracts benefits from all of the above mentioned protocol layers. The required inter-layer signalling challenge solved within this architecture is not only restricted to HFC networks, but could become important also in further shared access mediums in general. This paper described briefly the system architecture, the inter-layer signalling concept and finally presented some initial measured results. The open system architecture enables an interworking with other system architectures on the same network segment as well as an interworking with other access technologies and core network technologies.

This paper has shown that the architecture dynamically distributes available resources in a flexible manner to different customers according to agreed SLAs. A major goal of the architecture is the support of QoS for legacy IP applications which means that existing applications are supported with QoS without modifications.

References

1. J. Jaehnert, H. Fahner, et.al, " Interworking Technology and Applications - ATM-over-HFC-based Access to the Internet for Residential Users ", Proceedings of Intertworking98, Ottawa, 1998.
2. S. Wahl et. al, "Architecture and Experiences of a Multi-Service HFC Network", Proceedings of Conference on High Performance Switching and Routing, Heidelberg, 2000.
3. ITU-T Rec. J.112 Annex B DOCSIS: „Data over Cable Service Interface Specification" (Same as ANSI SCTE DSS97-2)
4. ETSI ETS 300 800 Digital Video Broadcasting (DVB); DVB interaction channel for Cable TV distribution systems (CATV); 1st edition, March 1998.
5. IEEE Project 802.14/a Draft 3/R2, „Cable-TV access method and physical layer specification", January 1999
6. IETF, Differentiated Services Working Group, RFC 2475 „Architecture for Differentiated Services", December 1998
7. J. D. Angelopoulos, Th. Orphanoudakis, "An ATM-friendly MAC for traffic concentration in HFC systems", *Computer Communications Journal*, Elsevier, Vol. 21, No. 6, 25 May 1998, pp. 516-529.
8. ATM Forum TM 4.1, Enhancements to Support IP Differentiated Services and IEEE 802.1D over ATM, BTD-TM-DIFF-01.02 Draft, Dec '99.
9. IETF, Differentiated Services Working Group, RFC 2474 „Definition of the Differentiated Services Field (DS Field) in the IPv4 and IPv6 Headers", December 1998.
10. Dinesh Verma, *"Supporting Service Level Agreements on IP Networks"*, Macmillan Technology Series, 1999.
11. IETF, Network working Group, RFC1483 " Multiprotocol Encapsulation over ATM Adaptation Layer 5", July 1993
12. J. D. Angelopoulos, N. Leligou, Th. Orphanoudakis, G. Pikrammenos, "The role of the MAC protocol in offering QoS to IP services over shared access systems", *Globecom'99Conference, Business applications session,* Rio de Janeiro 5-9, December 1999
13. http://www.athoc.de/AROMA/
14. IETF, Differentiated Services Working Group, RFC 2598 „Expedited Forwarding Behavior, June 1999.
15. IETF, Differentiated Services Working Group, RFC 2597 „Assured Forwarding Behavior, June 1999.
16. IETF, Network working Group, RFC 1577 "Classical IP and ARP over ATM", January 1994.
17. http://www.davic.org/speci.htm

TCP Conformance for Network-Based Control

Arata Koike

NTT Information Sharing Platform Laboratories
3-9-11 Midori-cho, Musashino-shi, Tokyo 180-8585, Japan
E-mail: koike.arata@lab.ntt.co.jp

Abstract. In this paper, we first review the potential roles of a network in improving end-to-end TCP control. We discuss problems associated with such network control when a network provider wants to provide a certain degree of service objective based on the control. When we use feedback control mechanism, we assume a specific behavior of end-hosts as a conformance condition. In case of TCP, there are a lot variants of implementation and the differences give big impacts on TCP performance. We evaluate such impacts for Ackadjust method by simulation and point out the necessity to clarify the relationship for the interworking between network-based control and TCP implementation.

1   Introduction

The Transmission Control Protocol (TCP) / Internet Protocol (IP) [1] is a basis for current computer communication networks. Various applications are realized on the top of TCP/IP. TCP assumes no reliability on underlying network. It features all control be done on an end-to-end basis. It deals with flow control and error control on data delivery.

The initial TCP protocol was specified in 19 years ago [2]. We could consider that choosing the window-based control scheme and the assigned values for control parameters for TCP were sufficient for the capabilities of networks and end-hosts at that time. We can see much improvement on the capabilities of network and end-hosts during last 10 years. For example, as for the network capabilities, we now have high-speed networks such as Asynchronous Transfer Mode (ATM) and Ethernet with the speed of 100Mbit/s or Gbit/s. Although we have gotten such great improvement on networks and end-hosts, we are still utilizing TCP as the transport layer protocol. Of course, there are many proposals on TCP algorithm to improve performance of data transmission to adapt the high-speed transmission. These enhancements, however, are mainly focused on the end-to-end control.

Let us again consider the characteristics of the TCP control mechanism:

- TCP is an end-to-end, window-based flow control mechanism.
- TCP indirectly detects packet loss within a network by observing a time-out or arrivals of duplicate Acknowledgement (Ack) packets and then limits the number of packets allowed entering the network.

Both of these are features of the end-to-end approach and TCP runs independent of network internal structure. With a low-speed network, this kind of end-to-end protocol works well. For high-speed networks, however, these features affect worse both for throughput performance and fairness among TCP connections. Fairness is not an important issue when the goal is merely end-to-end connectivity between different networks for TCP. However, with the evolution of commercial Internet service, many companies are using the IP network as their enterprise network. This raises fairness as an important issue in addition to end-to-end connectivity.

In addition to the improvement of TCP algorithms, several network-based controls for TCP flows are proposed to achieve better utilization of networks. Now, underlying networks, such as ATM, have many rich capabilities regarding to Quality of Service (QoS). End-users, however, only perceive QoS through end-to-end TCP performance and there is no explicit interworking between TCP and underlying networks. One of the objectives of these network-based controls is to provide capabilities of underlying networks to TCP end-users. For example, Available Bit Rate (ABR) service of ATM [3] performs rate-based control at ATM layer and thus we need an interworking mechanism to notify the change of lower layer characteristics to TCP. By providing lower layer capabilities to TCP layer, TCP can truly utilize the performance and QoS that are provided at the lower layer level.

Once a network begins to provide the above-mentioned network-based control as a network service, there arises an issue regarding to the conformity of users against the control. TCP was considered as an end-to-end protocol so that we did not need to consider conformance issues. But when a network provide various controls as a network service, the network must check the user behavior to keep the service level.

This paper is organized as follows. In the next section, we briefly overview interworking mechanisms of network-based control for TCP. In section 3, we describe the relationship between feedback control and associated conformance issues for ABR and TCP. We investigate TCP conformance for network-based control by simulation in section 4. We discuss the results in section 4 from the viewpoint of TCP conformance at section 5. We finally conclude our paper in section 6.

2 Interworking between TCP and Underlying Network

Since TCP is an end-to-end protocol, it means that there is limited room for intermediate network elements, such as routers, to contribute to the TCP control. In this section, we briefly review methods proposed to date in terms of network control. These methods can be categorized as either explicit or implicit. An explicit method is one in which a network element reports something explicitly to a TCP sender. An implicit method requires no modification of the algorithm to TCP senders.

2.1 Explicit TCP Control by Network

An explicit method can indicate the status of a network element directly. One such explicit contribution could be achieved by sending an ICMP (Internet Control Message Protocol) Source Quench message when a network element suffers severe congestion. According to the TCP/IP implementation, this will cause a TCP slow start. However, the use of a Source Quench message causes additional side effects. These disadvantages are summarized in [4].

A more explicit scheme is the study of Explicit Congestion Notification (ECN) [4][5]. ECN uses the TOS field of the TCP header and explicitly reports the network status to the TCP source. In contrast to the existing notification scheme, which utilizes packet loss, this scheme should work effectively by reporting network status directly without discarding packets. But this scheme requires modification of TCP software to accommodate ECN. As discussed in [4] and [5], in introducing a new scheme that affects the end-host algorithm, the coexisting environments of the existing and the new terminals must be taken into account.

2.2 Implicit TCP Control by Network

The other way to influence TCP control by a network is by implicit contribution. This kind of method does not require modifications of TCP software, yet the network has the capability of influencing TCP control, which effectively utilizes various TCP properties. One such idea that is widely employed is packet discarding. Random Early Detection (RED) [6] is the most well-known. It intentionally discards packets in a probabilistic manner when the number of stored packets in the buffer exceeds a certain threshold, indicating buffer congestion. It is reported that probabilistic discarding discards more packets of connections that share more bandwidths and is effective for recovering fairness. This scheme is based on packet discarding, but it will be a problem for high-speed networks due to the wait for the retransmission timeout.

In order to obviate packet discarding for high-speed networks, [7] proposes TCP-GATEWAY, which divides the TCP control segments. It maintains connectivity of the TCP connection at a LAN-to-WAN gateway point but it terminates the control loop at the gateway and uses a higher-speed native protocol for backbone segments. In terms of flow control, this method is especially effective for a network that has a large RTT, by reducing the length of each control loop and tuning control parameters appropriately for each segment. TCP, however, involves not only flow control but also error control in its feedback. Thus, in order to assure the reliability of TCP, huge amount of data must be maintained at the gateway.

In [8], [9] and [10], we proposed a rate adjustment function at a router to prevent packet loss and to influence TCP control (Fig. 1). This could be considered as a realization of transport middleware for TCP control. The key idea of this approach is based on the fact that TCP is a window-based control. This rate adjustment function could be done by scheduling forward and/or backward packets according to the current network status. Such scheduling causes a delay in response and limits the transmission of new packets from the TCP source, since the number of packets traveling without being acknowledged is determined by the TCP window size. We applied this scheme for interworking between legacy (e.g., Ethernet) networks and an ATM backbone network using ABR service. This approach enables us to relate rate-based control at the ATM layer and window-based control at the transmission layer. We translated the ATM layer rate information to an Ack packet interval, which controls the transmission speed of TCP packets. We want to emphasize that our approach does not require any modification of TCP end-hosts. This scheme is further explained in the next section.

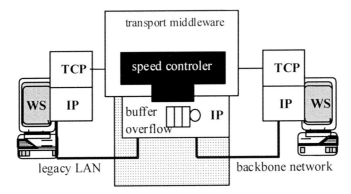

Fig. 1. Transport middleware for TCP rate adjustment

2.3   Interworking between TCP and ABR

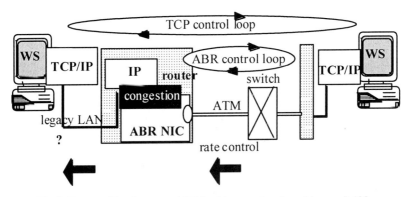

Fig. 2. Interworking between ABR backbone network and legacy LAN

We focus on the interworking between end-to-end TCP and backbone ABR network. ABR is a rate-based ATM layer protocol. It features a closed feedback control loop by using a special cell, called Resource Management (RM) cell. The allowed source transmission rate, ACR, is determined by feedback information that reflects the status of intermediate network nodes or the destination. By this nature, traffic characteristics of ABR services vary time to time. In case of backbone ABR network, the rate-based control is only done within the ATM-ABR segment. It can avoid congestion at the ATM network. On the other hand, when other networks are connected to the backbone ABR network through a router, congestion in ATM switches is simply moved to the router. And since the router has a limited buffer capacity, this congestion may cause buffer overflow at the router. This results in poor TCP performance even if we achieve good ATM layer performance (Fig. 2). As we described above, we proposed an interworking method to relate ACR information to intervals of Ack packets.

Our proposal achieves self-clocking of the interworking based on the queue length information of forward direction at a router. And we control intervals of Ack packet by creating buffers for Ack packets in a router. We call this scheme as 'AckAdjust' method (Fig. 3). The algorithm of our scheme is as follows. If the queue length l of a buffer for the forward direction in a router exceeds a certain threshold, say t, we schedule the Ack packet of the backward direction based on the rate ACR / H, where H is a parameter. If the queue length is less than or equal to the threshold t but greater than 0, we schedule the Ack packet based on the rate ACR / L, where L is also a parameter.

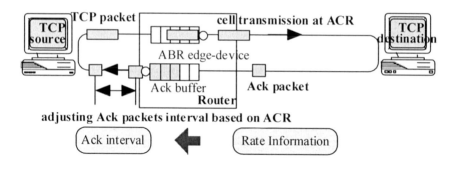

Fig. 3. Illustration of AckAdjust scheme

3   TCP Conformance

As described in the previous section, there have been made a lot of proposals on network-based TCP control mechanisms. An objective of these mechanisms includes providing network capabilities to end-users. Recent advanced networking technologies have potentials to provide high-performance communication. Unfortunately, TCP has no capability to utilize them directly. TCP end-users only perceive network performance and QoS on an end-to-end basis. Thus, using network-based controls makes it possible to implicitly enhance end-to-end TCP performance. TCP uses a feedback control. Therefore, when we initiate a network-based control, we assume a specific reaction of TCP end-hosts to the control. Let us consider two examples. In case of Source Quench, as an example of explicit control, a router that sent ICMP Source Quench packets expects initiation of slow-start at corresponding end-hosts. In case of RED, as an example of implicit control, source end-hosts are expected to shrink their congestion window size according to the packet discarding by a network.

If we use the network-based control only for an advisory purpose, we do not require any specific reaction of TCP end-hosts. But if we use it as a method to provide guaranteed performance and QoS to end-users, the correctness of a specific reaction of TCP end-hosts shall be one of the important issues. When we relay on a closed-loop feedback control, we always encounter this kind of requirement on the conformance of the behavior of end-users.

For example, ABR service of ATM also uses closed-loop feedback control. Switches can indicate allowed cell transmission rate (ACR) to an ABR source end-terminal by using resource management (RM) cells. Since ABR specification requires fairness among connections and minimum cell rate guarantee, it strictly specifies conformance for the feedback. The conformance requirement for feedback of ABR is a delay bound regarding to the processing delay of an end-terminal to react the indicated transmission cell rate by RM cells. If an end-terminal fails to meet the delay bound to adapt rate change, it is considered as non-compliant to the ABR conformance.

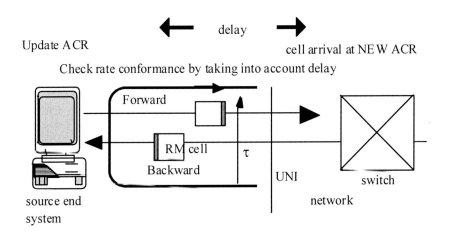

Fig. 4. ABR conformance for feedback

We consider the case for Explicit Rate (ER) switches (Fig. 4). In this case, we observe the following procedures. A backward RM cell passes through an ER switch with a new ER information. It reaches a source end system and ACR at the source was updated by the ER value contained in the backward RM cell. The source sends a new forward RM cell with the new ACR. Then the RM cell reaches the switch. A certain time, τ, is required in the course of these procedures. If the new ER value is less than the old one, a UPC function of the switch uses old ER value until a predetermined time τ for the policing threshold. After expiration of τ, the switch examine the cell arrival rate using the new ER value. If an end system continues sending cells at the old ER after the time τ, the UPC identifies the source as a violator of the conformance for feedback. We need this kind of requirement for end system when we provide services using feedback control. The requirement for the above example is delay in control. By using such requirement and the corresponding checking mechanisms, a network can provide services fairly among each end system.

In case of TCP, the time scale required to show the effect of TCP flow control is much longer than that of ABR control. Therefore, in practice, it has little meaning as a conformance requirement that takes into account delay bound regarding to the processing delay of an end-host. On the other hand, in case of TCP, differences in end-host mechanisms have impacts on the effectiveness of network-based control. In case ABR, end system behavior for the feedback is strictly specified. Therefore, we could expect the same reaction for all end systems for the same feedback. On the other hand, there exist various kinds of TCP implementations, variants and options according to the evolution of TCP. If we only consider connectivity as the role of TCP, we can assure interoperability among all different versions of TCPs. When we introduce network-based control to provide better performance for TCP end-users, however, such differences in TCP cause problems. This is because if we apply the same control mechanisms for different TCP flows, TCP end-hosts behave differently depending on their implementations. For example, in case of RED, we could expect different results for TCP with go-back-N and TCP with SACK even if a network discards packets in the same way. A network-based control assumes a specific end-host behavior but the control functions only when end-hosts support the expected behavior. Therefore, we must determine conforming TCP mechanisms to meet the requirement of network-based control.

4　　Simulation for TCP Conformance

Here, we investigate how TCP implementations affect the outcome of network-based controls. As an example of network-based control, we use Ackadjust scheme and consider interworking between end-to-end TCP and backbone ABR network [10]. This scheme translates ABR rate information to the interval of TCP's Ack packets at the edge of ABR connection (Fig. 3). As a result, data transmission rate of a TCP end-host is accommodated to that of ABR without changing the TCP source code. In this paper, we consider the effect of TCP implementation differences based on the AckAdjust scheme.

As an example of TCP implementation difference, we investigate the differences in the number of Ack packets sent from a destination. Since the Ackadjust scheme causes delay in Ack packets of each TCP flows based on ACR information, the differences in the number of Ack packets shall affect the outcome of the scheme. Some TCP implements so called 'delayed-Ack' method. The 'delayed-Ack' method allows acknowledging multiple data packets by a single Ack packet. Here, we assume for 'delayed-Ack' TCP that a destination end-host transmits an Ack packet when it receives five data packets or when 100 ms has passed after transmitting the last Ack packet.

250

4.1   Simulation Configuration

The simulation model is described in Fig. 5.

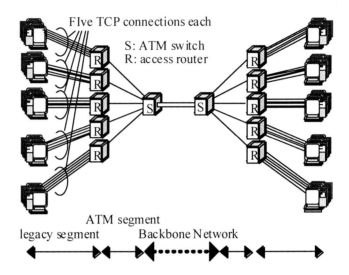

Fig. 5. Simulation model

There are five routers and five TCP sources at the source side. Each TCP source is connected to a router. The same configuration appears at the destination side. We call these segments legacy segments. Each router is assumed to be an ABR edge-device. Routers are connected to a switch and are aggregated to a backbone link. Then they are distributed to destination side routers at the opposite side switch. Router-switch (switch-router) segments are ATM segments. Each router terminates a Virtual Circuit (VC) and the five TCP connections are aggregated on a VC.

4.2   Simulation Parameters

We use the following simulation parameters.

4.2.1   Network Parameters
We assume Minimum Cell Rate (MCR) is 0 Mbit/s and Peak Cell Rates (PCR) of each VC is 150Mbit/s. Link rates for legacy segments are 100Mbit/s. Distances in both legacy segments and router-switch segments are 1 km. The backbone link length is 10 km. The light velocity is assumed to be 200,000 km/s. The ABR switch algorithm we used is CAPC2 [11] and we assume 95% utilization. Buffer size at ATM switches are 4096 cells and we did not observe any cell loss events at the ATM switch.

4.2.2 TCP Parameters

We use TCP-Reno, which includes fast retransmission and fast recovery. The TCP timer granularity is 200 ms. We assume the use of TCP window scale option and the maximum TCP window sizes is 128 Kbytes. Each packet size is 9188 bytes. We also assume the TCP destination packet processing delay obeys an exponential distribution with a 1-ms mean. Each TCP source transmits data until the end of simulation, and all sources start to transmit simultaneously.

4.2.3 Router Parameters

Data buffer size is 15 packets. The threshold of data queue is 3 packets and H (when $l > t$) = 2.08 and L (when $0 < l \leq t$) = 0.5 which is the same as in [10]. The Ack buffer size is unlimited. This final assumption is validated because Ack packets only contain a sequence number in their payload so the storage requirement is much smaller than for TCP datagram packets.

5   Discussion

We first describe the comparison of the TCP performance with and without the use of AckAdjust scheme for a non-delayed-Ack environment. In Fig. 6, we show the results for 25 TCP connections. Without using AckAdjust scheme, we can see the performance degradation both in throughput and fairness due to the packet discarding. In general, ABR achieves fairness among VCs but it is not expected to contribute fairness among TCP flows. But from this figure, by using AckAdjust scheme, we can see the realization of fairness among TCP flows in a VC.

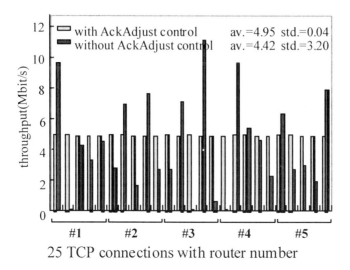

Fig. 6. Throughput characteristics with and without AckAdjust scheme

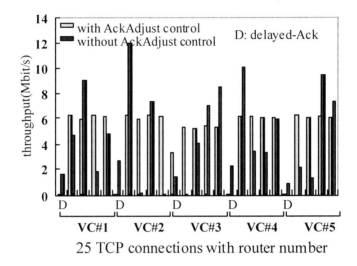

25 TCP connections with router number

Fig. 7. Throughput characteristics with delayed-Ack TCP connections

We next discuss the case for delayed-Ack environment. For each router, we assume one of the five TCP connections uses 'delayed-Ack' method. The other four TCP connections' destinations always return an Ack packet when they receive a data packet. Using this configuration, we show a result in Fig. 7. 'Delayed-Ack' TCP connections are indicated by a symbol 'D'. This figure shows that fairness among TCP connections are kept except for the connections using 'delayed-Ack' mechanism. 'Delayed-Ack' connections suffer unfairness in throughput since the mechanism causes delay in control and thus throughput degradation occurs. As a result, when we adopt the network-based control, i.e., Ackadjust scheme, to improve fairness, we can conclude that TCP implementation with 'delayed-Ack' is not a conforming implementation for that control.

6   Conclusion

In this paper, we briefly overview network-based control for TCP. We consider problems associated with such network-based control to achieve service objectives. Networks assume a certain end-hosts behavior as a conformance condition for feedback control. Differences in TCP implementation are an important conformance condition regarding to a network-based control. We investigate such example based on AckAdjust scheme, which controls Ack packets intervals and achieves TCP and ABR interworking. As a conclusion, we need to clarify the relationship between the performance and TCP implementation when we have performance objectives based on a network-based control.

Acknowledgement

Thanks are due to Ms. Kiyoka Takeishi and Ms. Mika Ishizuka for her support in the simulation and for useful discussions.

References

1. Stevens, W.R.: TCP/IP Illustrated Vol. 1, Addison Wesley (1994)
2. Postel, J. B.: Transmission Control Protocol, IETF RFC793, September (1981)
3. The ATM Forum: Traffic Management Specification 4.0, af-tm-0056.000, April (1996)
4. Floyd, S.: TCP and Explicit Congestion Notification, ACM Computer Communication Review, V. 24 N. 5, October (1994) 10-23
5. Ramakrishnan, K. K., Floyd, S.: A Proposal to add Explicit Congestion Notification (ECN) to IP, IETF RFC 2481, January (1999)
6. Floyd, S., Jacobson, V.: Random Early Detection gateways for Congestion Avoidance, IEEE/ACM Transactions on Networking, Vol. 1 No. 4, August (1993) 397-413
7. Hasegawa T., Hasegawa, T., Kato, T., Suzuki, K.: Implementation and Performance Evaluation of TCP Gateway for LAN Interconnection through Wide Area ATM Network, Trans. IEICE, B-I, Vol. J79-B-I, No. 5, (1996) 262-270 (in Japanese)
8. Koike, A.: TCP flow control with ACR information, ATM Forum/97-0758, September (1997)
9. Koike, A.: TCP/ABR interworking, ATM Forum/97-0998, December (1997)
10. Koike, A.: Interworking between end-to-end TCP flow control and ABR rate-based control for backbone network, Proc. of Interworking98, July (1998)
11. Barnhart, W.: Example Switch Algorithm for Section 5.4 of TM Spec.: ATM Forum/95-0195, February (1995)

Congestion Control Mechanism for Traffic Engineering within MPLS Networks

Felicia Holness, Chris Phillips

Department of Electronic Engineering – Queen Mary and Westfield College
University of London – London, UK, E1 4NS
Tel: +44 (0) 20 7882 3755, +44 (0) 20 7882 7989   Fax: +44 (0) 20 7882 7997
Email: f.holness@elec.qmw.ac.uk
Email: c.phillips@elec.qmw.ac.uk

Abstract: The transformation of the Internet into an important and ubiquitous commercial infrastructure has not only created rapidly rising bandwidth demands but also significantly changed consumer expectations in terms of performance, security and services. Consequentially as service providers attempt to encourage business and leisure applications on to the Internet, there has been a requirement for them to develop an improved IP network infrastructure in terms of reliability and performance [1]. Interest in congestion control through traffic engineering has arisen from the knowledge that although sensible provisioning of the network infrastructure is needed together with sufficient underlying capacity, these are not sufficient to deliver the QoS required [2]. This is due to dynamic variations in load. In operational IP networks, it has been difficult to incorporate effective traffic engineering due to the limited capabilities of the IP technology. In principle, Multiprotocol Label Switching (MPLS), a connection-oriented label swapping technology, offers new possibilities in addressing the limitations by allowing the operator to use sophisticated traffic control mechanisms.

However, as yet, the traffic engineering capabilities offered by MPLS have not been fully exploited. Once label switched paths (LSPs) have been provisioned through the service providers' network, there are currently no management facilities for dynamic re-optimisation of traffic flows. The service level agreements (SLAs) between the network operator and the customer are agreed in advance of the commencement of traffic flow, and these are mapped to particular paths throughout the provider's domain and may be maintained for the duration of the contract. During transient periods, the efficiency of resource allocation could be increased by routing traffic away from congested resources to relatively under-utilised links. Some means of restoring the LSPs to their original routes once the transient congestion has subsided is also desirable.

Today's network operators require the flexibility to dynamically renegotiate bandwidth once a connection has been set up [3] preferably using automated solutions to manage an access switch management algorithm and route connections. Although these services are already provided to some extent with provisioning, they tend to occur relatively infrequently (several times in a day) using prior knowledge and manual intervention. There are currently no mechanisms in place within the network to allow the operator to rapidly change the traffic paths in response to transient conditions.

This paper proposes a scheme called Fast Acting Traffic Engineering (FATE) [6][7] that dynamically manages traffic flows through the network by re-balancing streams during periods of congestion. It proposes management-based algorithms that will allow label switched routers (LSRs) in the network to utilise mechanisms within MPLS to indicate when flows may be about to experience possible frame/packet loss and to react to it. Based upon knowledge of the customers' SLAs, together with instantaneous flow information, the label edge routers (LERs) can then instigate changes to the LSP route to circumvent congestion that would hitherto violate the customer contracts.

Keywords: MPLS, Traffic Engineering, LDP, LSR, CR-LDP,FATE.

1.   Multi-Service Provisioning Environment

At present the Internet has a single class of service - "best effort". As a result of this single service all traffic flows are treated identically, there is no priority servicing regardless of the requirements of the traffic. A provisioning scheme that can be applied within an MPLS environment can be described as follows:

Consider **Fig. 1**, which shows a scheduler at each egress port of a LSR. The scheduler has been programmed to visit each class-based buffer at a rate commensurate with the loading of that particular buffer and its identified Quality of Service (QoS) constraint(s) i.e., a long-term guaranteed loss limit.

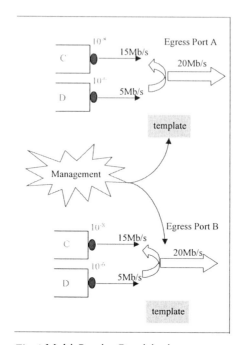

Fig. 1 Multi-Service Provisioning

The order and frequency with which the scheduler services each of the buffers is determined by a port template that may be programmed by the management module and read by the scheduler. In the scenario depicted, where buffer C would be serviced three times more than buffer D, the template could take the format shown below in **Fig. 2**.

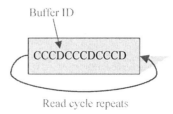

Fig. 2 Example Scheduling Template

A scheduling template is programmed according to a predetermined loss probability threshold for each buffer for a given anticipated load. In a situation where the traffic loading through a buffer stream increases, the management function has the ability to change the template of the scheduler if deemed appropriate. The management function uses its knowledge of the traffic characteristics and the current loading to determine whether a new constraint-based routed label switched path (CR-LSP) can be routed through a particular buffer, whilst attempting to maintain the buffer's traffic engineering constraints below the specified loss probability. Over a specified time period the loss probabilities through each buffer stream are recorded and in the event that the loss exceeds a predefined threshold the management function may decide to alter the template i.e. the scheduler's rate, to accommodate for the additional loading.

Scheduling reallocation is permissible provided the contracted QoS requirements of the LSPs traversing the buffer(s) are not violated.

However, as this method is based on predicted behaviour (estimated over the last time period) it does not cater for transient fluctuations in load. It is a provisioning mechanism. FATE on the other hand provides a means of dynamically redistributing existing CR-LSPs between buffers or via alternative paths in response to short term congestion events.

1.1. Traffic Flow Provisioning

So far basic provisioning has been considered i.e., mapping LSPs to buffers according to their particular QoS requirements and the long term loading situation. This mechanism allows for readjustment of the scheduling templates in response to predicted loading variations. It is relatively slow and operates on buffers – it does not provide granularity to respond to issues associated with individual LSPs.

FATE provides a fast acting mechanism, which augments the above scheme. It allows individual LSPs to be dynamically remapped to higher QoS buffers along a specified path in response to transient congestion situations.

2. Fate

This section describes the mechanisms and procedures that are employed within the FATE (Fast Acting Traffic Engineering) scheme. It proposes a fundamental set of novel mechanisms that can be employed in the effort to either pre-empt congestion or respond to its occurrence in a LSR along which a CR-LSP has already been established.

2.1 Congestion Detected in a CR-LSP

An ingress LER can determine the contribution it makes to the utilisation of the LSRs along each LSP, and can set up CR-LSPs[1] with that limited knowledge. However, it currently has no knowledge of how those same LSRs are being utilised by other LERs. It is this lack of information when deciding which LSP may meet an application's requirement that can lead to congestion occurring within a downstream LSR.

Assume over time that as a result of the increased load through a LSR, it starts to lose packets from a LSP. If the value exceeds a given threshold i.e., the loss probability assigned to that particular buffer, it is taken as a sign of possible congestion in that buffer stream, within that LSR.

Once the packet loss in the best effort LSP has risen above the predetermined threshold value, for an extended time period the LSR creates a LDP notification message containing the proposed *Congestion Indication* TLV. The objective of sending the *Congestion Indication* notification (CIN) message is to indicate to the ingress LER that there are packets being lost from a particular CR-LSP originating from it, allowing the ingress LER to either:

1. 1.Decide that the packet loss it is currently experiencing remains sufficiently low for it to continue to meet its SLA requirements, allowing/permitting no further action to be taken at this time.
2. 2.Renegotiate for new quality requirements along the existing LSP[2].
3. Negotiate for new quality requirements along an alternative LSP.

In order for the ingress LER to act on the received information, it needs to know the following:

1. The identity of the LSP that is experiencing congestion.
2. The current loss in the buffers the LSP is traversing.[3]
3. The LSRs this loss is occurring in.
4. The current loss the LSP is experiencing.

As a result of this information, the congested LSR generates a *CIN* message. This must contain the identity of the LSR that is experiencing loss, the identity of the CR-LSP along which packet loss was detected, and the packet loss the LSP and buffer are currently experiencing. The congested LSR uses the input port that the packet was received on and the input MPLS label, as an index into the Next Hop Label Forwarding Entry (NHLFE) table to obtain the CR-LSP ID. The CR-LSP ID identifies the *Ingress LSR Router ID* i.e., originating LER, and the local value assigned by that ingress LER to identify the CR-LSP initiated by it. The buffer the LSP traverses through at the congested LSR is obtained using the CR-LSP ID to index a separate *Buffer Table*[4] Each

---

[1] The terms CR-LSP and LSPs are used interchangeable, the main difference is that CR-LSPs are established based on constraints, e.g., explicit route constraints, QoS constraints, etc.

[2] Request that the LSP be promoted to pass through a higher priority buffer along the same path, and within the same LSR.

[3] Although each buffer and its servicing scheduling are dimensioned for a specific CLP, at any time due to traffic loading the current available packet loss within the buffer may have increased or decreased.

[4] The *Buffer Table* is to maintained by each LSR to a record of the availability of each buffer's resources.

LSR experiencing congestion records in its *Congestion Indication Table*[5] the CR-LSP *ID*, and the current LSP and buffer losses. A timer is set. If when the timer expires, the LSR is still suffering from congestion, the LSR will send another *CIN* message with the updated calculated loss values and reset the timer.

The LSR's own IP address is included in the message along with the current packet loss both the CR-LSP and the buffer are experiencing. The *CIN* message is then forwarded to the next hop LSR towards the ingress LER.

Rather than all congested LSRs always generating CIN messages, intermediate LSRs upon receipt of a *CIN* message may append relevant information to it concerning their status if they are also experiencing congestion. If a LSR receives a *CIN* message shortly after sending one, it checks the *Congestion Indication Table* to determine if the timer it has set has expired. If it has not expired, it will simply forward the message without appending its own information, otherwise it will include its information before forwarding.

Timers are used to control the responsiveness of the FATE scheme to traffic loading transients. For example, when a LSR is congested it can issue a *CIN* message. In doing so it sets a retransmission timer. It is not permitted to issue another message until the timer expires, thus avoiding signalling storms whilst improving the robustness of the protocol. Alternatively, if it receives a *CIN* message on route to the ingress LER from another congested LSR, it can simply append its own congestion information and set the timer accordingly. In doing so, avalanches of congestion notification messages towards the ingress LER are prevented. In addition, stability is improved by averaging the observed traffic parameters at each LSR and employing threshold triggers.

When the ingress LER receives a *CIN* message, it may do any of the actions previously outlined.

The motivation behind monitoring individual LSPs through a particular buffer stream stems from the ingress LER's need to ensure the SLAs between the customers and the MPLS network are maintained at all times. To enable it to do this, it needs to have knowledge of the loss encountered by the individual LSPs originating from it. Individual LSPs from a customer site are aggregated into LSPs that share class based buffer resources. As a result of this, the LSP loss rather the individual flow losses is reported back to the ingress LER, who has knowledge of which flows are affected via the flow/LSP binding information.

By monitoring both the losses experienced by individual LSPs and buffer streams, it gives the ingress LER two averages to consider when deciding whether to renegotiate QoS requirements along an existing path or a different path, or whether to accept the current condition.

For example, consider when an ingress LER receives an indication that the loss in a buffer its LSPs are passing through is experiencing a particularly poor loss probability (1 in 10^{-2}). However the loss probability experienced by the buffer it is traversing (1 in 10^{-5}) is acceptable. Or it may decide to set an *Optional Response Timer*; if it receives another CIN message before the timer expires it will take appropriate action. However, if the timer expires and no CIN message is received, it will assume the loss experienced by its flows has fallen within the negotiated value. Some means of averaging the loss statistics provides a useful dampening factor. To prevent an avalanche of CIN messages being sent to a single ingress LER, the congested LSR when it determines that more than one

---

[5] The *Congestion Indication Table* is maintained by each congested LSR, it contains information about the flow and buffer experiencing loss that has exceeded the predetermine thresholds.

CR-LSP traversing its buffers is experiencing a particularly poor loss probability, will aggregate the CIN messages for those individual buffers.

2.2   SCALABILITY

Monitoring losses in individual LSPs is not very scalable, even if those LSPs represent the aggregation of individual connections or flows from a customer site. It is quite possible that at any instance in time, a LSR could be expected to handle a very large number i.e., thousands of these LSPs. As a result of this scalability issue, detecting losses in individual LSPs described previously, may not be a viable option in an MPLS domain expected to maintain a large volume of LSPs[6]. This immediately poses two questions.

How is it possible for an autonomous MPLS network to apply congestion control mechanisms in a situation were it has numerous flows, some of which may be entering the domain just after exiting a customers premises, and others on route from or to another autonomous domain?

How can this service provider ensure the customers SLA is met whilst traversing this network?

In monitoring a single LSP or a number of LSPs that connect between a specific source and destination, connected within a single autonomous system, it is quite easy to identify the ingress and egress LERs, and the exchange of messages can be easily handled under the control of the operator.

Consider the case when the source and destination are not within the same domain, or where the MPLS domain is as an intermediary transport 'pipe'. It is not possible or desirable for the operator to determine the absolute source and destination of each LSP.

The author proposes assigning a *Virtual Source/Virtual Destination* (VS/VD) [4] pair for the aggregation of LSPs entering the domain at one point and exiting at another point, using label stacking or tunnelling within the autonomous MPLS domain of interest.

All LSPs arriving at a particular ingress LER and exiting at a particular egress LER are assigned to a FEC. The ingress LER also known as the virtual source, is the entry point to the MPLS domain and it is at this point that an additional label is 'pushed' onto the label stack, and used to 'tunnel' the packet across the network. On arriving at the egress LER, also known as the virtual destination, the label is 'popped' and the remaining label used to forward the packet.

By employing label stacking within the domain and assigning VS/VD pairs, the issue of scalability is removed whilst allowing the operator control of the LSPs traversing its network. It allows for efficient utilisation of the limited network resources and the additional capability of controlling congestion.   With the VS/VD paradigm, the congestion control message need only propagate along as far as the virtual source for the ingress LER and to the virtual destination for the egress LER. With the virtual endpoints of the LSP defined, aggregation of many LSPs can be treated as an individual LSP as described previously.

---

6 However [4] explains how operation and maintenance (OAM) cells are used in ATM for fault management and network performance on a point to point connection basis, thus implying it is possible to monitor a large number i.e., thousands of flows or connections.

2.3 Renegotiation Procedures

On receiving a *CIN* message the ingress LER extracts the following information: CR-LSP ID that encodes the *Ingress LSR Router ID* and a locally assigned value *Local CR-LSP*, from the LSP-ID TLV. The LSR Router ID experiencing loss, and the value of packet loss the LSP and buffer are currently experiencing, from the *Congestion Modificaiton* TLV. The *Ingress LSR Router ID* along with the *Local CR-LSP* identifies that this message has been received by the correct ingress LER. With this information the ingress LER is able to identify the particular LSP and its traffic parameters.

The ingress LER needs to determine whether it should renegotiate along an existing LSP for a higher buffer stream offering improved servicing or whether it should negotiate for a new LSP route. The decision depends on information gathered from Statistical Control messages explained later.

2.3.1 Renegotiation Along an Existing LSP Within a Higher QOS Buffer

If the ingress LER decides to renegotiate along an existing path for a higher service class, it will carry out the following procedure: The ingress LER formulates a Label Request message with the ActFlag set, to indicate that this is an existing CR-LSP along which the traffic parameters need to be modified. The Label Request message contains the newly requested modified traffic parameters along with the service class it requires. When each LSR receives a Label Request message it uses the globally unique CR-LSP ID as an index into the *Buffer Table* to determine which buffer stream the CR-LSP traverses, the amount of bandwidth initially reserved and the loss probability assigned to that CR-LSP identified by the CR-LSP ID.

The LSR then chooses a higher buffer stream to the one the CR-LSP currently traverses. It then determines whether it can allocate the bandwidth and the minimum loss probability requested within one of the alternative buffer streams. If it can, it temporarily assigns that amount in the new buffer stream, whilst maintaining the original entry. It alters the available bandwidth within the *Buffer Requirement Table* and forwards the Label Request message to the next hop.

If all the LSRs along the CR-LSP are able to meet the requirement on receipt the egress LER will create a LDP notification message containing a *RenegSuccess* TLV indicating the resources have been reserved and send it to the upstream LSR towards the ingress LER of the CR-LSP.

On receiving a *RenegSuccess* notification message each LSR will permanently assign the resources to the path. The *RenegSuccess* notification message is then passed upstream. On receipt of a *RenegSuccess* message the ingress LER updates the FEC/label binding to reflect the higher buffer stream through which the CR-LSP will now be routed.

The *Reneg Success* notification message includes the CR-LSP ID, along with the parameters agreed[7] on, in terms of bandwidth required and minimum LSP loss probability

If a LSR cannot allocate the additional resource it will send a proposed *RenegFailure* TLV within a notification message to the message source and not propagate the Label Request message any further. The LSR will append to the *RenegFailure* notification message the maximum current available bandwidth it can allocate within each of its buffer streams that are also capable of meeting the minimum loss probability requested.

---

[7] This document assumes the bandwidth is controlled by the operator by possibly using policing and shaping mechanisms, but these mechanisms are beyond the scope of this document.

On receipt of a *RenegFailure* notification message, the LSR will deduce that another LSR further upstream has been unable to allocate resources for a LSP which traverses one of its own buffers.

The ingress LER on receiving a *RenegFailure* notification message will have enclosed a single value representing the lowest currently available bandwidth that can be offered along that CR-LSP, whilst realising that renegotiation along the existing path has failed for that CR-LSP and decides on remedial action. The protocol supports a "crank back" mechanism. For instance, when the ingress LER receives a *RenegFailure* notification message it can select an alternative path either by referring to a topological link cost database maintained by a separate routing protocol or the decision is made by the network management module. It then sends a Label Request message along the revised path. When it receives a Label Mapping confirming a new path has been set up, it replaces the old Label Mapping with the newly received Label Mapping, it can then delete the original label or keep it to send other data along the path it represents. If the decision is to delete the original label, the ingress LER will send a Label Release message [5] including the newly replaced Label along the LSP to the egress LER. This procedure results in the label being removed from the pool of "in use" labels. This Label Release message should be sent a few seconds after the last packet is forwarded along that path to ensure the egress LER receives the last packet before it removes the label from forwarding use[8].

2.4   Monitoring Procedures

Proposed *Statistical Control* TLVs contained within LDP notification messages, known as *Status Requests*, are sent into the network periodically from the ingress LER or when the ingress LER receives a *CIN* message.

When the ingress LER chooses to issue a Status Request, it uses the CR-LSP ID to determine which CR-LSP it refers to. It then formulates the Status Request message with the explicit route and CR-LSP ID included and transmits it to the next hop in the ER.

As each LSR receives it, it appends its own statistical information to the message. This includes the loss probability experienced by this CR-LSP. It also includes the current losses of all the alternative class-based buffers the CR-LSP could pass through at this LSR along the specified path[9]. It then forwards the *Status Request* to the next LSR. When the message reaches the egress LER, it is sent back to the ingress LER. Upon receipt of a *Status Request* message that it earlier issued, the ingress LER extracts the CR-LSP ID, and records for each LSR along that path the loss experienced both by this CR-LSP and the loss currently being experienced by all the relevant buffers at each LSR. This information is recorded in a *Statistical Buffer Table* for monitoring purposes.

The *Status Request* messages provide an overall view of the status of the links and LSRs along a particular CR-LSP. It includes the available bandwidth and loss probabilities within every buffer stream within a LSR, as well as the loss experienced by a CR-LSP.

---

[8] Alternatively a 'flushing' mechanism could be used to ensure all data sent along the former path has reached its destination prior to forwarding more data along the new path [4].

[9] In this thesis loss is used as an example statistical parameter, however, this could be easily generalised to a variety of traffic engineering performance metrics.

The CIN messages only return status information about the CR-LSP suffering unacceptable loss and the particular buffer it traverses in the congested LSRs between the ingress LER and the initiator of the message i.e., not the entire CR-LSP.

Subsequently, if the ingress LER receives a *CIN* message, it examines the information held in its *Statistical Buffer Table* to help determine whether it should renegotiate along the existing path, as the higher buffer streams seem capable of meeting its QoS requirements. Alternatively, it can choose to negotiate for an alternative path.

2.5   Simulation Results

One scenario considered for the simulation involved a network of two LSRs and two LERs as shown in Figure 3. A CR-LSP between the ingress and egress LERs is established, with flows passing through all the buffer streams within each LSR/LER ☺. After a specified simulation time the number of flows are increased to cause the loss probability within LSR 1 to rise above a predetermined threshold. The moment at which that point is detected the FATE congestion mechanism responds. In this particular scenario, the flows are switched onto a higher buffer stream along the same LSP ☺.

The associated graph shows the point at which congestion is detected in a buffer and within a LSP. It also shows the operation of congestion indication mechanisms as witnessed by the subsequent reduction in packet loss once the flows are transferred to a higher buffer stream.

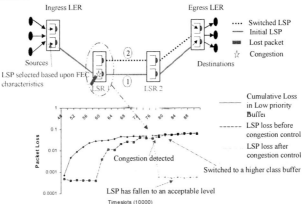

Fig. 3 Loss detected before and after FATE Mechanism

2.6   Discussion

Preliminary results from the FATE scheme illustrate the benefit of dynamic renegotiation of LSPs to control congestion within an MPLS domain. Without this control, network operators have no viable means of redistributing traffic flows at short notice onto under-utilised links / LSRs. The issue of scalability has been addressed by exploiting LSP aggregation to provide tunnels between VS/VD pairs. This mechanism also allows the operator to exercise full traffic engineering control within their domain without affecting the content of the received flows.

Timers are used to control the responsiveness of the FATE scheme to traffic loading transients. For example, when a LSR is congested it can issue a Congestion Indication notification message. In doing so it sets a retransmission timer. It is not permitted to issue another message until the timer expires, thus avoiding signalling storms whilst improving the robustness of the protocol. Alternatively, if it receives a Congestion Indication notification message on route to the ingress LER from another congested LSR, it can simply append its own congestion information and set the timer accordingly. In doing so, avalanches of congestion notification messages towards the ingress LER are prevented. In addition, stability is improved by averaging the observed traffic parameters at each LSR and employing threshold triggers. Although the averaging window has so far been set to 5 seconds, current research is examining the stability of the system as this parameter is adjusted.

A further area of ongoing research is the extension of FATE, called FATE+ where the decision in situations of congestion is the responsibility of the congested LSR and is not passed to the ingress LER. FATE+ is suitably employed along 'loosely' routed CR-LSPs.

References

1. Coombs. S, Nortel Networks Initiates Major Step in MPLS Multivendor Interoperability, http://www.newswire.ca/releases/March1999/16/c4402.html, March 1999.
2. Borthick.S, 'Router Startups Betting on MPLS', http://www.bcr.com/bcrmag/11/nov98p14.htm.
3. Semeria.C, 'Traffic Engineering for the New Public Network', http://www.juniper.net/techcenter/techpapers/TE_NPN.html, 22/02/00, 25/02/00.
4. McDyson, Spohn, 'ATM Theory and Applications' ISBN 0-07-645356-2
5. Andersson,L Doolan,P et al, LDP Specification, http://search.ietf.org/internet-drafts/draft-ietf-mpls-ldp-08.txt.
6. Holness,F Phillips,C 'Dynamic QoS for MPLS Networks', accepted at the 16th UK Teletraffic Symposium, (UKTS), Nortel Networks, Harlow, England, May 22-24, 2000.
7. Holness,F Phillips,C 'Dynamic Traffic Engineering within MPLS Networks',accepted at the 17th World Telecommunications Congress (WTC/ISS2000) Birmingham, England, May 7-12,2000.

Software Switch Extensions for Portable Deployment of Traffic Control Algorithms

G. Kormentzas[1], and K. Kontovasilis[1]

[1] National Center for Scientific Research "DEMOKRITOS",
Institute for Informatics & Telecommunications,
GR-15310 AG. PARASKEVI, POB 60228, GREECE
e-mail: {gkorm, kkont}@iit.demokritos.gr
Tel: +301 650 3167
Fax: +301 653 2175

Abstract. ATM networks employ traffic control as an important mechanism for assuring QoS levels, while also achieving economical resource usage. Current ATM networking equipment incorporates rather elementary Traffic Control Algorithms (TCAs), hardwired into the software controlling the devices. This arrangement does not allow upgrading to more advanced control schemes, as these become available. Considering this problem, the paper proposes a software infrastructure for portably and transparently embedding advanced traffic control functionality in ATM switches. In the proposed infrastructure, there are abstract software entities and programmable interfaces between entities, both complied with the emerged P1520 reference model. Given that current signaling standards do not support externally defined traffic control functionality, the paper discusses in detail an appropriate for the presented infrastructure signaling protocol, called Virtual Signaling Protocol (VSP). The examination of VSP messages that run through the L interface (of the P1520 model) constitutes one of the main objectives of this paper. The functionality of the presented infrastructure (as well as VSP) is tested by means of an implemented prototype system.

1 Introduction

ATM, a particularly pervasive BISDN technology provides explicitly and by design the foundations for deploying Traffic Control Algorithms (TCAs). These algorithms are necessary for balancing the tradeoff between efficient utilization of network resources and assurance of particular QoS (Quality of Service) levels required by the users.

Given the significance of traffic control, current ATM switches come equipped with a suite of TCAs (such as routing, Call Admission Control (CAC), resource configuration/allocation, etc.), specified and implemented by the manufacturer. In most cases, internal details of the algorithms are not released and access to the switch control software is not provided. Even in those cases where some access is allowed through an API (Application Programming Interface), it is usually limited to configuration tasks and does not allow modifications to the core of TCAs.

As long as ATM switches are closed boxes that execute a restricted set of manufacturer TCAs, it is difficult for network operators and/or value-added service providers to: (a) dynamically reprogram ATM switches in order to, e.g., extend their functionality and (b) implement customized traffic control policies and/or protocols. Addressing the inflexible architecture of current ATM switches, the paper proposes an open software infrastructure in which specific TCAs can be portably installed and transparently instantiated on demand. With this architecture, the network operators/service providers of a shared ATM network infrastructure:

a) are freed from manufacturer-dependent traffic control mechanisms,
b) can develop signaling protocols and control programs richer than standard ones,
c) can quickly introduce new (advanced) TCAs to support QoS and
d) can implement different traffic control policies according to their particular needs and/or purposes.

The software infrastructure presented herein includes abstract software entities and generic interfaces between entities, both complying with the P1520 reference model (see Figure 1) [1]. The said model has been developed in the context of the IEEE P1520 standards development project [2], with the aim of standardizing software abstractions of network resources and APIs for manipulation of these resources. The P1520 reference model follows a layered view of a programmable network[1] that is similar to the abstraction of an operating system in a personal computer [4]. An operating system provides programming interfaces for applications to make use of the physical resources of the computer. With an analogous software layer over a programmable network, different TCAs (including customized signaling protocols) can operate on the same network.

The rest of the paper is organized as follows: Section 2 presents the proposed flexible and extensible software architecture through which drastic changes in the traffic control functionality of an ATM switching system can be achieved. The next Section

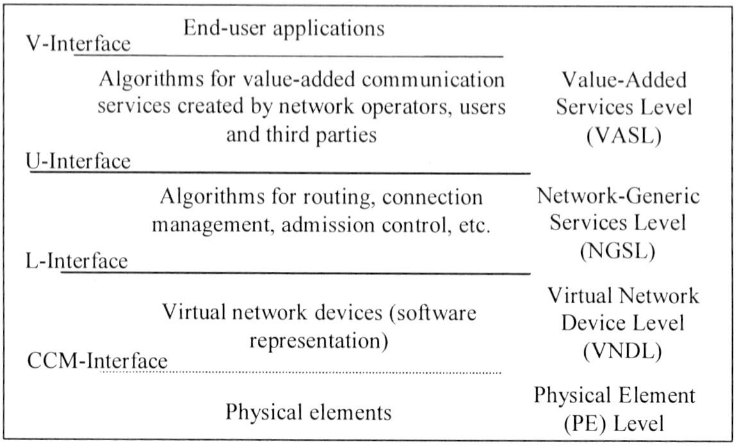

V-Interface	End-user applications	
	Algorithms for value-added communication services created by network operators, users and third parties	Value-Added Services Level (VASL)
U-Interface		
	Algorithms for routing, connection management, admission control, etc.	Network-Generic Services Level (NGSL)
L-Interface		
	Virtual network devices (software representation)	Virtual Network Device Level (VNDL)
CCM-Interface		
	Physical elements	Physical Element (PE) Level

Figure 1. The P1520 reference model

---

[1] According to [3], programmable calls a network of which the functionality can be extended by dynamically installing software extensions on the networking nodes.

3 discusses, in the form of a set of messages, a Virtual Signaling Protocol (VSP) suitable for the proposed architecture. A prototype system that implements the abovementioned software infrastructure is presented in Section 4. Lastly, Section 5 concludes the paper, outlining also some plans for future exploitation.

2 A Software Infrastructure for Portably and Transparently Embedding Traffic Control Functionality in ATM Switches

In this section, we propose a software infrastructure through which the control plane functionality of an ATM switch can be extended.

Figure 2 presents the proposed open infrastructure for portably and transparently embedding TCAs in an ATM switch. As it may be seen in this figure, the heart of the proposed software infrastructure constitutes an *abstract information model* (called Switch-Independent Management Information Base (SI-MIB)) that is placed on top of the target ATM switching system. The SI-MIB provides a virtual environment for portably deploying TCAs by separating the functioning of these algorithms from the details of the switch hardware and low-level management/control software. This device independence is achieved through generic traffic control constructs (in other words, logical switch-independent objects appropriate for traffic control schemes) that constitute software abstractions of the resources and the traffic load conditions within the corresponding ATM switch. (An in-depth description of the SI-MIB can be found elsewhere [5].)

Generally speaking, an abstract information model: (a) has the ability to semantically represent the different MIB trees of various controlled/managed with a small yet comprehensive number of constructs, and (b) allows for high-level, device- and manufacturer-independent control/management actions [6].

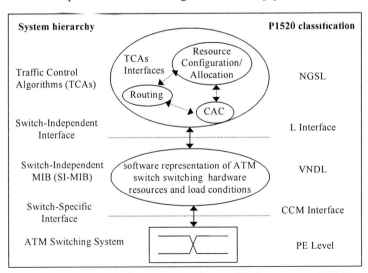

Figure 2. Software switch extensions for portable deployment of TCAS

Returning to the proposed open control architecture, in terms of the P1520 initiative, the TCAs belong to the Network Generic Service Level (NGSL), the switching hardware and low-level software at the Physical Element (PE) level, while the SI-MIB standing in between, corresponds to the Virtual Network Device Level (VNDL).

The software entities implementing a TCA communicate with the SI-MIB at a virtual level through a switch-independent interface, called an L-Interface, in terms of P1520. The L interface is programmable and allows a TCA to access the abstract software version of the switch representation within the SI-MIB. Since this interface deals only with the generic traffic control constructs of the SI-MIB, different TCAs can be applied, without requiring any additional effort for customizing the L interface to the functioning of the particular algorithm. Moreover, as the L interface is uncoupled from the equipment details, the abstract software implementing a particular TCA can be used as is on switches of different manufacturers.

Matching the real status of the switch to the abstract counterpart maintained within the SI-MIB is achieved through a switch-specific interface (a CCM-Interface, in terms of P1520). The CCM interface is *not* programmable, but a collection of switch-specific protocols or software drivers that enable exchange of state (and management/control information in general) at a very low level between a switch and the respective SI-MIB.

Given the proposed open architecture, the exchange of control information between switches can not take place through control paths/channels established by fixed standards signaling protocols, as these protocols do not support externally defined traffic control functionality. Therefore, it is necessary an open signaling protocol, which will ensure that the enhanced control functionality supported by the presented software infrastructure, can be communicated between different switches. Addressing this issue, the next section presents a signaling protocol appropriate for the proposed architecture.

3. A Virtual Signaling Protocol

As mentioned in the previous Section 2, the proposed software architecture overriding the standards-based signaling protocols requires the definition of a customized one. Towards this direction, this section discusses in detail how the software entities of the proposed infrastructure handle a call request/call release in the context of a Virtual Signaling Protocol (VSP) suitable for this infrastructure.

3.1 Call Set-up

When a user requests a new call, one of the following things may happen:
1. All CAC modules involved to the new connection(s) of the call, as well as the destination accept the call request, thus the new connection(s) is(are) set-up successfully;

2. One of the involved CAC modules rejects the new call, then an alternative (if it exists) route is examined and only if all routes fail, the new call is rejected;

3. The destination rejects the call request and a posteriori the new call is rejected.

For the first case (i.e., an accepted call request), the interactions between the end terminals (Source and Destination respectively) and the routing modules, CAC modules and SI-MIBs that are placed on top of the switches are depicted in Figure 3. (The number associated with each message in Figure 3 indicates the message's relevant position in a chronological sequence.)

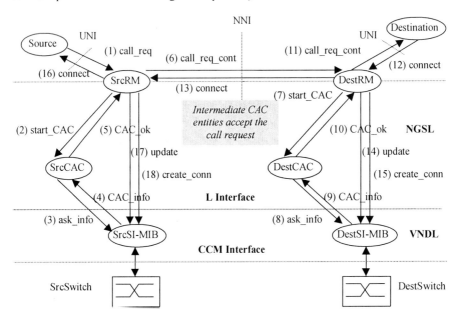

Figure 3. Interactions between the proposed software entities in case of an accepted call request

Elaborating on Figure 3, a user (Source) sends a message call_req to the source routing module (SrcRM). This message includes as parameters the addresses of the Source-Destination pair (parameter *S-D*), an indication of the desired QoS level (parameter *QoS*), in terms of delay or cell loss, that the new call may tolerate, a set of parameters describing the traffic profile for this call (parameter *TrProf*) and a unique call reference identifier (parameter *RefNum*).

Based on the parameters *S-D* and *QoS*, the SrcRM finds an appropriate route for the new call. Specifically, the SrcRM defines the relevant output port (parameter *portOut*) of the source switch (SrcSwitch) and the next switch (parameter *NextSw*). The routing in the rest of switches is based on a unique route number (parameter *routeNum*) identifying the selected route. Additionally, the SrcRM reserves an input and an output VCI (parameters *VCIn* and *VCIout* respectively) in order to be used for the establishment of the appropriate VC crossconnection in case of call acceptance.

Subsequently, the SrcRM sends a message start_CAC to the source CAC module (SrcCAC) in order to initialize a CAC process. This message contains the parameters *portOut*, *QoS*, *TrProf* and *RefNum*. The SrcRM uses the first two of these parameters in a message ask_info in order to retrieve from the source SI-MIB

(SrcSI-MIB), through a message CAC_info, appropriate information for the application of a CAC scheme. This information consists in available resources (parameter *res*) and the load conditions (parameter *bgTraffic*) within the (specified by the parameters *portOut* and *QoS*) Virtual Multiplexing Unit (VMU)[2] of the source switch (SrcSwitch).

Using the information of the parameters *QoS*, *TrProf*, *res* and *bgTraffic*, the SrcCAC performs a CAC scheme in order to accept or reject the new call request. Given that the SrcCAC accepts the call (see Figure 3), it proceeds to compute the new value *updatedbgTraffic* of the parameter *bgTraffic*. Then, it informs the SrcRM for the acceptance of the new call request, through a message CAC_ok with parameters *RefNum* and *updatedbgTraffic*.

Receiving the message CAC_ok, the SrcRM continues the per hop CAC process in each one of the ATM switches of the selected route. In this context, it sends a message call_req_cont to the routing module of the next ATM switch. The said module operates as described above. The parameters of the message call_req_cont are the same as the message call_req, plus the parameters *routeNum* and *VCIout*. As mentioned, the parameter *VCIout* will be used in case of an accepted call for the appropriate call set-up.

According to the scenario depicted in Figure 3, the message call_req_cont (following the abovementioned per hop process) reaches at the destination routing module (DestRM). Again, the described CAC process is repeated and finally, given that the destination CAC module (DestCAC) accepts the new call request, a message CAC_ok returns from the DestCAC to the DestRM. Then, the DestRM (through the message call_req_cont) informs the destination for the new call request.

As the destination accepts the call request, it sends to the DestRM a message connect with the parameter *RefNum*. The DestRM transfers this message to the previous routing module, which passes it to its own previous routing module, etc. Eventually, the message connect reaches at the SrcRM and from there it goes to the source.

Besides passing a message connect to the previous routing module, each routing module initializes a message update (with parameters *portOut*, *Qos* and *updatedbgTraffic*) and a message create_conn (with parameters *PortIn*, *VPIin*, *VCIin*, *PortOut*, *VPIout* and *VCIout*). The first message updates the traffic load of the VMU (defined by the parameters *portOut* and *Qos*) in order to reflect the newly established call. The second message incurs the set-up of the crossconnection defined by the parameters of the message. At the final stage, the successful set-up of a new call is the result of the application of messages update from all the involved to the new call ATM switches.

We proceed now to examine the two cases where the new call request is rejected, either by the network, or by the destination.

At the fist case (see Figure 4a), the call request is rejected as there is no route where all the involved CAC modules accept the call. By comparing Figures 3 and

---

[2] VMU is a virtual notion that may correspond to only a part of an actual ATM multiplexer (e.g., a VP equipped with part of the buffer and link capacity of some output port of an ATM switch).

4(a), it is apparent their concurrence as long as the CAC modules of a selected (by the SrcRM) route accept the new call request. But, when a CAC module (the DestCAC according to the scenario of Figure 4a) rejects the call request, there is a slight variation. Instead of message CAC_ok, the DestCAC sends to DestRM a message CAC_reject with parameter *RefNum*. Subsequently, the DestRM passes a message call_reject with parameter *RefNum* to the previous routing module, which transfers it to its own previous routing module, etc. Eventually, the message call_reject reaches at the SrcRM.

Receiving the message call_reject (or a message CAC_reject in case where the SrcCAC rejects the new call request), the SrcRM examines alternative routing paths. If it finds a route where all the involved CAC modules accept the call request (Figure 3), the call is accepted, otherwise (Figure 4a) it is rejected. The SrcRM informs the source for the rejection of the call request through the message call_reject.

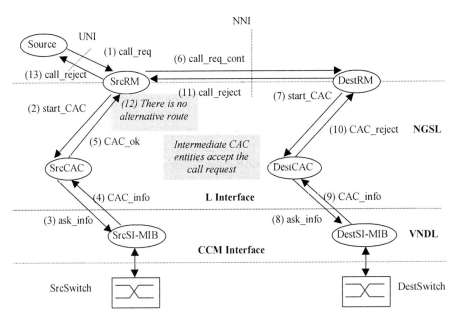

Figure 4a. Call rejection by the network

A different scenario for the rejection of a new call request is depicted in Figure 4b. According to this scenario, the call request for a selected (by the SrcRM) route is accepted from all the involved CAC modules. Thus, the DestRM sends to the destination a message `call_req_cont`. But, the destination, instead of a message `connect`, answers to DestRM with a message `call_dest_reject`. The parameter of this message is the *RefNum*. Receiving the message `call_dest_reject`, the DestRM passes it to its previous routing module, which transfers it to its own previous routing module, etc. Eventually, the message `call_dest_reject` reaches at the SrcRM, and from there it goes to the source of the new call request.

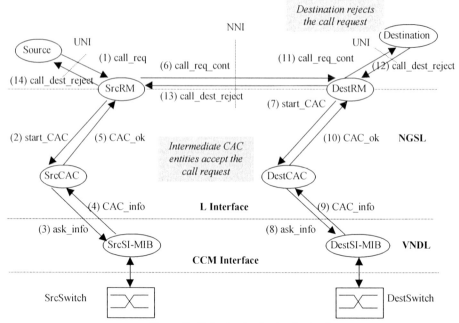

Figure 4b. Call rejection by the destination

3.2 Call Release

According to the call release scenario depicted in Figure 5, the call release is decided by the source. In this case, the source sends a message `call_release` (with parameter *RefNum*) to the SrcRM. The SrcRM passes this message (via the SrcCAC) to the SrcSI-MIB, in order to receive through a message `CAC_info_release` (with parameters *QoS*, *TrProf*, *res* and *bgTraffic*) the required information for the computation of the new value *updatedbgTraffic* of the parameter *bgTraffic*. After this computation, the SrcCAC sends a message `CAC_release` (with parameters *RefNum* and *updatedbgTraffic*) to the SrcRM.

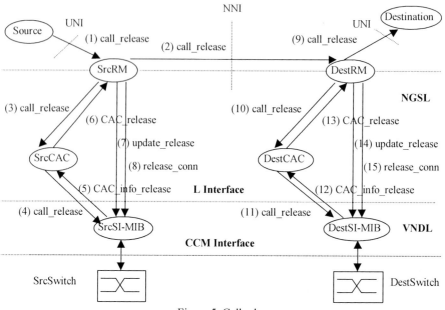

Figure 5. Call release

Receiving the message CAC_release, the SrcRM forwards a message update_release (with parameters *portOut*, *QoS* and *updatedbgTraffic*) to the SrcSI-MIB, in order to update the traffic load of the VMU defined by the parameters *portOut* and *Qos*. Additionally, the SrcRM sends a message release_conn (with parameters *PortIn*, *VPIin*, *VCIin*, *PortOut*, *VPIout* and *VCIout*) to the SrcSI-MIB, in order to incur the release of the crossconnection defined by the parameters of the message. As this process is repeated for all the switches that are involved to the route of the discussed call, the DestSI-MIB releases the last relevant VC crossconnection of the outgoing call.

By examining Figures 3, 4 and 5, it is apparent that some of the messages (e.g., ask_info, CAC_info, update, etc.) run through the L interface. Table 1 includes all messages that are relevant to the L interface. The set of these messages constitute (to the best knowledge of the authors) the first effort in the literature to employ the L interface, as specified in P1520, in a standardised and systematic way for the purposes of hardware independent traffic control.

Table 1. A proposal for applying the L interface structure to portable provision of traffic control

Message	Parameters	Short description
ask_info	portOut, Qos	Through this message NGSL asks from VNDL the traffic control information that corresponds to the VMU defined by the parameters of the message.
CAC_info	res, bgTraffic	By this message, VNDL returns the information requested from the message ask_info (network resources and traffic load conditions within the particular VMU).
update (update_release)	portOut, Qos, updatedbgTraffic	Through the update message NGSL updates the traffic load of the VMU defined by the parameters portOut and Qos.
create_conn	PortIn, VPIin, VCIin, PortOut, VPIout, VCIout	By the message create_conn, NGSL requests the set-up of the cross-connection defined by the parameters of the message.
call_release	RefNum	By the message call_release, NGSL asks from VNDL traffic control information related to the connection defined by the parameter RefNum.
CAC_info_release	Qos, TrProf, res, bgTraffic	Through this message VNDL returns the information requested from the message call_release (network resources, traffic load conditions and QoS of the VMU within traverse the connection defined by the parameter RefNum, as well as the traffic characteristics of the flow that produces the said connection).
Release_conn	PortIn, VPIin, VCIin, PortOut, VPIout, VCIout	By the message release_conn, NGSL requests the release of the cross-connection defined by the parameters of the message.

4. A Prototype Implementation

The section discusses the implementation of a prototype system that follows the general architectural framework of Sections 2 and 3 and makes use of intelligent software agents.

Although not mandated by the design, and for the purpose of facilitating the implementation process, the software entities of the prototype infrastructure have been built on top of an agent platform [7] that supports communication primitives according to the ontology specified by the FIPA organization [8]; therefore communication between the various distributed software entities in the prototype is achieved by means of an Agent Communication Language (ACL). The underlying agent platform used in the implementation was based on background work, conducted in the scope of the IMPACT[3] research project [9]. Again, this choice was made for achieving a faster development cycle, without being forced by design or other constraints.

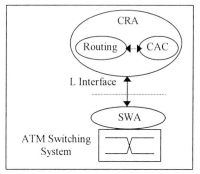

Figure 6. Structure of the implemented software infrastructure

Figure 6 outlines the structure of the implemented software infrastructure. As shown in this figure, the implemented infrastructure consists of distributed software entities that communicate through the facilities of the agent platform just mentioned. Two different types of agents may be identified: the *Switch Wrapper Agent* (SWA) that furnishes an abstract software environment for performing management and control operations on an ATM switch; and the *CAC & Routing Agent* (CRA), which is the software entity expressing in abstract and portable terms the actual traffic control algorithms. Comparing with Figure 2, the SWA corresponds to the VNDL of an ATM switching system, while the CRA plays the role of NGSL. The remaining of this section elaborates on implementation details for each of the two agent-types. Each topic is treated in an individual subsection. (An in depth description of the said agents can be found elsewhere [10].)

4.1   CAC & Routing Agent

The CRA encapsulates the core of traffic control functionality available to the implemented prototype. This type of agent should normally has been called "Traffic Control Agent", but the name CRA was chosen to reflect the fact that this prototype implementation emphasizes CAC procedures and, necessarily, routing actions, which are intimately interwoven with CAC decisions.

The main function of the current CRA implementation is to receive call requests, to determine if they can be accepted through the corresponding switch (without violating the QoS constraint of both the incoming call and all other established connections) and, if yes, forward the call to the next node along the route being checked.

A CRA is implemented as multithreaded software. This is because CRA should be capable of handling/processing simultaneous incoming calls. Call requests arrive at

[3] The IMPACT project (Implementation of Agents for CAC on an ATM Testbed—AC324) was funded by the European Commission under the Advanced Communications Technologies and Services (ACTS) Research Programme.

the CRA through an appropriate message of the agent communication language supported by the underlying agent platform. Upon arrival, the request is placed in a queue of incoming calls, which is served according to a FIFO policy from a thread named `CRAWorkerThread`. For each item removed from the queue, a new instance of a thread named `CallProcessThread` is spawned for processing the call, simultaneously with other pending requests.

The `CallProcessThread` invokes a routing module that selects an appropriate route for the call, this choice depending upon the source-destination pair and the desired degree of QoS. The current implementation supports a simple form of static routing, using fixed universal routing tables. Route selection occurs only at the first network node (the one at the UNI boundary); in subsequent nodes, identification of the selected route is through a corresponding 'label'.

Returning to the operation of CRA, once the call process request is picked by `CallProcessThread` for consideration, the appropriate output port of the corresponding switch is determined, and a software object, called `PortInfo`, is invoked for interrogating the port. (Each output port of the switch corresponds to a separate instance of this object.) `PortInfo` determines the appropriate VMU to check, by consulting the QoS specification of the call request. Information about the current status of this VMU (in terms of the traffic load conditions and the resources available) is maintained (in portable, abstract form) within the SI-MIB of the switch and `PortInfo` retrieves this information by communicating with the SWA. Then, on the basis of the VMU status, and following the rules of the CAC scheme, `PortInfo` decides to either accept or reject the call and relays this decision back to the `CallProcessThread`. If the call was accepted, the CRA signals the next switch along the selected route for the incoming call and the CAC procedure is continued in that next hop.

If, during this process, all switches along the route accept the call, then each involved CRA sends appropriate commands to the SWA attached to the corresponding switch, so that the SI-MIB will update its information to reflect the newly established connection. If, on the other hand, the call is rejected at some node, the fact is back-propagated to the originating node, where an alternate route is selected (if available) and the whole procedure is started over. If no alternate route may be found, the call is rejected by the network.

The above description makes clear that in the implemented prototype, a `CallProcessThread` corresponds to a routing module of Figures 3, 4 and 5, while a `PortInfo` plays the role of a CAC module of the said figures.

4.2 Switch Wrapper Agent

A Switch Wrapper Agent (SWA) is the software component that provides a virtual environment for the portable deployment of control and management operations on the underlying ATM switch. As noted in the previous subsection, during the course of a CAC procedure the SWA attached to an involved switch interacts with the CRA controlling this switch. Specifically, the CRA retrieves from the SWA information about the current status of the involved output port (actually, the appropriate VMU within that port). Based on this information, the CRA performs the CAC check, and, in case of acceptance, informs the SWA, so that the latter will update the status of the switch to include the new connection. Furthermore, the SWA issues the appropriate low-level management commands to the switch for establishing the physical connection between the input and output ports involved.

The core components of a SWA are:

- SI-MIB, i.e., the switch-independent MIB providing an abstract virtual representation of the resources and traffic load conditions within the switch; and
- a library, consisting of a set of software drivers, for accessing the low-level management facilities of the ATM switch.

In the prototype, the low-level library has been implemented using the Java Management API (JMAPI). Currently there are vendor-specific versions for three models of ATM switches, specifically: FORE ASX-200, CISCO LS1010 and Flextel WSS 1200. Porting the library to other devices is easy, since the task involves only the mapping of the abstract objects in SI-MIB to the counterparts in the specific MIBs of the target switch.

5. Conclusions

Following recent proposals for programmable network infrastructures, the paper proposes a flexible and extensible architecture for portably and transparently embedding traffic control functionality in ATM switches. The proposed architecture includes abstract software entities and generic interfaces between entities, both complying with the P1520 reference model.

The paper focuses on the functionality of the L interface. In this context, it discusses how the presented open architecture (consisting of a software infrastructure and a virtual signaling protocol) provides (to the best knowledge of the authors) a first effort in the literature for applying the L interface structure to portable provision of traffic control.

It should be noted that, although the software infrastructure presented in this paper is tailored to ATM networking equipment, it is fairly general and can be extended for covering network nodes of a different technology (provided that this technology supports the potential for QoS concept and at least some notion of "connection" or, more generally, "traffic flow"). This extensibility is highly desirable, since ATM is not regarded anymore as the sole QoS-aware networking technology. (For recent efforts towards injecting QoS capabilities over IP-based networks, see e.g., [11].)

We plan to extend our work in the directions of applying the presented architecture to non-ATM networking technologies (for a programmable IP router architecture supporting control plane extensibility, see e.g., [12]), performance evaluation and security issues.

References

1. J. Biswas, A. Lazar, S. Mahjoub, L.-F. Pau, M. Suzuki, S. Torstensson, W. Wang and S. Weinstein, "The IEEE P1520 Standards Initiative for Programmable Network Interfaces", *IEEE Communications Magazine*, pp. 64-70, October 1998.
2. Information electronically available from `http://www.ieee-pin.org/` and `http://stdsbbs.ieee.org/groups/index.html`.
3. D. Tennenhouse and D. Wetherall,, "Towards an Active Network", *Computer Communications Review*, Vol. 26, No.2, pp. 5-18, April 1996.
4. T. Chen, "Evolution to the Programmable Internet", *IEEE Communications Magazine*, pp. 124-128, March 2000.
5. G. Kormentzas, J. Soldatos, E. Vayias, K. Kontovasilis, and N. Mitrou, "An ATM Switch-Independent MIB for Portable Deployment of Traffic Control Algorithms", In Proc. *7th COMCON Conference*, July 1999, Athens, Greece.
6. K. Kontovasilis, G. Kormentzas, N. Mitrou, J. Soldatos, and E. Vayias, "A Framework for Designing ATM Management Systems by way of Abstract Information Models and Distributed Object Architectures", to appear in *Computer Communications*, 2000.
7. J. Bigham, L.G. Cuthbert, A.L.G. Hayzelden, Z. Luo and H. Almiladi, "Agent Interaction for Network Resource Management". In Proc. *Intelligence in Services and Networks '99 (IS&N99) Conference*, Barcelona, April 1999.
8. Foundation for Intelligent Physical Agents, *FIPA 97 Specification, Part 2: Agent Communication Language (ACL)*, 1997. Electronically available from `http://www.fipa.org/` or `http://drogo.cselt.stet.it/fipa`.
9. MPACT AC324, Technical Annex, March 1998. See also `http://www.acts-impact.org/`.
10. J.Soldatos, G. Kormentzas, E.Vayias, K.Kontovassilis, N.Mitrou, "An Intelligent Agents-Based Prototype Implementation of an Open Platform Supporting Portable Deployment of Traffic Control Algorithms in ATM networks", In Proc. of the *7th COMCON Conference*, July '99, Athens, Greece.
11. X. Xiao and L.M. Ni, "Internet QoS: A Big Picture", *IEEE Network Magazine*, Vol.13, No 2, March/April 1999.
12. J. Gao, P. Steenkiste, E. Takahashi, and A. Fisher, "A Programmable Router Architecture Supporting Control Plane Extensibility", *IEEE Communications Magazine*, pp. 152-159, March 2000.

Capacity Management for Internet Traffic

S. O. Larsson and A. A. Nilsson

Dept. of Telecommunications and Signal Processing
University of Karlskrona/Ronneby
Sweden

Abstract. We describe methods to guarantee a certain level of service for Internet traffic by reserving capacity along fixed logical paths. The amount of needed capacity is calculated by using a trade-off between connection rejection probability and the utilization of the capacity. This can be used in the process of dimensioning capacity on a long term basis. Two dynamic allocation methods are proposed, which periodically reallocates capacity according to measured traffic loads. Call admission control is used and automatic connection retrials are studied. Each method is designed for a particular scenario. In the first one, a LAN reserves capacity to get a certain transmission speed for every connection. Capacity is paid for according to how much of it that is reserved. The second one considers an aggregated traffic generated by users having limited transmission speed in their connection to the Internet (*e.g.* modem- and mobile-users) and the operator of the network manages the capacity. The Internet traffic is modeled as Web page fetches from the World Wide Web.

Keywords: Internet Traffic, QOS, Capacity Dimensioning, Resource Allocation.

1 Introduction

As the global Internet grows so does the demand for capacity. There is also a desire to incorporate real-time services and thereby extending the best effort class of service to multiple classes of services. For near real-time applications (*e.g.* IP Telephony) some level of quality of service (QoS) is needed. Reservation of capacity can accomplish this. To be able to guarantee a maximum delay variation and maximum packet loss probability one should not setup more connections than can be handled. A connection admission control (CAC) algorithm controls that the QoS will be met and subsequently rejecting connection at some times when the load is high. The reservation is best done along a fixed path. Internet has traditionally only supported a single best-effort traffic. It is based on a connection-less packet-networking infrastructure. Packets can always be sent but no guarantees are given that they will arrive to the destination. This depends on the stateless Internet protocol (IP) infrastructure and an inadequate support from network controls. By building a state-based infrastructure and using a connection-oriented approach, support can be given for QoS guarantees.

Lots of effort has been put into developing protocols to support this. ATM Forum has developed the Multiprotocol over ATM (MPOA), while IETF has developed the Multiprotocol label switching (MPLS). MPLS enables labelled switched paths (LSPs) over Internet. Compared to virtual path connections in an ATM network, the LSPs can be used ``inside'' other LSPs, *i.e.* used in hierarchical levels or layers [12]. The logical paths also enables fast switching of the data packets since the routing already has been done. With new services for mobile users with limited capacity, reservation of capacity will give the users value for their money. The amount of reserved capacity should also be adjusted according to actual traffic load. The reservations can be achieved with the proposed integrated services (IntServ) framework [13]. It will keep track of the flow-state (with additional regular updates) for each connection by the use of the resource reservation protocol (RSVP) [14]. At the edges of a core network, individual buffers ensure that connections are protected against each other, but the resources should be reserved for an aggregate of users in the core network. Considering the scaling problem with IntServ, IETF has proposed a different framework called differentiated services (DiffServ) [16]. In this case, packets are tagged with priority according to delay requirements and by value of importance. The tag is used by core routers for scheduling, handling congestion, and so on, on a hop-by-hop basis. This will work as long as enough capacity is available. To decrease the packet loss for low priority packets in high load situations or when other traffics misbehave, some sort of intelligent scheduling mechanism must be used. Weighted fair queuing (WFQ) scheduling has been proposed for these purposes. This management system supports several classes of services between best-effort and guaranteed services. IETF's controlled-load services [15] is a guideline about how to support real-time applications with a degree of service that corresponds to a moderate loaded network. A CAC algorithm is also introduced. To be able to guarantee QoS, our proposed method for capacity reservation will be useful for ensuring that not more traffic, than can be handled, is injected into the network.

We assume that users have limited bandwidth for the transmission of data (*e.g.* by the modem speed) and that LSPs (or similar) for which capacity are reserved, are used for these users. This means that we use constant transmission speeds. It has been assumed that the transmission speed is controlled by some mechanism on the sending side [6]. For IP Telephony this is obvious but it can be used for the most data transfers and with this approach we do not suffer from slow start, round-trip delay and other problems as compared to TCP. Capacity sharing and utilization accomplished with TCP is an important aspect that should be used for some services. In the following, a lower limit for the transmission speed can be used instead of a constant transmission speed.

Section 2 describes our model and assumptions of the traffic. It is difficult to dimension networks to cope with Internet traffic, since this traffic is fundamentally different from traditional voice traffic. Over-provisioning of capacities is one way to postpone the problem. In Section 3 a strategy for the dimensioning problem is proposed. Even if a proper design can be found for steady traffic, the ever-changing traffic should be handled efficiently. If capacity is reallocated dynamically the network will be better utilized. In Section 4 we propose an allocation function for Web traffic destined to a certain LAN, while in Section 5 we propose a management of the capacity needed for an aggregated stream of Web traffic to be routed through the Internet.

2 Traffic Model

The traffic model used is a model of how Web pages are fetched through a Web browser. Characterization of Web traffic involves both the structure and size of the pages. Each page is designed around a main ``anchor" page and a possible additional set of embedded items that appear within this anchor page. The embedded items are fetched immediately after the anchor page and are shown together with it, on the user screen.

The Web page contents are classified into two groups: HTML and non-HTML. This is only natural because of the nature and intent of the participating items. The embedded items are mostly non-text in nature and are glued together using HTML text. The modelling process of the file sizes, corresponding to the two classes, also brings out their inherent characteristic diff-erences. The non-HTML type is more variable than the HTML type. Examples of embedded items can be either: Java scripts, frames, maps etc. for HTML and Java applets, postscripts, images etc. for non-HTML. The probability of fetching HTML- and non HTML-items is set to 50% for both.

The distributions for the file sizes and the number of embedded items have been computed in [1] from data collected by probing a broad spectrum of Web servers (*e.g.* newspapers, universities, search engines, commercial sites etc.). The contents from the servers were successively downloaded and analyzed for their structural information. (Only unique items were selected.) We assume that the file contents and what is actually transferred through the network is very much the same. No regards has been taken to the cache used in Web browsers.

Some of the items can be very large. The file size (X) is modelled with a Pareto distribution in [2,3]. In the following, we consider the case of a combination of a uniform distribution and a truncated Pareto distribution for the file sizes, which corresponds to measurements done on commercial servers [4]. In our simulation experiments we have used the following distribution:

$$P\{X \le x\} = \begin{cases} \dfrac{(x-a)}{(b-a)}\,pu, & a \le x \le b \\ \left(\dfrac{1-(b/x)^\alpha}{1-(b/N)^\alpha}\right)(1-pa)+pa, & b \le N \end{cases} \tag{1}$$

According to [1] the parameters for HTML items are: a=100, b=4200, α=1.36, and the mean file size is 12888 bytes. The probability $_{pa}$ denotes the probability that $a{\le}b$, and it is calculated from the total mean file size and is 0.215. For non-HTML items a=100, b=1024, α=1.06, the mean file size is 12000 bytes, and pa is almost zero.

Measurements show that the number of embedded items Z in a Web page can be modelled by a negative binomial distribution [1]:

$$P\{Z = x\} = \binom{\delta + x - 1}{\delta - 1}\left(\frac{\beta}{1+\beta}\right)^x\left(1-\frac{\beta}{1+\beta}\right)^\delta, \tag{2}$$

where $\delta=1.386$, $\beta=6.985$. The probability to have no embedded items is set to 12% and the mean number of embedded items is ten. We assume that for a certain Web page the number of embedded items and their file sizes are independent.

The evaluation of the capacity management methods is done by simulations. Traffic is generated by Web page fetches where the embedded items are transferred in parallel after the reception of the ``anchor" page. If a new connection request is rejected, the application will try again a maximum of R_{max} times. This is only natural since an upper limit of connection setup delay is wanted. R_{max} could be made to depend on the type of service. The waiting time to the next connection attempt is doubled each retrial. The delay has been set to one second for the first retrial. This means that if $R_{max}=6$, then if a transfer of an item has not started within one minute, it will not be fetched (unless the user retry manually). The truncation of the Pareto distribution by N can not be smaller than $\approx 8 \cdot 10^{10}$ if the mean value for the non-HTML file sizes is to be kept. This still means that for the simulations, a large ``warm up" period is needed to measure the correct value of the mean number of connections. This is caused by the very slow addition of very long connections. As an example, assume a constant traffic intensity that suddenly changes to a new level. This will result in a long settling time to reach a new steady state, *i.e.* when the mean number of connections changes very slowly. By adding a couple of persistent connections at start up, the warm up period can be decreased while still having a correct mean value of the number of connections. This is particularly useful when dimensioning the capacity for steady state traffic. In the sequel, 95% confidence intervals are presented for the simulated and measured parameters.

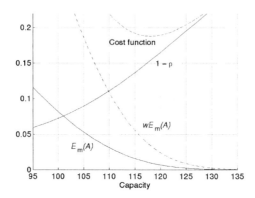

Fig. 1. Cost function and its components for voice traffic.

3 Dimensioning

In order to simplify the evaluation, *i.e.* to reduce the needed simulation time, we have not limit our investigation to a particular transport protocol. The scope is to dimension the transmission capacity of the links and the logical paths. To dimension resources, the distribution of the number of simultaneous connections has to be known. We denote the probability for a certain number of ongoing connections (*i.e.* the used capacity), as the occupation distribution. The variance of the occupation distribution when having batch arrivals is larger compared to the case with single arrivals. In order to dimension the links, we first evaluate the case with voice traffic. A trade-off is done between the utilization of the capacity and the connection setup blocking probability (CBP). By doing this we can find a relationship between the utilization and CBP, which we can apply on any type of traffic. We can also state that this will take into consideration both the network provider and the users point of views. For simplicity, we assume that the provider wants to maximize the utilization. Alternatively, let us consider that, by having this view, we get a minimum capacity at the end that can be used when dimensioning a network to meet a certain service level. Increasing the utilization will also increase the CBP. The users want a low CBP but this on the other hand will give a low utilization. It should be noted that best-effort traffic will increase the utilization further but we do not take this into account. We get the following cost function (C) to minimize:

$$C = 1 - \rho + wE_m(A), \tag{3}$$

where ρ is the utilization defined as:

$$A (1 - E_m(A))/m. \tag{4}$$

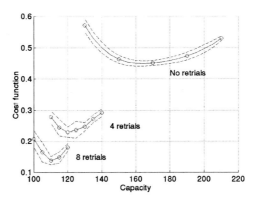

Fig. 2. Cost function for Internet traffic.

The Erlang-B formula is used, where $E_m(A)$ is the CBP given a traffic load of A and a capacity of m (maximal number of simultaneous connections). For a given w, the m that minimizes the cost function, is denoted as the target capacity and the resulting CBP as the target CBP. To obtain the value of w, we study the case of a voice traffic where the offered traffic is Poissonian. If we let calls be blocked with a probability of $\approx 1\%$ for a selected traffic load of 100, then a weighting factor w can be found that optimizes (1). A weight of ~3.5 gives the results shown in Figure 1. The cost function, 1-ρ, and the CBP are shown. By using this weight, the needed capacity can finally be found for our model of Internet traffic. The optimum for different number of retrials has been found by simulations. Figure 2 shows the result with a traffic of 100. The capacity is in this case expressed as units of 56 kb/s. The dashed lines show the confidence intervals.

The capacity needed to optimize C when allowing up to eight retrials is for a traffic of 100, 200, and 300: 110, 213, and 315 respectively. The target blocking probabilities are: 0.8%-1.1%, 0.3%-0.6%, and 0.4%-0.9% respectively. The corresponding target capacities are shown in Figures 3 and 4. Figure 4 shows the dependence to the amount of traffic. Figure 3 shows the dependence on the number of retrials for the traffics 100, 200, and 300. For this traffic range, using a maximum of about five retrials, the target capacities are almost equal to the capacity giving 1% blocking (calculated with Erlang-B formula) for a Poisson traffic without retrials and using a negative exponential service time of one. These capacities are shown as solid horizontal lines in Figure 3.

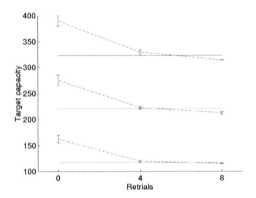

Fig. 3. Target capacity for different number of retrials.

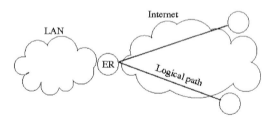

Fig. 4. Model of the network configuration.

4   LAN - Internet

4.1   Allocation Function

Figure 5 shows the case when a LAN is connected to the Internet via an edge router (ER) [7] or some sort of a label-switching router (LSR). Our goal is to calculate the needed capacity for the logical paths (*e.g.* LSPs) so that they can accommodate the traffic destined for the LAN. In this case, only Web browsing is considered. One can also see the reservation as a possible way to manage capacity for a virtual private network (VPN), where the owner of the VPN pay in relation to the capacity reserved. The ER is used for CAC, which means that connections can only be established if there is enough reserved capacity left to accommodate them. One way to introduce explicit requests that can be counted, is by letting the end-terminals send requests to the ER before connection establishment. This could correspond to the proposal in [7] for setting up IP Telephony connections by asking the ER. It can seem as a very costly solution to make connection requests and CAC for every fetch of a Web page item, however, this can and should be simplified compared to an IP Telephony connection setup. In [8] computers monitors the traffic in a campus LAN. A connection rejection is implemented by issuing a TCP RST messages during the TCP connection establishment phase. This way they can control how the capacity on the Internet access link is to be divided between the students and the staff.

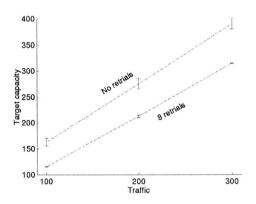

Fig. 5. Target capacity for different traffics.

The connection requests are used for estimating the traffic load, which in turn is used for the reservation. The capacity reallocations are done periodically with a period of T_u and the actual traffic load is estimated each minute.

Given the transfer rate as well as a knowledge of the mean file sizes, we can calculate the mean connection time $\bar{x}$. The mean number of simultaneous connections is $A \cdot \bar{x} \cdot 10$, since the mean number of embedded items per page is ten.

An important aspect of Web traffic is the occurrence of very long lasting connections. When reallocating the capacity periodically, these connections will have very little impact on the performance of the reallocations. Simulations have shown that in a smaller time scale and allowing for retrials, the needed capacity that minimizes the cost function is almost the same as the traffic load. The problem then is to estimate the traffic from measuring the total number of connection requests (with retrials included).

To model the connection requests we add automatic retrials when connection requests are rejected by the CAC. Figure 6 shows fresh new connection requests (A or ``first traffic") and the retrials (A_R). To get an estimate $\hat{A}$ of the fresh traffic A we use:

$$p = E_m \left(A_{total} \right) \tag{5}$$

$$A_{total} = A\left(1 + p + p^2 + \ldots + p^{R_{max}}\right) \tag{6}$$

$$\hat{A} = A_{total} \frac{1-p}{1-p^{R_{max}+1}} \tag{7}$$

$$A_R = A(p + p^2 + \ldots + p^{R_{max}})$$

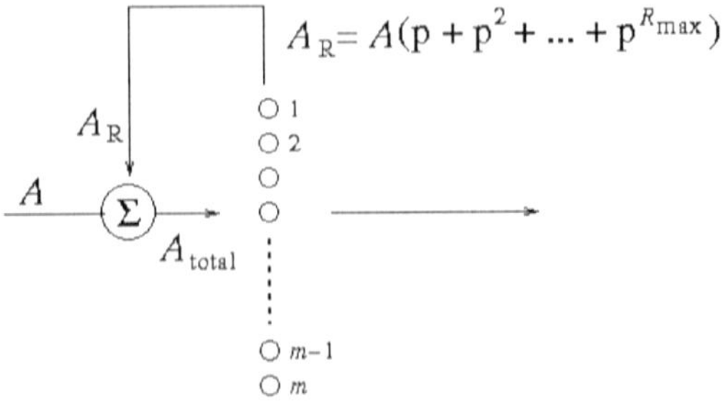

Fig. 6. Model of the retrials.

Table 1. Cost function comparisons.

Expected service time [secs.]	Utilization [%]	CBP [%]	Cost-function
13.5	92.08 ± 1.38	0.25 ± 0.03	8.8
12.5	93.07 ± 1.38	0.44 ± 0.04	8.5
11.5	93.99 ± 1.44	0.75 ± 0.06	8.6
10.5	94.89 ± 1.44	1.19 ± 0.06	9.3

Table 2. Performance comparisons.

Method	Capacity	CBP [%]	Utilization [%]	Mean service time
WWW	220	0	90.04 ± 1.06	14.94 ± 1.45
CAC	220	0.13 ± 0.06	89.70 ± 0.80	13.45 ± 0.98
WWW	195	0	98.66 ± 0.55	19.67 ± 2.22
CAC	195	4.52 ± 0.82	97.11 ± 0.35	12.91 ± 1.18
WWW-DYN	212.9 ± 1.09	0	91.23 ± 1.44	15.63 ± 1.63
CAC-DYN	213.6 ± 1.92	0.46 ± 0.04	92.81 ± 1.86	13.57 ± 0.84

The reservation is never less than the actual number of allocated connections n. The allocation formula is:

$$m = \max\left(n, \quad \hat{A}\right)$$ (8)

where n is the actual number of allocated connections. It should be noted that if the number of retrials are limited or set to zero, there should be an offset of capacity in the allocation formula (3). The extra capacity needed to reach the target CBP is almost zero for about eight retrials and can be estimated as a linear function of the traffic. According to the results in Figure 4 this function is approximately $45 + A \cdot 1.15$, when not allowing any retrials. The actual maximum number of retrials can be estimated based on the application. In case of ``surfing" the Web, R_{max} in formula (2) could for example be set to six, which is about one minute.

In formula (2) there is an assumption that the retrials meets the system, where the occupation distribution will be equal to the mean distribution. Although there is an exponential back off, this is not a correct assumption. However, the difference between the estimated traffic without retrials (first traffic) and the real one, is small. Having a capacity of 190 (A_{total} is 699.9-768.7), $\hat{A}$ is estimated to 201-206. For a capacity of 200 and 210 (A_{total} is 556.3-615.4 and 409.6-468.2), $\hat{A}$ gets 203-205 and 209-210, respectively.

4.2 Evaluation

The problem of allowing too many users into the Internet often results in what has been called ``World Wide Wait." This effect is a result of the sharing of capacity as well as capacity shortage. When the traffic load reaches a limit where the capacity (of routers or servers) is not enough to handle the total stream of packets, some packets will get lost and TCP connections decrease the transfer rate and will occupy the resources for a longer time. New connections will meet an already congested network, which results in even longer delays.

For fixed rate connections (8 kb/s) the time to load one embedded item is 12.44 seconds (*i.e.* the service time). Table 1 shows the cost function for different settings of an expected service time when using the dynamic allocation (DYN) (3) with CAC and allowing up to eight retrials (traffic=200). The cost function has a minimum at the theoretical value, which shows that the allocation function works. Table 2 shows a comparison between an ideal fair share allocation without CAC (denoted as WWW), and the use of CAC with up to eight retrials.

We let the method WWW have a maximum of 8 kb/s for each connection and if the new calculated capacity is less than the maximal capacity that can be used by the current number of established connections, then the new capacity will be set to the mean value of these capacities. The other methods always enable 8 kb/s and capacity is always allocated so that the ongoing connections can be accommodated. The dynamic method uses a reallocation period (T_u) of one minute. No traffic prediction is used but the estimated traffic is smoothed by taking the mean of the previous estimate and the current measurement. The offered first traffic is fixed at 200. The performance is shown for different amount of allocated capacities. When the available capacity decreases the utilization for WWW increases but so does the service time. Having a dynamic allocation even for WWW the service time will be ``acceptable." Although, the utilization for WWW and CAC are similar, the

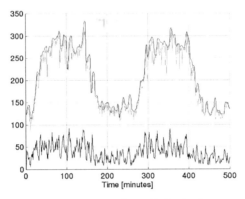

Fig. 7. Dynamic reservation with $T_u=1$ minute.

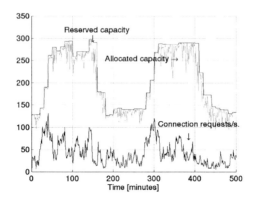

Fig. 8. Dynamic reservation with $T_u=10$ minutes.

distribution of used capacity is different. WWW will have a peak at the maximum capacity, which depends on the fact that no blocking occurs since all connections share the capacity equally. Using retrials, the connections will be delayed and the occupation distribution will increase gradually and reach its peak at the maximum capacity so that the utilization stays high all the time. In the case of having TCP connections, slow start and retransmissions are involved that would have decreased the utilization compared to our results. By using CAC, the overall throughput will increase [8]. Having a control of the transfer rates as well as having proper queuing mechanisms in the switches, the minimum transfer rate is guaranteed.

Evaluations of the performance of CAC-DYN is given in Figures 7 and 8. In this case, the traffic intensity has been varied linearly up and down between steady states that last 100 minutes. T_u has been set to one minute and to ten minutes. The bottom curve is the total arrival rate of connection requests as indicated in Figure 8. The resulting blocking probability for $T_u=1$ minute is $0.72\% \pm 0.05\%$ with a mean allocated capacity of 214.8 ± 2.3. For $T_u=10$ minutes, the blocking is $2.71\% \pm 0.22\%$ and a mean allocated capacity of 211.8 ± 2.0.

The evaluation shows that the allocation function works and that sufficient capacity is reserved. By changing the reservation dynamically, the utilization of the capacity does not suffer from the capacity reservation.

5   Aggregated Traffic

5.1   Allocation Function

In this case we assume that users have limited transfer rates (*e.g.* by the modem speed) and that LSPs (or similar) with reserved capacity are used for these users. Having different LSPs for different classes of traffic, we can in this case group users, having a certain modem speed, into one class. Our model has similarities to the model shown in Figure 5, except that the LAN is replaced with an Internet subnetwork and the ER with an ``aggregation router" [7]. The ER could also correspond to a ``bandwidth broker" as in a DiffServ environment. This router, which allocates capacity on LSPs, counts the total packet rate for each LSP. Dividing this bitrate with the user transfer rate for this LSP gives an estimate of the total number of simultaneous Web page fetches. The occupation distribution is modelled as Poissonian and no retrials are permitted here. The needed capacity is calculated with a formula given in [9,10]. The allocation can be seen as a confidence interval:

$$m = \bar{n} + K\sqrt{\bar{n}}, \tag{9}$$

where $\bar{n}$ is the number of simultaneous Web page fetches estimated by the mean of several samples. The transmission speed is not needed in (4) as long as all the users use the same speed (for the LSP in question). It is important to note that (4) can be used in the previous scenario if there is only one connection established per Web page fetch, as proposed for HTML version 1.1 or if the traffic is IP Telephony.

5.2   Management

There are basically two methods for reserving capacity. First, if no cost is related to the capacity, all of it should be allocated or reserved (except for some capacity left for best effort services). This can be done by trying to mini-mize the total CBP. When doing this one should take into account the selected T_u and the number of LSPs that share the total capacity on an average link. The available capacity will also affect the result. The setting of K should in this case be done adaptively to take into account the multiplexing effect between LSPs when reallocating frequently. Secondly, if there is some cost related to the reservation of capacity, the cost function has to be optimized, (*i.e.* just enough capacity is to be used). The setting of K can be made to depend on the cost of allocating the capacity. Another strategy is to have a central network management centre that supports a global control. There are many proposals in the literature for this and for large networks the number of control messages is

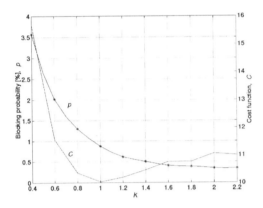

Fig. 9. Blocking probability and cost function for different Ks.

minimized. There are, however, situations when a central management is not logical, *e.g.* in diverse Internet scenarios.

5.3 Evaluation

Figure 9 shows the CBP (p) and the value of the cost function when there are four LSPs for 56kb/s-modem users. T_u is set to one minute. The mean transfer time of one object is 1.78 seconds. In the sequel, we denote with λ the number of connection arrivals per second. The capacity needed for the LSP in the case of $\lambda=10$ (offered traffic=178) is about 11 Mb/s. The link capacity is set to $4 \cdot 198$ (where $E_{198}(178)=0.01$).

As seen in the figure the CBP is lower for higher values of K. In the case when all the capacity should be utilized, there is a value of K that minimizes the total CBP, which is a little larger than 2.2 in this case. If just enough capacity is to be allocated, then the K is set to 1.0.

With our allocation function, the CBP decreases for higher traffic. When the traffic intensity increases to $\lambda=20$ the blocking decreases from $0.88\% \pm 0.02\%$ to $0.64\% \pm 0.01\%$. For $\lambda=30$ the blocking is $0.57\% \pm 0.02\%$. These values are very similar to the target blocking probabilities. It is possible to adjust the allocation function to equalize the blocking for different traffics. Blocking probabilities that maximize the total amount of handled connections have greater difference. Setting K to a fixed value can be seen as a trade-off between these extremes of blocking probabilities and further indicate the usefulness of the target blocking probabilities.

6 Conclusion

We have described methods to guarantee a certain level of service for Internet traffic by reserving capacity along fixed logical paths. A trade-off between utilization and CBP is made to enable a method for capacity dimensioning for Internet traffic. If consideration is taken to limit the mean service time and increase the throughput, there is an advantage of having CAC. This has also been proposed in [11]. The cost of this is a small CBP. If near real-time services are to be guaranteed some QoS, especially for mobile users, they probably will prefer not to be connected if the service would be less than acceptable. Two allocation functions have been proposed. Each one is tailored for a certain situation of capacity management. In one situation, the network provider manages the capacity. In the other case, capacity is paid for according to how much of it that is reserved.

7   Future Work

The problem is to find a cost for reallocations and a cost for not doing them frequent enough. Performance of TCP connections in a realistic network model together with capacity realloc-ations will show how users perceive the performance, which also will be a parameter for the calculation of this cost.

References

1. P. Pruthi, A. K. Jena, and A. Popescu, ``HTTP Interactions with TCP,'' 11'th ITC Specialist Seminar, Japan, pp. 195-204, 1998.
2. M. F. Arlitt and C. L. Williamson, ``Web Server Workload Character-ization: The Search for Invariants,'' IEEE/ACM Trans. on Netw., Vol. 5, No. 5, pp. 631-645, 1997.
3. P. Barford and M. E. Crovella, ``Generating Representative Web Workloads for Network and Server Performance Evaluation'', Proc. on Performance'98 / SIGMETRICS'98, pp. 151-160, 1998.
4. A. K. Jena, ``Modelling and Evaluation of Internet Applications'', *Doctoral thesis*, ISSN 1101-3931, Lund 2000.
5. R. Guérin and H. Ahmadi, ``Equivalent Capacity and Its Application to Bandwidth Allocation in High-Speed Networks,'' IEEE J. Sel. Areas in Commun., Vol. 9, No. 7, pp. 968-981, Sept. 1991.
6. N. Bhatti and R. Friedrich, ``Web Server Support for Tiered Services,'' IEEE Network, Vol. 13, No. 5, pp. 64-71, 1999.
7. P. Goyal, *et al.*, ``Integration of Call Signalling and Resource Management for IP Telephony,'' IEEE Network, Vol. 13, No. 3, pp. 24-32, 1999.
8. A. Kumar, *et al.*, ``Nonintrusive TCP Connection Admission Control for Bandwidth Management of an Internet Access Link,'' IEEE Commun. Magazine, pp. 160-167, May 2000.
9. J. Virtamo and S. Aalto, `Remarks on the effectiveness of dynamic VP bandwidth management,'' COST 257 TD(97)15, Jan. 1997.
10. S. O. Larsson and Å. Arvidsson, ``Performance Evaluation of a Local Approach for VPC Capacity Management,'' IEICE Trans. on Com., Vol. E81-B, No. 5, pp. 870-876, 1998.
11. J. W. Roberts and L. Massoulié, ``Bandwidth sharing and admission control for elastic traffic,'' 11'th ITC Specialist Seminar, Japan, pp. 263-271, 1998.
12. X. Xiao, *et al.*, ``Traffic Engineering with MPLS in the Internet,'' IEEE Network, Vol. 14, No. 2, pp. 28-33, 2000.
13. S. Shenker, R. Braden, and D. Clark, ``*Integrated Service in the Internet Architecture: An Overview*,'' IETF   RFC1633, July 1994.
14. R. Braden, *et al.*, ``*Resource ReServation Protocol (RSVP) - Version 1 Functional Specification*,'' IETF RFC2205, Sept. 1997.

15. J. Wroclawski, ``*Specification of the Controlled-Load Network Element Service*,'' IETF RFC2211, Sept. 1997.
16. S. Shenker, R. Braden, and D. Clark, ``*An Architecture for Differentiated Services*,'' IETF RFC2475, Dec. 1998.

Building MPLS VPNs with QoS Routing Capability [1]

Peng Zhang, Raimo Kantola

Laboratory of Telecommunication Technology,
Helsinki University of Technology
Otakaari 5A, Espoo, FIN-02015, Finland
Tel: +358 9 4515454   Fax: +358 9 4512474
Email: {pgzhang@tct.hut.fi, raimo.kantola@tct.hut.fi}

Abstract. Recently MPLS is used for building up VPNs in IP backbone, called MPLS VPNs. In this paper, we discuss issues on finding routes with QoS requirements (i.e., QoS routing) in MPLS VPNs. We first present background on MPLS VPNs as well as QoS routing. Then we discuss both the benefits and problems resulted from introducing QoS routing into MPLS VPNs. We particularly present an architecture of MPLS VPNs with QoS routing capability, on which we discuss some important issues on running QoS routing in MPLS VPNs.

1. Introduction

With the rapid development of the Internet, there arise great interests in the deployment of Virtual Private Networks (VPNs) across IP networks. Many preliminary works have been done in this area. For example, a framework for IP based VPNs is proposed in [1], in which various types of VPNs, their respective requirements and mechanisms for implementations are discussed; An approach for building core VPN services in a service provider's MPLS backbone is presented in [2]; An extension to CR-LDP for VPNs is proposed in [3] by adding an optional VPN-ID TLV to CR-LDP label request message to identify the VPN that the request is meant for. In these documents, MPLS is believed to be a key technology for building up VPNs (i.e., MPLS VPNs) due to a number of reasons as follows.

- MPLS offers fast forwarding capability;
- MPLS connects sites through setting up label switch paths (LSPs) on which traffic engineering can be applied;
- MPLS provides supports for various L2 protocols, e.g., ATM, Frame Relay, etc.;
- MPLS supports signaling protocols, which can facilitate fast configurations of VPNs;
- MPLS is capable of scaling into very large networks.

Meanwhile, QoS is regarded as a key element of any VPN services. For example, services with stable and good qualities in terms of bandwidth and delay are

[1] This work is supported by IPANA project which is carried out in Helsinki University of Technology.

expectedly offered in VPNs. Among various mechanisms of traffic engineering (e.g., traffic scheduling, resource management), QoS routing is one of the enhancing mechanisms for deploying quality classes into the IP networks[4]. The general objective of QoS routing is to improve the efficient utilization of network resources and to provide flexibility in support for various services. Therefore, QoS routing is expectedly used in MPLS VPNs. However, there still lacks insensitive study on this topic.

In this paper, we investigate the issues of QoS routing in MPLS VPNs. In particular, we present an architecture of MPLS VPNs with QoS routing capability as well as some methods for operating QoS routing in MPLS VPNs.

The remainder of this paper is organized as follows. In section 2, we describe the background on MPLS VPNs and QoS routing. In section 3, we discuss the benefits and problems resulted from introducing QoS routing into MPLS VPNs. We present and describe an architecture of MPLS VPNs with QoS routing capability in section 4. In section 5, we present and discuss some issues on operating QoS routing in MPLS VPNs. Some conclusions are given in the final section.

2.   Background on MPLS VPNs and QoS Routing

In this section, we give the general information on MPLS VPNs and QoS routing. We describe the definitions and current status of some components as follows.

2.1   VPNs

A VPN is a set of sites which are attached to a common network (i.e., backbone), applying a set of specific policies (e.g., addressing, security, etc). VPN services are widely used for interconnecting sub-divisions of an organization or a company in multiple areas. VPNs are meant for sharing resources within VPNs.

Although VPN services have appeared for a few years, constructing VPNs across IP backbone is a relatively new topic [1]. There are two different methods to construct VPNs across IP backbone, i.e., CPE (Custom Premises Equipment) based and network based. Most current VPN implementations are based on CPE equipment. VPN capabilities are being integrated into a wide variety of CPE devices, ranging from firewalls to WAN edge routers. On the other hand, there is also significant interest in 'network based VPNs', where the operation of the VPN is outsourced to an Internet Service Provider (ISP), and is implemented on network as opposed to CPE equipment. This method attracts both customers seeking to reduce support costs and ISPs seeking new revenue sources. In this paper, we discuss QoS routing in network based VPNs. However, most of the methods presented in this paper can also apply to CPE based VPNs.

2.2   MPLS

MPLS integrates a label swapping framework with network layer routing [5]. Its basic idea involves assigning short fixed length labels to packets at the ingress to an MPLS cloud (based on the concept of forwarding equivalence classes) and making forwarding decisions accroding to the labels attached to packets throughout the interior of the MPLS domain. Thousands of papers on MPLS have been presented in various aspects including traffic engineering and implementations. MPLS is regarded as a key technology for realizing Differentiated Services (DiffServ) networks.

2.3   MPLS VPNs

VPNs are built up by using MPLS. A MPLS VPN can consists of that are from the same enterprise or from different enterprises and these sites may attach to the same service provider or to different service provider. If more than one different service providers are used, the bilateral or multilateral agreements should be pre-determined.
    Moreover, MPLS based VPNs provide the following benefits [6].
- A platform for rapid deployment of additional value-added IP services, including intranets, extranets, voice, multimedia, and network commerce;
- Privacy and security are equal to layer-2 VPNs by constraining the distribution of a VPN's routes to only those routers that are members of that VPN, and by using MPLS for forwarding;
- Easy management of VPN membership and rapid deployment of new VPNs;
- Increased scalability with thousands of sits per VPN and hundreds of VPNs per service provider;
- Scalable any-to-any connectivity for extended intranets and extranets that encompass multiple businesses.

2.4   QoS

The QoS requirements for a service are generally clarified by a set of parameters such as bandwidth, delay and so on. Offering QoS guaranteed or assured services in the Internet is becoming more and more attractive. Great efforts have been devoted to this field in various aspects, e.g., traffic scheduling, resource management, QoS routing, etc[7].

2.5   QoS Routing

Constraint based routing, a general term of QoS routing, selects routes according to not just a single metric (e.g., hop count) but also additional routing metrics (e.g., bandwidth and delay) and administrative policies (e.g., access authentication). In particular, QoS routing provides support for alternate routing, for instance, if the best existing path cannot admit a new flow, the associated traffic can be forwarded in an adequate alternate path. QoS routing algorithms can prevent traffic shifting from one path to another "better" path only if the current path meets the service requirements of the existing traffic. A framework for QoS routing in the Internet is presented in [8]. QoS routing has been introduced into OSPF as described in [9]. A large number of routing algorithms are summarized in [4]. Some mechanisms for operating inter-domain QoS routing in DiffServ networks are presented in [10].

2.6 QoS Routing in MPLS VPNs

MPLS supports explicit paths and alternative paths so that QoS routing can be naturally used in MPLS VPNs. QoS routing might be used in such cases as finding routes for connecting a number of sites into a VPN or setting up paths for sessions within VPNs. QoS routing is also believed to be one of the key components for supporting QoS in MPLS VPNs.

3. Benefits and Problems of QoS Routing in MPLS VPNs

QoS routing determines routes under the knowledge of network resource availability, as well as the requirements of flows. As a result, the performance of applications is guaranteed or improved in comparison with that without QoS routing. Meanwhile, QoS routing optimizes the resource usage in the network by improving the total network throughput. QoS routing is likely used for constructing an efficient and high performance MPLS VPNs. These benefits might be achieved in a number of ways as follows.
- QoS routing selects feasible routes by avoiding congested nodes or links;
- If workload exceeds the limit of existing paths, QoS routing offers multiple paths for transferring additional traffic;
- If a link or node failure occurs, QoS routing selects alternative paths for quick recovery without seriously degrading the quality.

However, these benefits of QoS routing also incur the cost of developing new routing protocols or extending the existing ones. Moreover, it potentially increases higher communication, processing and storage overheads. It brings out a number of problems as follows[8]:
- What kinds of resource information can be used for determining feasible routes?
- Which protocols are suitable for distributing route and resource information within domain or across multiple domains?
- How to select routes across multiple domains?
- How to balance the complexities and benefits of introducing QoS routing into the real networks?
- In which ways the cost of running QoS routing in MPLS VPN networks can be minimized?

Currently, there lacks deep and broad investigations on these problems although some work have already been carried on[9].

4. An Architecture for QoS Routing in MPLS VPNs

4.1 Architecture

We present the architecture as shown in Figure 1. A MPLS VPN is built up by connecting MPLS sites through tunnels across IP backbone. Each MPLS site has a Bandwidth Broker (BB), which is to exchange route and signaling information and to manage and maintain VPN networks.

A Central Bandwidth Broker (CBB) in IP backbone is likely used, however, not necessarily. If the IP backbone can provide QoS support, CBB performs similar functions as BBs in each MPLS site. BBs of each MPLS site can negotiate with the CBB in order to setup QoS guaranteed tunnels or sessions. CBB performs VPN management in a central way, for example, CBB determines the acceptance of a MPLS site into the MPLS VPN. CBB can be implemented in any router in IP backbone, or virtually in BB of a MPLS site.

Both BB and CBB have two major tasks related to route management:
- Finding routes for connecting a number of sites into a VPN;
- Setting up paths for sessions within VPNs.

The first task has a longer time scale than the second task. In this paper, we intend to focus on the second task.

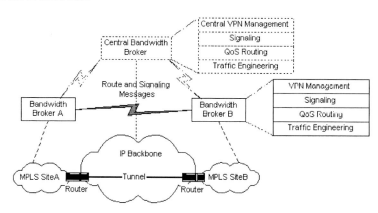

Fig. 1. An Architecture of MPLS VPNs with QoS Routing

Each bandwidth broker consists of a number of components, i.e., VPN Management, Signaling Protocol, QoS Routing and Traffic Engineering. VPN Management performs functions of management and administrative policies, e.g., addressing, access authentication, tunneling management, etc. Signaling Protocol is needed to setup tunnels between MPLS sites or sessions between applications of different MPLS sites. QoS routing is used for finding feasible routes for tunnels or sessions and for maintaining topology of MPLS VPNs. Traffic Engineering includes a number of mechanisms (e.g., classifying, marking, shaping and queuing) for forwarding packets.

In practice, there are several candidates for implementing these components. For VPN Management, SNMP might be used; For Signaling Protocol, CR-LDP or Extended RSVP can be used; For QoS Routing, QOSPF or inter-domain QoS routing might be used; For traffic engineering, Integrated Service or Differentiated Service might be used.

The functions of these components can be understood by depicting the process of setting up a path for a flow with quantitative QoS requirements.

1. When a BB (or CBB) receives a request for a flow, it determines a set of possible routes and then selects a feasible route;
2. Once a path has been found, the BB (or CBB) assures that the flow follows the path;
3. The BB updates its local resource database and broadcasts the route and resource information to other nodes;
4. The BB marks the flow packets and polices the flow;
5. The BB monitors the link state to detect a link failure and performs rerouting in case link failure occurs.

4.2 QoS routing model

Since this paper focuses on QoS routing, we present an implementation of QoS routing component in Figure 2.

As shown in this figure, this model consists of three functional blocks (i.e., Policy Control, Route Computation & Selection, and Routing Information Advertise and Update) and three tables (i.e., VPN topology database, tunnels & sessions table, and tunnels & sessions routing table).

Policy Control exerts specified policies on finding routes and exchanging routing information. Route Computation & Selection determines routes based on the knowledge of topology information and policy constraints.

Routes are computed and saved into tunnels & sessions table for data forwarding. The tunnels & sessions table is used to store information related to specific flows, in terms of traffic parameters, requirements for QoS, etc. Routing Information

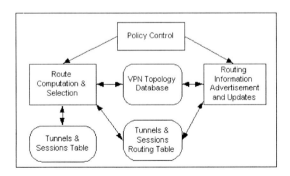

Fig. 2. An Implementation of QoS Routing

Advertisement and Update is in charge of broadcasting routing information (e.g., resource information, policy constraints, routes selected, etc) and updating local database when receiving VPN routing information from other sites.

Here, we introduce two simple routing algorithms: Lowest Cost (LC) algorithm and Widest Bandwidth (WB) algorithm[11].

Consider a directed graph $G=(N, E)$ with numbers of nodes N and numbers of edges E, in which each edge (i, j) is weighted by two parameters, b_{ij} as the available bandwidth and c_{ij} as the cost. The cost is an additive parameter, e.g., hop number, delay, etc. Let $b_{ij} = 0$ and $c_{ij} = \infty$ if edge (i, j) does not exist in the graph.

Given any directed path $p= (i,j,k,\cdots, l, m)$, define $b(p)$ as the bottleneck bandwidth of the path, i.e., $b(p) = \min[b_{ij}, b_{jk}\cdots, b_{lm}]$, and define $c(p)$ as the sum of the cost, i.e., $c(p) = c_{ij} + c_{jk}+\cdots+ c_{lm}$. Given two nodes i and m of the graph and two constraints B and C. To LC algorithm, the QoS routing problem is to find a path p^* between i and m so that $b(p) \geq B$ and $c(p) \leq C$. To WB algorithm, the QoS routing problem is then to find a path p^* between i and m so that $b(p) \geq B$ and the path has the widest bandwidth, and if there are more than one widest paths the path with the lowest cost is selected. Let C_i be the estimated cost of the path from source node s to destination node t. Let B_i be the estimated bandwidth of the path from source node s to destination node t.

- LC algorithm
  Step 1: Set $c_{ij} = \infty$, if $b_{ij} < B$;
  Step 2: Set $L= \{s\}$, $C_i = c_{si}$ for all $i \neq s$;
  Step 3: Find $k \notin L$ so that $C_k = \min_{i \notin L} C_i$;
      If $C_k > C$, no such a path can be found and the algorithm terminates,
      If L contains node t, a path is found and the algorithm terminates.
      $L:=L\cup\{k\}$.
  Step 4: For all $i \notin L$, set $C_i:=\min[C_i, C_k+c_{ki}]$;
  Step 5: Go to Step 3.
- WB algorithm:
  Step 1: Set $b_{ij} = 0$, if $b_{ij} < B$;
  Step 2: Set $L = \{s\}$, $B_i = b_{si}$ and $C_i = c_{si}$ for all $i \neq s$;
  Step 3: Find set $K \notin L$ so that $B_K = \max_{i \notin L} B_i$;
  Step 4: If K has more than one element, find $k \in K$ so that $C_i (s, \cdots, k, t) = \min_{i \in K} [C_{(s, \cdots, k, t)}]$. $L:= L\cup\{k\}$. If L contains all nodes, the algorithm is completed.
  Step 5: For all $i \notin L$, set $B_i:= \max[B_i, \min[B_k, b_{ki}]]$;
  Step 6: Go to Step 3.

Both algorithms first eliminate the link whose available bandwidth is below the required bandwidth and produces a new graph. Then, the former calculates path with the lowest cost by using Dijkstra's algorithm while the latter calculates the path with the widest bandwidth by using a variation of Dijkstra's algorithm.

The other important topic of QoS routing is cost. The cost of QoS routing includes three parts, that is, storage cost, computation cost and distribution cost. Usually, it mainly depends on the distribution cost. Therefore, the updating algorithm of route and resource information is very important. Here, we briefly present two updating algorithms as follows.

- The first algorithm, called Period Based algorithm (PB), performs update periodically.

- The second algorithm, called Threshold Based algorithm (TB), performs update when the variation of available bandwidth of the link exceeds a configured threshold.

5. Issues on running QoS Routing in MPLS VPNs

In this section, we present and discuss some issues on running QoS routing in MPLS VPNs in the following subtopics.

- Distributing label and VPN attributes

In MPLS VPNs, labels and VPNs attributes (e.g., label ID, VPN ID, etc) can be distributed and maintained by using QoS routing protocols. Extensions to BGP for carrying label and VPN attributes in MPLS VPN are proposed in [2, 12]. One can construct different kinds of VPNs, by setting up the Target and Origin VPN attributes.

For example, label distribution can be piggybacked in the BGP Update message by using the BGP-4 Multiprotocol Extensions attribute[13]. The label is encoded into the NLRI field of the attribute. Label mapping information is carried as part of the Network Layer Reachability Information (NLRI) in the Multiprotocol Extensions attributes.

Fig. 3. Format of NLRI for label distribution in BGP-4

The Network Layer Reachability information is encoded as one or more triples of the form <label, length, prefix> as shown in Figure 3. The Length field indicates the length in bits of the address prefix plus the label(s); The Label field carries one or more labels; The Prefix field contains address prefixes followed by enough trailing bits to make the end of the field fall on an octet boundary.

The other alternative uses signaling protocol for distributing label and VPN attributes.
- Distributing route and topology information
QoS routing can be used for maintaining VPN topology within VPN. It is used for understanding not only the topology information but also resource states in VPNs, in which deliberate control and management can be applied. The resource states can be clarified with a number of parameters, e.g., bandwidth, delay, etc.

For example, BGP-4 can be extended for supporting traffic engineering[14]. The BGP update message will contain a new Optional Transitive attribute called TE Weight. The traffic engineering weights act as a cost or distance function, describing the quality of a path to a destination network in traffic engineering terms. Each TE Weight type could be:
– Maximum Bandwidth Available
– Maximum Number of IGP Hops

– Maximum Transit Delay
– Color
– Etc.

Therefore, the routes with quality information are distributed, then BGP Route Selection process is extended to select routes on the basis of the TE weights.

- Finding feasible routes

There are a number of algorithms for finding QoS routes in a single domain[4]. The two routing algorithms presented in section 4.2 are very promising and expectedly used in the real network because of their simplicities. On the other hand, both algorithm use *bandwidth* as the key parameter because in many cases *bandwidth* dominates the quality of service.

Moreover, some mechanisms for operating inter-domain QoS routing are proposed in [10]. In this case, Figure 4 illustrates the main functions and the procedures for setting up paths across domains. Signaling entity (SE) is a signaling agent of a MPLS site, while routing entity (RE) is a routing agent of a MPLS site running inter-domain QoS routing protocols. SE's functions include outgoing and incoming parts. The outgoing part collects QoS requests from interior routers and determine to initiate path setup requests; The incoming part processes path setup requests from other SEs. SE queries its local RE for external routes, and RE replies SE with next hops or whole routes. Note that the *path setup request* message usually contains the specifications of the flow and the requirements for QoS.

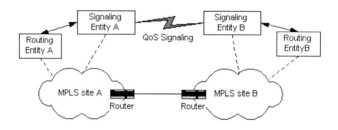

Fig. 4. Setting up paths across domain

We present five mechanisms for operating QoS routing across domains [10]:
1. SE based - crankback
2. SE based – flooding
3. Cache based Routing
4. *D*-hop resource routing
5. RE based source routing

For brevity, we just describe the first mechanism as follows.

When SE receives a *path setup request* message from an upstream SE, it first requests its local RE for next hop. If RE replies a non-blank next hop, SE checks if there is enough available resource on the link to that hop. If yes, SE adds itself to route list of the path and sends a request to that hop. If no, it requests the local RE for next hop again. If SE has queried RE for K times, SE sends a *path setup failure* message upstream. Here, K is a given constant. If SE receives a *path setup failure* message from downstream SE, it also requests its local RE for next hop again. A feasible route will be found until the request reaches the destination. In this case, resource reservation is proceeded downstream.

This mechanism does not require RE to understand the global resource information, that is, there is no need for global topology and resource information database. As a result, advertising and updating resource information can be avoided. The current inter-domain routing protocol (i.e., BGP) can be directly used, except minor modifications on interface with SE.

6. Conclusions

MPLS is likely used in VPNs due to its distinguished merits, e.g., fast forwarding, tunneling, etc. QoS routing is naturally used in MPLS VPNs for providing feasible routes with considerations on QoS constraints. QoS routing is beneficial for developing QoS guaranteed MPLS VPNs across IP networks. In this paper, we investigate both benefits and problems when introducing QoS routing into MPLS VPNs. Particularly, we present an architecture of MPLS VPNs with QoS routing capability and discuss some issues on running QoS routing in MPLS VPNs. However, there are still a great number of open research problems concerning QoS routing in MPLS VPNs, e.g., methods of advertising and updating resource information, algorithms of computing routes, etc.

References

1. B. Gleeson, et al: A Framework for IP Based Virtual Private Networks. IETF RFC2764 (2000)
2. K. Muthukrishnan, et al: Core MPLS IP VPN Architecture. IETF Draft (2000)
3. P. Houlik, et al: Extensions to CR-LDP for VPNs. IETF Draft (2000)
4. Chen, S., Nahrstedt, K.: An Overview of Quality of Service Routing for Next-Generation High-Speed Networks: Problems and Solutions. IEEE Networks, Vol. 12, No. 6 (1998) 64-79
5. R. Callon, et al: A Framework for MPLS. IETF Draft (1999)
6. Cisco VPN Solution Center: MPLS Solution User Guide. Chapter 1 (1999) page 3-4
7. S. Blake, et al: An Architecture for Differentiated Services. IETF RFC2475 (1998)
8. E. Crawley, et al: A Framework for QoS-based Routing in the Internet. IETF RFC2386 (1998)
9. G. Apostolopoulos, et al: QoS Routing Mechanisms and OSPF Extensions. IETF RFC2676 (1999)
10. P. Zhang, R. Kantola: Mechanisms for Inter-Domain QoS Routing in Differentiated Service Networks. Accepted by QoS of future Internet Services (QofIS'2000). Berlin (2000)
11. Z. Wang and J. Crowcroft: Quality of Service Routing for Supporting Multimedia Applications. IEEE JSAC, Vol.14, No.7 (1996) 1228-1234
12. P. Houlik, et al: Carrying Label Information in BGP-4. IETF Draft (2000)
13. T. Bates, et al: Multiple Extensions for BGP-4. IETF RFC2283. (1998)
14. B. Abarbanel, S. Venkatachalam: BGP-4 Support for Traffic Engineering. IETF Draft (2000)

A Linux Implementation of a Differentiated Services Router

Torsten Braun, Hans Joachim Einsiedler[1], Matthias Scheidegger, Günther
Stattenberger, Karl Jonas[2], Heinrich J. Stüttgen[2]

Institute of Computer Science and Applied Mathematics, University of Berne
Email: [braun|mscheid|stattenb]@iam.unibe.ch
[1]now: T-Nova Deutsche Telekom Innovationsgesellschaft mbH, Berkom
Email: einsiedler@berkom.de
[2]Computer & Communications Network Product Development Laboratories Heidelberg, NEC
Europe Ltd.
Email: [karl.jonas|stuttgen]@ccrle.nec.de

Abstract. The Internet Engineering Task Force (IETF) is currently working on the
development of Differentiated Services (DiffServ). DiffServ seems to be a
promising technology for next-generation IP networks supporting Quality-of-
Services (QoS). Emerging applications such as IP telephony and time-critical
business applications can benefit significantly from the DiffServ approach since the
current Internet often can not provide the required QoS. This paper describes an
implementation of Differentiated Services for Linux routers and end systems. The
implementation is based on the Linux traffic control package and is, therefore, very
flexible. It can be used in different network environments as first-hop, boundary or
interior router for Differentiated Services. In addition to the implementation
architecture, the paper describes performance results demonstrating the usefulness
of the DiffServ concept in general and the implementation in particular.

Keywords: Quality-Of-Service (QOS), Internet Protocol (IP), Differentiated Services
(Diifserv), Assured Service, Assured Forwarding (AF), Premium Service, Expedited
Forwarding (EF), Linux

1 Differentiated Services

For scalable QoS support in the Internet, the IETF is developing the Differentiated
Services Architecture [2]. The IETF focuses on two services - *Assured Service* and
Premium Service [3]. For discrimination of these different services from the currently
used *Best Effort* service, the IETF proposed a special byte in the Internet Protocol (IP)
header, the so-called Differentiated Services Byte (DiffServ byte). This byte contains 6
bits called DiffServ code-point (DSCP). DSCPs describe the so-called per-hop behavior
(PHB), which is the externally observable forwarding behavior applied to a DiffServ flow
at DiffServ capable nodes [5].

Premium Service (Expedited Forwarding, EF [7]) is understood as a Virtual Leased
Line service where users cannot exceed the bandwidth. Premium Service is designed in
order to achieve low queuing delay, e.g. for real-time applications like IP telephony.
Assured Service (Assured Forwarding, AF [8]) assures the customer a certain amount of
bandwidth. The bandwidth cannot be guaranteed but packets are labeled with higher
priority for transmission over the network. Four *Assured Service* classes have been
defined [6] with three dropping precedence levels (low, medium and high) each. The

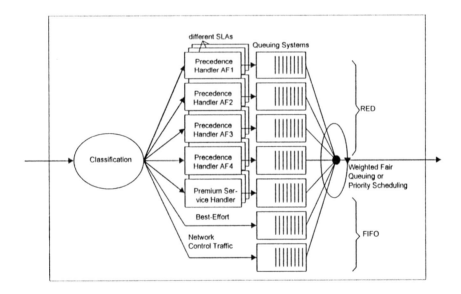

Fig. 1. Implementation Architecture of a DiffServ boundary router

dropping precedence levels might be increased but the packet should stay in the same class. The different classes are handled independently from each other, but packets of one micro-flow are mapped to the same class.

At a DiffServ node an incoming packet is first classified by a classifier, which identifies the service to be supported. A *Behavior Aggregate* (BA) classifier selects packets based on the DiffServ code-points only. The *Multi-Field* (MF) classifier looks also into other IP or higher layer header fields. The classifier forwards the packet to the service-dependent traffic conditioner which may include a meter, a marker, a shaper and a dropper [10]. With these components several kinds of routers can be built. A Differentiated Services network requires four different kinds of routers:

- The first hop router is placed adjacent to the sender host. Packets are classified (BA or MF) and marked per flow according to a user profile. Service handlers have to ensure conformance of the flows with the predefined profiles.
- Egress routers are located at the border between two DiffServ domains such as Internet Service Providers (ISPs). They have to make sure that the leaving traffic behaves according to the Service Level Agreement (SLA) negotiated with the adjacent domain.
- Ingress routers are also located at the entry points of a DiffServ domain and perform BA/MF classification. Their policing mechanisms restrict the incoming traffic according to the SLAs. Very often, routers at the boundary of a domain (boundary router) work as ingress routers for incoming traffic and as egress routers for outgoing traffic.
- Interior routers within DiffServ domains are responsible for forwarding according to the service requirements of the packets. They have to give higher priority to Differentiated Services packets than to *Best Effort* ones. Interior routers consider aggregated Differentiated Service flows only.

2   DiffServ Router Implementation Architecture

Figure 1 shows the developed implementation architecture of our DiffServ boundary router. After classification, the traffic is processed by the corresponding service handlers such as *Premium Service* shapers, policers, or *Assured Service* dropping precedence handlers and then forwarded to the associated queuing systems. For *Premium Service* traffic, network control traffic and *Best Effort* traffic, the queuing system can be a simple FIFO queue, while *Random Early Detection* (RED) is proposed for *Assured Service*. An output scheduling mechanism such as Priority Scheduling or Weighted Fair Queuing is required for selection of packets to be sent via the outgoing interface.

After classification at interior routers the packets are immediately directed to the outgoing queuing systems. While in a first hop router and in an ingress router each connected customer needs a BA/MF classifier and a conditioner, we only have BA classification in an egress router. There is only the need for one classifier and one conditioner per class.

Table 1. Structure of the DiffServ table

source address	source address mask	source port	destination address	destination address mask	destination port	protocol	input DSCP	output DSCP	Bandwidth AF medium dropping precedence	Bandwidth EF / AF low dropping precedence
32 bits	32 bits	16 bits	32 bits	32 bits	16 bits	8 bits	8 bits	8 bits	32 bits	32 bits

A DiffServ profile can be implemented based on the IP header fields such as the IP version, source /destination address including network masks, protocol type, source/destination port and DSCP. In addition to the IP header fields, a profile table (Figure 2) includes the parameters such as bandwidth values. The profile influences marking and dropping packets within DiffServ nodes. A *Premium Service* (EF) profile requires a peak bandwidth value which is used to compute the amount of tokens that are left for the service. For *Assured Service* (AF) bandwidth values for low and medium dropping precedence are necessary.

Figure 3 shows the logical structure of an egress router or a first hop router for *Premium Service*. After classification, the packets are stored in a queue until tokens become available. Then, the packet can be sent to the outgoing queue. In the case that more packets arrive in the buffer than packets can be sent, packets must be discarded. In egress routers, it is expected that the senders (or the upstream routers) will send with the agreed rate so that the queue can be kept small [13]. In first hop routers, the buffers have to be bigger because of bursty traffic from non-DiffServ clients.

In a Premium Service ingress router, the packets are discarded as shown in Figure 4. There, we have no buffer (queue) in the traffic conditioner. The arriving packets are stored in the output queuing system, if tokens are available. Otherwise, they are discarded immediately after classification.

Another important difference between the various DiffServ router types is that in an egress router, we usually have BA classification only while first hop or ingress routers have BA/MF classification because they are connected to external customers, which have negotiated SLAs with the provider of the DiffServ domain [11].

Figure 5 shows the architecture of the Assured Service Handler which is an implementation of the three color marking concept [1]. Token buckets support the decision, whether the dropping precedence of a packet must be modified before forwarding to the *Assured Service* queuing system. The *High Dropping Precedence* packets are handled as *Best Effort* traffic but are still marked as *Assured Service* traffic. The restriction of the high dropping precedence bandwidth will be done in the outgoing queuing system. This has to provide some kind of policing functionality. The *Assured Service* queuing mechanism is an extended RED mechanism [9] with three dropping curves for each dropping precedence as depicted in Figure 6. The dropping probability is calculated by the following formula:

$$
dp^{drop} = \begin{cases} 0 & for \quad ql < th_{min}^{drop} \\ \dfrac{ql - th_{min}^{drop}}{th_{max}^{drop} - th_{min}^{drop}} & for \quad th_{min}^{drop} \leq ql \leq th_{max}^{drop} \\ 1 & for \quad ql > th_{max}^{drop} \end{cases} \tag{1}
$$

dp	Dropping probability with $dp^{highDrop} \leq dp^{mediumDrop} \leq dp^{lowDrop}$
$drop$	Dropping precedences (low, medium and high)
th_{min}	Minimum threshold of the queue
th_{max}	Maximum threshold of the queue

$$
th_{Start-drop} = th_{min}^{HighDrop}
$$

$$
th_{Hard-drop} = th_{max}^{LowDrop}
$$

In our implementation, we try to protect the network control traffic against dropping in the way that its priority is just below the priority of *Premium Service*. This is valid for both Priority Scheduling or Weighted Fair Queuing output queuing, which are the two options implemented for output queuing (see Section 3.2)

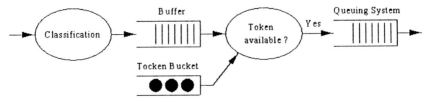

Fig. 2. Egress or first hop *Premium Service* route

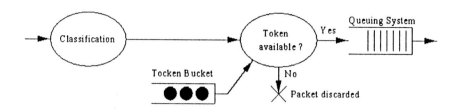

Fig. 3. Ingress *Premium Service* router

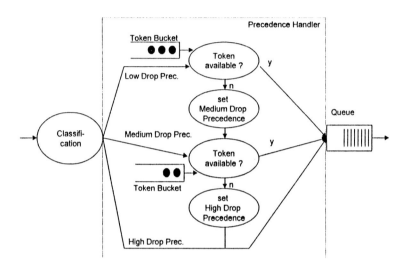

Fig. 4. Functionality of the precedence handler

Titel:
TrioQueue1.eps
Erstellt von:
fig2dev Version 3.2 Patchlevel 0-beta3
Vorschau:
Diese EPS-Grafik wurde nicht gespeichert
mit einer enthaltenen Vorschau.
Kommentar:
Diese EPS-Grafik wird an einen
PostScript-Drucker gedruckt, aber nicht
an andere Druckertypen.

Fig. 5. RED queue for three dropping precedence

3   Linux Implementation of DiffServ Routers

Linux kernels allow a wide variety of traffic control functions [13]. Several DiffServ modules have been made available for Linux [14, 15]. The traffic control functions are either hard-coded during compiling the kernel or they can be loaded dynamically after

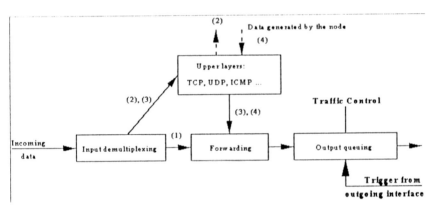

Fig. 6. Network packet processing

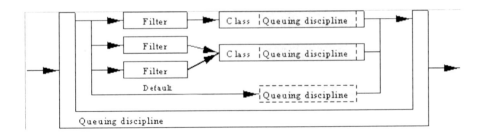

Fig. 7. Queuing discipline with filters and classes

system initialization or during run-time. Our implementation takes advantage of the second option and allows to change the configuration without resetting the node. New modules such as shapers, markers, meters and droppers can be added or removed via the command line. The traffic control allows to compile a single kernel for a node, which can be configured as DiffServ first hop, ingress, egress and interior routers, supporting flexible SLAs.

3.1 Linux Networking Support

Figure 7 shows IP packet processing in the Linux kernel. A router forwards received packets directly to the network, e.g. to another interface (1). If the node is also an end system (server, workstation, etc.) or an application level gateway, the packets destined to it are passed to higher layers of the protocol stack for further processing (2). This can also include manipulation of fields and then forwarding to the network (3) again. An end system can generate packets by itself, which then will be sent through the protocol stack to the forwarding block (4). The forwarding component does not only include the selection of the output interface but also the selection of the next hop, encapsulation, etc. From there, the packet is queued for the particular interface. This is the point of *traffic control* execution. Manipulation, such as delaying packets, changing header fields, dropping etc. can be done there. After traffic control has released the packet, the particular network device can pick it up for transmission. The output queuing block is triggered by the output interface. For processing the next packet, the interface sends a start signal to the output queuing block.

After compilation of t c and loading the modules, the code components can be added via the command line or a management tool (a Shell or Perl script) to the outgoing queuing block. The code consists of queuing disciplines, classes (the identification of a queuing discipline), filters, and policing functions (within filters and classes). Figure 8 shows an example of a queuing discipline. Packets, which are forwarded over the same interface, may desire different treatment. They have to be enqueued into different queuing disciplines. For an enqueued packet the called queuing discipline runs one filter after the other until there is a match with a class. Otherwise, the default queuing discipline is used. In the case of a match, the packet is enqueued in the queuing discipline related to the class for further manipulation of the packet. Different filters can point to the same class. Policing functions are required in the queuing disciplines to ensure that traffic does not exceed certain bounds. For example, for a new packet to be enqueued, the policing component can decide to drop the currently processed packet or it can refuse the enqueuing of the new one. Each network device has an associated queuing discipline, in which the packets are stored in the order they have been enqueued. The packets are taken from the queue as fast as the device can transmit them.

3.2   DiffServ Modules for Linux

The framework of our implementation mainly focuses on the enqueue and dequeue components of the queuing discipline structure because these components are the right place for the Differentiated Services implementation. The implementation of six queuing disciplines was necessary to cover all four kinds of DiffServ nodes. Some queuing disciplines work with profiles. During initialization of these queuing disciplines, the information is copied from a file into the memory, from where the queuing discipline can access it.

The *DiffServ Service Handler* sets the Class Selector Codepoint according to a profile. Packets that do not match with the profile are forwarded as *Best Effort* packets. Network control traffic is forwarded untouched as well. During enqueuing a function is called, which compares the packet header with the profile and then marks the packet with the respective service.

The *DiffServ Classifier* splits the traffic into the seven service branches (*Premium Service*, network control traffic, four *Assured Service* classes and *Best Effort* service) according to the DSCP. For dequeuing packets from the different queues, priority scheduling or a weighted fair queuing variant can be used.
- For priority scheduling, the packets are dequeued depending on the priority parameters given to the queuing discipline during initialization. The queue with the highest priority is the first queue that will be accessed. A queue can send only if all queues with a higher priority are empty. The recommended default priority sequence is Premium Service, network control traffic, Assured Services Classes 1-4, Best-Effort.
- The Weighted Fair Queuing (WFQ) variant gives always highest priority to Premium Service packets over network control packets. The remaining bandwidth is shared among the Assured Service and Best-Effort packets according to the configured parameters.

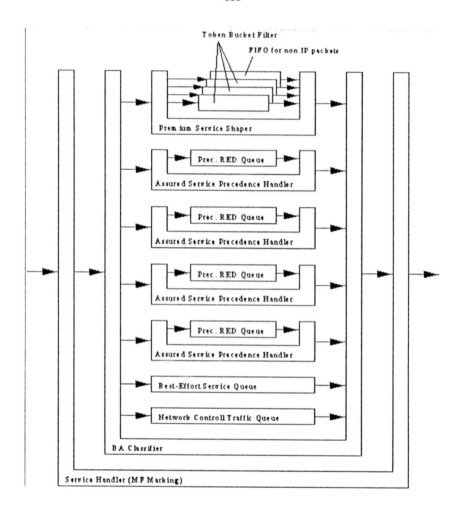

Fig. 8. First hop router

The *Assured Service Precedence Handler* checks incoming packets whether they are in-profile by measuring the packet size (in bytes) against a token bucket. An in-profile packet is forwarded and the tokens are decremented by the size of the packets. Otherwise, the packet is re-marked with a higher dropping precedence. In the case of medium dropping precedence, the packet is measured against the respective token bucket. If no tokens are available, the packet is forwarded directly into the outgoing queue and marked with high dropping precedence. The marked packets might then be discarded in the following Three Way RED queue.

The *Three Way RED Queue* drops packets according to the RED parameter values. This is done in the enqueue component. The dropping probability is calculated depending on the dropping precedence of incoming packets. The packet will then either be dropped or forwarded to the following FIFO queue in which the packet is stored for dequeuing. Two parameter sets are required. A limit defines the maximum number of packets, the

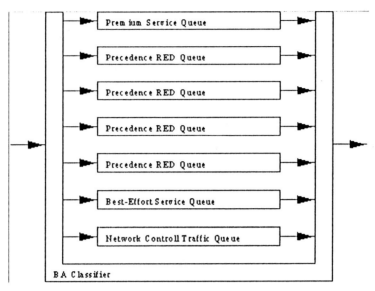

Fig. 9. Interior router

Three Way RED queue can buffer, and floating-point numbers define the thresholds of the discarding curves.

The **Premium Service Policer Handler** polices the *Premium Service* traffic according to *Premium Service* profiles. An arriving packet will be forwarded to the FIFO queue, if there is a match with the profile and if there are enough tokens available in the respective token bucket. Otherwise, the packet is not conforming to the profile and is discarded.

Since we have to use a shaper for each *Premium Service* flow, there is the need for a **Premium Service Shaper Handler** that forwards the packet to these shapers. IP packets matching a profile are forwarded to the respective shaper and are discarded otherwise. This shaper module is part of the standard Linux kernel distribution [20, 22].

3.3 Structure of DiffServ Routers

Figure 9 shows the structure of a first-hop router with MF and BA classification. For *Premium Service*, shaping is used, the Three Way RED queue has been selected for the different AF classes. An egress router looks quite similar but does not need a service handler for MF marking. An ingress router differs from the egress router in having policing instead of shaping for *Premium Service* and in having a service handler for BA/MF re-marking.

Figure 10 shows the simplicity of an interior router. The packets are classified in the BA classifier and forwarded to the corresponding queue of each service.

4 Performance Measurements

For performance measurements we used the `ttcp` tool and a self-written UDP socket program for the generation of aggressive bursty flows to several destinations. We configured one router as first hop and ingress router in order to test the interoperation between the queuing disciplines. For the measurements we used `tcpdump` and shell scripts processing the `tcpdump` traces.

The following setup has been chosen for the measurements with two flows: The host `elmer` has been the source of the DiffServ flows (*Assured* or *Premium Service*) and `weasel4` has been the destination. The *Best Effort* background traffic has been sent from `elmer` to `weasel`. The background traffic has been generated for each measurement and filled up the rest of the available bandwidth on the 10 Mbps link. In the sessions with three flows, elmer has been the source of the third flow and `weasel5` has been the destination. `weasel`, `weasel4` and `weasel5` were different logical interfaces using the same Ethernet interface of one host. The traffic source `elmer` was connected via an 100 Mbps link. The router had an incoming 100 Mbps interface and formed by the 10 Mbps outgoing interface a bottleneck.

Table 1 shows measurements of a *Best Effort* background flow and a *Premium Service* flow with different bandwidth values for shaping and for policing.

Table 2 shows the measurements of two *Premium Service* flows through two shapers together with a *Best Effort* background flow.

Table 3 shows measurements with an *Assured Service* flow and a *Best Effort* background flow. The *Assured Service* parameters have been as follows:

- Queue length: 10 packets
- Low dropping precedence, Start: 0.9, End: 1.0
- Medium dropping precedence, Start: 0.1, End: 0.5
- High dropping precedence, Start: 0.0, End: 0.1

Table 4 shows measurements with constant Three Way RED queue parameters but different bandwidth values for the Precedence Handler. We had an *Assured Service* flow and a *Best Effort* background flow.

Assured Service parameters:

- Queue length: 20 packets
- Low dropping precedence, Start: 0.9, End: 1.0
- Medium dropping precedence, Start: 0.2, End: 0.5
- High dropping precedence, Start: 0.0, End: 0.2

Finally, we mixed *Assured Service* traffic with *Best Effort* traffic and *Premium Service* traffic, the last one was policed (ingress router setup). We had an *Assured Service* flow, a *Premium Service* flow, and a *Best Effort* background flow.

Assured Service parameters:

- Queue length: 20 packets
- Low dropping precedence, Start: 0.9, End: 1.0
- Medium dropping precedence, Start: 0.1, End: 0.5

High dropping precedence, Start: 0.0, End: 0.1

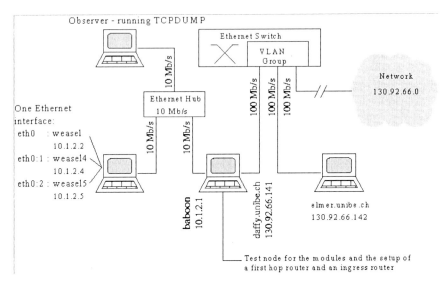

Fig. 10. Test network

Table 2. *Premium Service* shaping and policing with different bandwidth values

	Bandwidth (setup)	Number of Packets	Achieved Bandwidth	Time Period
Shaper	64 kb/s	3570	65.7 kb/s	628.39 s
		4708	65.6 kb/s	830.29 s
	128 kb/s	6372	131.3 kb/s	561.78 s
		6350	131.4 kb/s	559.35 s
Policer	800 kb/s	41408	799.8 kb/s	599.35 s
		32229	799.2 kb/s	466.80 s
	1.28 Mb/s	133105	1.280 Mb/s	1203.74 s
		126654	1.279 Mb/s	1146.25 s

Table 3. Two parallel *Premium Service* shapers in parallel

	Bandwidth (setup)	Number of Packets	Achieved Bandwidth	Time Period
Shaper	128 kb/s	8404	130.8 kb/s	743.78 s
	64 kb/s	4217	65.6 kb/s	

Table 4. Assured Service Flows

Dropping Precedence	Bandwidth (setup)	Number of Packets	Achieved Bandwidth	Time Period
low	800 kb/s	58201	799.5 kb/s	842.73 s
medium	640 kb/s	46559	639.5 kb/s	
high	-	0	0	

Table 5. Different precedence handler bandwidths

Dropping Precedence	Bandwidth (setup)	Number of Packets	Computed Bandwidth	Time Period
low	800 kb/s	63850	798.9 kb/s	
medium	640 kb/s	51006	638.2 kb/s	925.17 s
high	-	190059	2378.1 kb/s	
low	1280 kb/s	70555	1276.6 kb/s	
medium	960 kb/s	52321	945.2 kb/s	640.80 s
high	-	87669	1583.7 kb/s	

Table 6. Ingress router with *Premium, Assured* and *Best Effort* Service

Service	Dropping Precedence	Bandwidth (setup)	Number of Packets	Achieved Bandwidth	Time Period
Assured Class I	Low	1280 kb/s	44812	1276.8 kb/s	406.29 s
	Medium	960 kb/s	32212	917.8 kb/s	
	High	-	13454	383.3 kb/s	
Premium	Policer	1280 kb/s	44813	1276.8 kb/s	
Assured Class I	Low	1280 kb/s	72544	1278.8 kb/s	656.70 s
	Medium	960 kb/s	52886	932.2 kb/s	
	High	-	23740	418.5 kb/s	
Premium	Policer	1280 kb/ s	71349	1257.7 kb/s	

5   Related Work

There are currently several DiffServ implementations under Linux being developed, e.g., the KIDS implementation from University of Karlsruhe [16]. The most similar one to our implementation is the implementation described in [14] which we will call the EPFL implementation hereafter. This and our implementation are both based on the Linux traffic control package. While for our implementation sophisticated DiffServ queuing and scheduling components such as the Three-Color-Marking for Assured Service have been developed, the EPFL implementation tries to use more general components not tailored to DiffServ. The classification of the EPFL is more flexible but required to modify core

Linux data structures while our implementation avoided this. For output queuing, we developed a WFQ variant based on the bad performance behavior experienced from other available output scheduling mechanisms. Another significant difference is the kind of configuration of both implementations. The EPFL implementation requires rather long and more complex tc configuration scripts. By using ASCII configuration tables we believe that our approach simplifies the configuration of a DiffServ router by human users. Our implementation also allowed to integrate a layer-4 flow detection mechanism [12]. In addition, special queuing disciplines for ATM have been implemented that allow to replace software shaping and policing by ATM hardware [17].

6 Conclusions and Outlook

This paper described a DiffServ implementation for Linux performing DiffServ processing at the egress point of a router or an end system. The measurements clearly show the usefulness of our implementation architecture. In addition, some Differentiated Service processing such as policing could be located at the ingress interface of a router. This would allow to perform traffic conditioning functions on the whole traffic received from a single DiffServ domain, e.g. from a single customer. Otherwise, if the traffic is spread over several output interfaces, the aggregate traffic can not be policed correctly. More detailed performance measurements allowing the comparison with the KIDS implementation will be published in a subsequent paper.

Acknowledgements

The implementation platform used at the University of Berne has been funded by NEC Europe Ltd. the SNF R'Equip project no. 2160-053299.98/1, and the foundation "Förderung der wissenschaftlichen For-schung an der Universität Bern". NEC Europe Ltd. funded several persons involved in this project at University of Berne. The authors are grateful to Werner Almesberger (EPF Lausanne) for constructive discussions on Linux related issues.

References

1. J. Heinanen, R. Guerin: "A Single Rate Three Color Marker", Internet Draft draft-heinanen-diffserv-srtcm-01.txt, May 1999.
2. IETF-DiffServ-Working Group, "Differentiated Services for the Internet." http://www.ietf.org/html.charters/diffserv-charter.html/.
3. K. Nichols, V. Jacobson, and L.Zhang, "A Two-bit Differentiated Services Architecture for the Internet", Internet RFC 2638, July 1999.
4. K. Nichols, S. Blake, F. Baker, and D. L. Black, "Definition of the Differentiated Service Field (DS Field) in the IPv4 and IPv6 headers," Internet RFC 2474, December 1998.
5. M. Borden and C. White, "Management of PHBs", Internet Draft: draft-ietf-diffserv-phb-mgmt-00.txt, September 1998.
6. J. Heinanen, F. Baker, W. Weiss, and J. Wroclawski, "Assured Forwarding PHB Group", Internet RFC 2597, June 1999.
7. V. Jacobson, K. Nichols, and K. Poduri, "An Expedited Forwarding PHB", Internet RFC 2598, June 1999.

8. D. Clark and J. Wroclawski, "An Approach to Service Allocation in the Internet", Internet Draft draft-clark-diff-svc-alloc-00.txt, July 1997.

9. S. Floyd and V. Jacobson, "Random Early Detection Gateways for Congestion Avoidance" in IEEE/ACM Transactions on Networking, Vol.1 N.4, pp. 397-413, August 1993.

10.S. Blake, D. Black, M. Carlson, E. Davies, Z. Wang, and W. Weiss, "An Architecture for Differentiated Services", Internet RFC 2475, December 1998.

11.Y. Bernet, D. Durham, and F. Reichmeyer, "Requirements of Diff-Serv Boundary Routers", Internet Draft draft-bernet-diffedge-01.txt, Nov. 1998.

12.T. Harbaum, M. Zitterbart, F. Griffoul, J. Röthig, S. Schaller, H. J. Stüttgen: Layer 4+ Switching with QoS support for RTP and HTTP, Proceedings of IEEE Globecom Conference, Rio de Janeiro, Brazil, December 1999

13.A. Kuznetsov, "Traffic Control Software Package for Linux." ftp://ftp.inr.ac.ru/ip-routing/.

14.W. Almesberger, J. H. Salim, and A. Kuznetsov, "Differentiated Services on Linux", Internet Draft: draft-almesberger-wajhak-diffserv- linux-01.txt, June 1999.

15.W. Almesberger, "Linux Traffic Control - Implementation Overview." ftp://lrcftp.epfl.ch/pub/people/almesber/pub/tcio-current.ps.gz, Nov. 1998.

16.K. Wehrle, R. Bless: "Evaluation of Differentiated Services using an Implementation under Linux", Proceedings of the International Workshop on Quality of Service (IWQOS'99), London, UK, May 31 - June 4, 1999.

17] T. Braun, A. Dasen, M. Scheidegger, K. Jonas, H. Stüttgen: Implementation of Differentiated Services over ATM, IEEE Conference on High-Performance Switching and Routing, Heidelberg, June 26-29, 2000.

Design of a Multi-layer Bandwidth Broker Architecture

George A. Politis, Petros Sampatakos, Dr. Iakovos ,S. Venieris

National Technical University of Athens
Greece

Abstract. Internet is widely known for lacking any kind of mechanism for the provisioning of Quality of Service guarantees. The Internet community concentrates its efforts on the Bandwidth Broker architecture towards this problem. This paper presents a design model of a multi-layer Bandwidth Broker architecture that introduces a Resource Control Layer, which is divided into two sub-layers. The upper one is responsible for the overall network administration, while the lower one performs per-flow policy-based admission control. The design models, the mechanisms, and algorithms adopted in this architecture will be delineated.

Keywords: Bandwidth Broker, Quality of Service, Integrated Services, Differentiated Services, Resource Control Point, Resource Control Agent, Application Middleware.

1. Introduction

Internet is the technology that has become part of our every-day life over the past years and gains significant momentum day by day. Although it started as an experiment [1,2], nowadays, it is a serious business and it aims to be the integrated infrastructure that will concentrate most or even all of the services, existing and future ones. However, the protocols and mechanisms of the current Internet technology seem to be insufficient for delivering the traffic of the arising and demanding multimedia applications with the appropriate Quality of Service (QoS) characteristics, and thus enhanced mechanisms have to be deployed to provide a QoS-enabled Internet infrastructure.

Despite the notion that many have adopted, QoS is not solved merely by increasing the capacity of the links, since there are always merging points in the network that inevitably lead to congestion situations. But first, it should be clarified what this term means; a working definition states: "*IP QoS enables a network to deliver a traffic flow end to end with the guaranteed maximum delay and guaranteed rate required by the user process, within agreed error boundaries*" [3]. In order to bring QoS into the network, three components have to be deployed: traffic handling, signalling, and provisioning and configuration [3,4,5]. The first refers to the classification of data packets into separate flows, the scheduling and the buffer management algorithms performed on each flow at the network devices. The second component allows the end-user to signal specific flow requirements and enables the end-to-end co-ordination of QoS between the network nodes. Finally, the third component decides which network device performs which specific traffic handling mechanism based on the policies of the network operator. Moreover, it refers to monitoring, measurement and traffic engineering mechanisms needed for evaluating the QoS guarantees, fixing overloaded links, measuring the characteristics of traffic.

There are several initiatives from the Internet community to resolve this problem, in principle the *Integrated Services* (IntServ) [6,7] and the *Differentiated Services* (DiffServ) [8,9] approach. The first approach, which uses explicit resource reservations, is considered as rather difficult to scale up to a worldwide network, while the second one does not yet provide all required mechanisms for end-to-end QoS provisioning. Although they are two independent models where the DiffServ model was introduced as a rather simple and easily deployable model that came to replace the IntServ model and overcome the scalability issues that follow it, finally it is realised that they are not competitive but rather complementary [4,10,11].

The concept of the Bandwidth Broker (BB) that has been introduced from the early stages of the DiffServ model [9] is responsible for performing policy-based admission control, managing network resources, configuring specific network nodes, among others. Nowadays, the Internet community directs its efforts towards the specification and standardisation of the mechanisms of the BB, as well as the development of a prototype [12,13,14]. This paper presents the architecture and design decisions of a multi-layer BB, which is currently under development [15].

This paper is organised as follows: Section 2 gives a short presentation of existing approaches for QoS provisioning over IP. Section 3 discusses the basic concepts of the proposed architecture and specially the Resource Control Layer. Section 4 describes the design model of the Resource Control Point and its mechanisms. Finally, the work to be done in the future is delineated in Section 5, while a summary of the main topics of this paper are given in Section 6.

2. Existing Approaches

Both telecommunications industry and research community have spent a lot of effort on investigating and developing new technologies that could provide QoS over IP-based networks, during the last years. The first attempts focused on providing an automatic optimisation of IP traffic over switched-based networks, such as ATM (e.g. MPOA, IP switching). However, the disadvantage of those approaches is that the application software does not have an interface which can control the specific capabilities of the underlying network.

A different approach, coming from the Internet Engineering Task Force (IETF), is the IntServ architecture, which provides a starting point to establish the necessary infrastructure for advanced multimedia services on top of the IP protocol suite. Integrated Services architectures have been defined using protocols which are being implemented for IP routers (e.g. RSVP [16]). The basic concept of the IntServ model is the enhancement of the existing IP router with tasks traditionally executed in switch-based networks and thus giving Internet a connection-oriented character. Hence, operations like policing, shaping, admission control and QoS management must be provided by all of the RSVP routers on a per IP flow basis. However, in a large scale network with millions of connected users, the number of IP sessions handled by a core RSVP router can be very large. Therefore, the execution of the above functions for every active IP flow in a core IP router leads to pure performance and to a non-scalable network architecture. Furthermore, many important issues remain unsolved, in particular appropriate charging and admission control mechanisms in order to make an Integrated Services architecture economically viable.

The above considerations have forced the Internet community to define a new model for QoS provisioning over IP networks. The new model, known as DiffServ model defines a set of traffic classes each of which is designed to serve applications with similar QoS demands. A traffic class describes the Per Hop Behaviour (PHB) that the packets of this class should receive in each network node. The per hop behaviour determines the priority, the maximum delay in the transmission queues, the link-sharing bandwidth and the probability of a packet to be dropped. The DiffServ model ensures high scalability by separating the operations performed in the borders of the network from those accomplished in the core network. Border routers perform more complex tasks such as traffic policing and shaping, marking and prioritised routing. Marking is the process of classifying IP packets belonging to a specific IP flow and assigning them to the appropriate traffic class. All of the above operations are performed on per flow basis as in the IntServ model.

The small number of active IP flows handled by a border router does not cause the scalability problems that exist in the IntServ architecture. On the other hand, core routers carry out only one simple task that is prioritised routing. DiffServ core routers do not keep any information for the established IP flows. On the contrary, they simply serve packets according to the traffic class that the Ingress border router has assigned to. Hence, each DiffServ core router has to know only the number of traffic classes and the corresponding per hop behaviour of each class. However, in the DiffServ model the functions that would allow end users to request network resources in a dynamic manner are missing. In other words, the *signalling* interface between users and border routers is still not defined.

The framework, which takes advantage of both models and harmonises their different qualities, introduces a logical entity, the so-called Bandwidth Broker (BB) [9,12,13,14]. The main components of this architecture, as well as their functionality are depicted in Figure 1. It can be seen that this model attacks the problem of QoS provisioning in the three aspects mentioned above i.e. signalling, traffic handling and management.

In this new model, Internet is assumed to be separated into various Administrative Domains or Internet Service Providers (ISPs), where each core network is based on the DiffServ model forwarding the aggregate IP traffic based on the DiffServ Code Points (DSCPs) of the traffic flows. Nevertheless, this traffic should be policed, shaped and marked on a per-flow basis at the ingress points of a DiffServ domain, and this is performed by the Edge Devices (EDs). Moreover, since adjacent ISPs have contracts between them, the so-called Service Level Agreements (SLAs), that specify, among others, the traffic characteristics that one domain injects to the other, the egress points of a domain has the responsibility to shape the aggregate traffic sent downstream to the neighbouring domain, in order not to violate the contracts. This is the task of the Border Router (BR). Therefore, the routers are configured appropriately with traffic conditioning mechanisms, as well as scheduling and buffer management modules that specify the PHB of the router according to the network services offered by each domain.

Furthermore, the user may signal her/his QoS requirements to the BB of the domain via a mechanism that could be based on the IntServ/RSVP model or another mechanism such as COPS [17], CORBA [18] etc. The BB, which is responsible for monitoring and controlling the available bandwidth within the DiffServ domain, identifies the path of the new

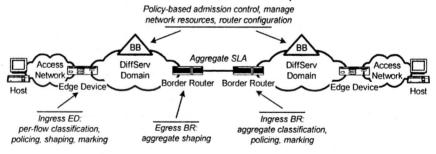

Fig. 1. Bandwidth Broker Principle Architecture

flow and checks whether there are enough available resources in the DiffServ routers that belong to this path. If the path goes beyond the set of DiffServ routers controlled by a specific BB, then the request is forwarded to the appropriate adjacent BB until the BB, which handles the destination ED is reached. Finally, if all of the DiffServ routers, which are involved in the data path of the user's request, have enough available resources, the request is accepted.

Moreover, all involved BBs update their resources database in order to reflect the new established flow. After the successful establishment of the flow, user's packets belonging to the specific IP flow are policed, shaped, and classified (according to the traffic profile sent initially by the user in the reservation message) as mentioned above.

3. Resource control point (RCP)

This section presents the main principles, components and mechanisms of the multi-layer BB architecture. Moreover, special attention is given to the structure and functionality of a specific logical entity, the Resource Control Point (RCP).

3.1 Architectural Principles

This architecture is highly related to the general BB architectural framework described in the previous section. In order to comprehend the decisions made for the design of this specific architecture, it would be helpful to give some fundamental assumptions. First, it is assumed that the data-plane consists of DiffServ-aware routers, while there is no intention in developing any new technology in this field. Therefore, this plane is used as it is and the focus is on the design of an overlay Resource Control Layer (RCL) that manages the resources of the underlying DiffServ data-plane. Second, this architecture is limited to the single-domain case, thus no inter-domain mechanisms are discussed; although most of the ideas presented apply to both cases.

Last but not least, there is a compromise that has to be accepted between provisioning hard QoS guarantees or simplifying the design by sacrificing a percentage of the network resource utilisation. The resulting architecture is rather simplified, but of course, such a decision can only be taken by also considering the network services to be offered. Therefore, the target network services for this model address to applications that exhibit a "light" DiffServ behaviour:
– Delay and jitter sensitivity with small IP packet lenghts e.g. Voice over IP.
– Delay sensitivity (looser than the above) and high-bandwidth requirements e.g. video-conferencing.
– Packet loss sensitivity, security requirements and low duration sessions e.g. SAP.
– No need for guarantees i.e. best-effort.

3.2 Resource Control Layer (RCL)

The Resource Control Layer (RCL) is separated into three logical entities that have been assigned distinct tasks. First, the *Resource Control Point* (RCP) is responsible for managing and distributing the network resources to the corresponding elements. The initial values of the network resources come from the network administrator during the start-up configuration. Second, the Resource Control Agents (RCAs) are assigned the task of performing policy-based admission control so that each reservation request is accepted after ensuring that the customer has administrative rights and there are sufficient resources in the network. In order to perform admission control, the RCAs are assigned by the RCP an amount of resources for which they are responsible. Moreover, each RCA controls an Edge Device or a Border Router, configures the traffic conditioning parameters, allocates the bandwidth resources received by the RCP and handles the users reservation requests. Third, the Application Middleware (AMW) provides an interface to the end-user applications that enables the end-user to signal her/his requirements to the QoS infrastructure. Figure 2 depicts the logical components of the architecture, their associations, as well as the relation to the underlying data-plane.

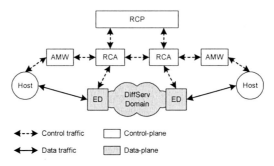

Fig. 2: Resource Control Layer Architecture

Each RCA is associated to a single ED and handles the reservation requests that come from the hosts attached to the ED. The RCA performs local admission control checking whether the local "area" is capable to handle the new traffic flow. In order to make this decision independent of any central entity (e.g. the RCP), the RCP should allocate to the RCA a resource share that would represent the resources of the nearby network. Therefore, the load from the signalling processing of a reservation request is distributed to the RCAs, while the RCP redistributes the resources among the RCAs whenever one or more of them runs out of them.

Depending on the nature of the requested network service and the degree of its guarantees, the RCL may allocate resources either at the ingress RCA or the egress one or both. In case that both RCAs make the reservation, then the ingress RCA should be able to locate the egress one and forward the reservation request. The mechanism of RCA identification is the task of another entity, not shown in the model, which is responsible for the mapping of a host IP address to the corresponding address of the RCA.

Although the admission control is restricted to the edges of the network (ingress, egress or both), this model promises to provide QoS guarantees. The key is to apply an efficient mechanism to the RCP, so that the resources distributed to the RCAs reflects the QoS traffic that can be handled by the core network without violating the requested QoS guarantees

3.3 Hierarchical Structure of the RCP

The functionality of an RCP includes the start-up configuration of the network, the distribution of the resources to the RCAs and the reconfiguration of the available resources according to the variations of the traffic load. In order to reduce the interactions between the RCAs and the RCP and at the same time, provide an efficient resource management entity, the RCP is structured in a hierarchy of RCPs, as illustrated in Figure 3.

Each RCP has the responsibility of its "children" RCPs. Initially, following a top-down approach, the available bandwidth, which is primarily determined by the capacity of the backbone network, is distributed from each RCP to their children according to the initial amounts defined by the network administrator. In addition, after the start up process of the network and the initialisation of the resources, the RCPs are the managers of the resources that have been assigned to them. In other words, the distribution of the resources is rather a dynamic procedure than a static configuration. Thus, in case that the resources that have been assigned to an RCP are insufficient, the RCP will request more from his "parent" RCP, in order to take advantage of any unused resources. The request can be propagated to the upper levels if necessary. In this way the redistribution of the resources takes place as closest to the requester of the resources as possible and the process load is reduced.

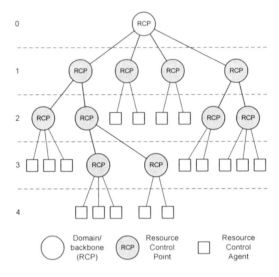

Fig. 3: Hierarchy of the RCPs

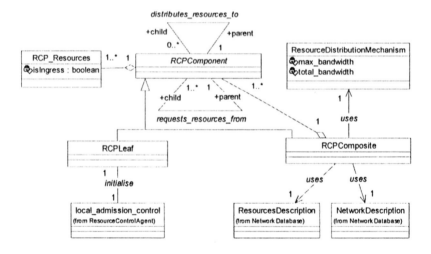

Fig. 4: Resource Control Point Coarse Design Model

The structure of the RCPs should reflect the structure of the underlying network. Therefore, it is required to take into account a number of variables in order to manage to make the best possible mapping. The more accurate this mapping the more efficient the hierarchy will become. Knowledge about network topology and routing as well as information about the expected SLAs of the customers should be helpful in solving this problem.

There are some basic principles that can be used as guidelines for the formation of the hierarchy. Firstly, the RCP should represent a set of physical links that are topologically related. They can represent the links of a sub-area or sub-network, e.g. the network of a university laboratory. When two or more sub-areas are connected to the same router then a new RCP could be formed including the two RCPs that represented those sub-areas. As it is obvious, in a network that uses a fully meshed (or nearly fully meshed) topology the concept of the hierarchy could not be applied. Furthermore, it is not allowed to an RCP to include a link that is already member of another RCP of the same level; the level of an RCP should be taken into account because the parent RCP will always include links that are already members of its children RCPs. Also, the routing information could be an additional input for the case. The sub-areas of the same level of hierarchy should not be directly linked. The local traffic for each sub-area should not use links that are members of another RCP, otherwise this will result in a leak of resources. The above guidelines are not intending to produce an optimal mapping between the network topology and the hierarchical structure of the RCPs, but provide a relatively easy guide for the network administrator.

4. Design of the RCP

4.1 Coarse Design Model

The coarse design model of the RCP is depicted on figure 4 using the Unified Modelling Language (UML) notation [19].

In order to depict the tree structure of the resource control points the composite pattern is used [20], where the following classes are defined: *RCPComponent*, *RCPComposite* and *RCPLeaf*.

The tree of the resource control points is created using the information retrieved by the network database concerning the *NetworkDescription* and *ResourcesDescription*. This information is used to create the resource control points and assign initial resources to each of them. The network database is managed by the network administrator.

Each RCP has a set of objects *RCPResources* that represent the resources that are assigned to each traffic class provided by the network. Each RCP has 5 values that are related to the resources that it possess. The *max_bandwidth* represents the maximum amount of bandwidth that could ever be assigned to the RCP. This value is restricted principally by the capacity of the physical link. The *total_bandwidth* value is the amount of bandwidth that is actually assigned to the RCP for distribution among its children RCPs, initially is the same as the initial amount of resources. The *spent_bandwidth* value apparently depicts the amount of bandwidth that is already distributed to the child RCPs or in case of an *RCPLeaf* the amount of bandwidth spent for the reservation requests of the corresponding RCA. Obviously the subtraction *total_bandwidth – spent_bandwidth* represents the available bandwidth of each RCP.

4.2 Resource (Re-) Distribution

A static approach of the resource distribution mechanism will eventually result in bottlenecks and in unfair and inefficient management of the network resources. Thus, an adaptive mechanism should be defined that will adjust the distribution of resources accordingly to the demand. For this purpose the watermarks mechanism is introduced. Two watermarks are defined, one low and one high watermark for each RCP and RCA.

When a new reservation request has been received and the already spent bandwidth plus the bandwidth required by the new request exceed the high watermark, a request for more resources is made to the parent RCP. The amount of resources that will be redistributed to the requester is decided by the parent and depends on the amount of the available bandwidth of the parent RCP and the amount of the original request. There are two obvious solutions, the parent could distribute exactly the amount that has been requested or can give all the available bandwidth. To find the optimal solution is not an easy task and also depends on the network topology. The algorithm should minimise the communication overhead but should also be able to distribute the resources without permitting greedy components to dominate. A first approach of such an algorithm is described in pseudocode.

```
if (3 * req ≤ 25% of the available resources)

give 3*req;

elseif (2 * req ≤ 25% of the available resources)

give 2*req;

elseif (req ≤ available bandwidth)

give req;

else (request_resources from parent);
```

The algorithm described could be adaptive by altering the factor that will be multiplied with the *req* value (requested bandwidth) or the percentage of the available bandwidth according to the network needs. The concept is to distribute more resources than requested in order to avoid a frequent communication with the specific child, but to save also enough resources for the other children.

The low watermark is an indication that there are unused resources which should be released in order to be used be other children. When the low watermark is crossed the child calls the *release_resources* of his parent. The amount of resources that is going to be released is determined by the high watermark, e.g. in an RCP that has 1Mbps of total bandwidth, a high watermark at 800Kbps, a low watermark at 200Kbps and the spent bandwidth is 128Kbps there should be released 200Kbps so the new total bandwidth is 800Kbps. Obviously the watermarks have to be reconfigured based upon the new total bandwidth value, so the new high watermark is (800*80%) = 640Kbps and the low watermark (800*20%) = 160Kbps.

4.3 Software and Hardware Platform

Since, the functions of the particular RCL comprises actions on various platforms of an IP network and the hosts connected to this network, it is reasonable to use a platform-independent system. Thus, the RCL is implemented as a distributed software system, where the OMG's Common Object Request Broker Architecture (CORBA) is applied. Therefore, the interfaces between distributed components are described using OMG's Interface Description Language (IDL), and that an object request broker (ORB) is applied to allow communication of the distributed components. As the system is implemented using Java (SDK 1.2.2), the IDL-to-Java-compiler as well as the ORB included in this SDK is used.

5. Future Work

The architecture presented in this paper is restricted to the case with one ISP and thus the inter-domain mechanisms between adjacent RCPs and adjacent RCAs and how they provide QoS guarantees are not investigated, yet. However, the algorithms described in the single-domain scenario should be examined and tested that the QoS guarantees they offer, at least justify the simplicity of the model.

Moreover, the RCL should be enhanced with some additional mechanisms necessary for the provisioning of hard guarantees requested by advanced multimedia applications. Such mechanisms include taking advantage of the routing information, monitoring the core DiffServ domain and developing a measurement platform that enable the RCL estimate and foresee the traffic loads, and therefore, take more effective decisions. Last but not least, a powerful technology that will be investigated in the context of this work, is the Multiprotocol Label Switching [21,22].

6. Conclusions

The overall architecture presented in this paper addresses the problem of QoS provisioning in IP networks, in a complete and consistent manner. It introduces a multi-layer Resource Control Layer that is responsible for the handling of the reservation requests, performing policy-based admission control, provisioning and configuring the network in a top-down approach, managing the network resources and dynamically redistributing them among the network elements.

This paper presents the specific design model of the Resource Control Point which is structured in a hierarchical manner in order to manage effectively the network resources and control the Resource Control Agents which are distributed at the edges of the network. Two field trials are expected to take place in the near future (for the intra- and inter-domain, correspondingly) that will provide useful information about the efficiency and the level of QoS provisioning this model can offer.

Acknowledgements

This paper is partly funded by the European research project AQUILA, Information Societies Technology (IST) programme, IST-1999-10077. The authors are currently engaged in the implementation of the concept presented in this paper.

Appendix

AMW	Application Middleware
ATM	Asynchronous Transfer Mode
BB	Bandwidth Broker
BR	Border Router
COPS	Common Open Policy Service Protocol
CORBA	Common Object Request Broker Architecture
DiffServ	Differentiated Services
DSCP	DiffServ Code Point
ED	Edge Device
IDL	Interface Description Language
IntServ	Integrated Services

IP	Internet Protocol
ISP	Internet Service Provider
MPOA	Multiprotocol over ATM
PHB	Per-Hop Behaviour
QoS	Quality of Service
RCA	Resource Control Agent
RCL	Resource Control Layer
RCP	Resource Control Point
RSVP	Resource Reservation Protocol
SLA	Service Level Agreement
UML	Unified Modelling Language

References

1. Krol, E.; Hoffman, E.: FYI on "What is the Internet?". RFC 1462, May 1993.
2. McQuillan, J.; Walden, D.: The ARPA Network Design Decisions. Computer Networks, 1, pp. 243-289, 1977.
3. Roberts, Lawrence.: Quality IP. April 21, 1999. http://www.data.com/issue/990421/roberts.html
4. Bernet, Y.: The Complementary Roles of RSVP and Differentiated Services in the Full-Service QoS Network. IEEE Commun. Mag., Vol. 38, No. 2, Feb. 2000.
5. Cisco: Introduction: Quality of Service Overview. http://www.cisco.com/univercd/cc/td/doc/product/software/ios120/12cgcr/qos_c/qcintro.htm
6. Clark, D.; Shenker, S.; Zhang, L.: Supporting real-time applications in an integrated services packet network: architecture and mechanism. Proc. ACM SIGCOMM, pp. 14-26, August 1992.
7. Braden, R.; Clark, D.; Shenker, S.: Integrated services in the Internet architecture: an overview. RFC 1633, 1994.
8. Black, D.; Blake, S.; Carlson, M.; Davies, E.; Wang, Z.; Weiss, W.: An Architecture for Differentiated Services. RFC 2475, 1998.
9. Nichols, K.; Jacobson, V.; Zhang, L.: A Two-bit Differentiated Services Architecture for the Internet. RFC 2638, 1999.
10. Eichler, G.; Hussmann, H.; Mamais, G.; Venieris, I.; Prehofer, C.; Salsano, S.: Implementing Integrated and Differentiated Services for the Internet with ATM Networks: A Practical Approach. IEEE Communications, January 2000.
11. Bernet, Y.; Yavatkar, R.; Ford, P.; Baker, F.; Zhang, L.: A Framework for End-to-End QoS Combining RSVP/Intserv and Differentiated Services. Internet Draft, IETF, 1998.
12. Neilson, R.; Wheeler, J.; Reichmeyer, F.; Hares, S.: A Discussion of Bandwidth Broker Requirements for Internet2 Qbone Deployment. Internet2 Qbone BB advisory Council, August 1999.
13. Report from the First Internet2 Joint Applications/ Engineering QoS Workshop. http://www.internet2.edu/qos/may98Workshop/9805-Proceedings.pdf, May 1998.
14. Terzis, A.; Ogawa, J.; Tsui, S.; Wang, L.; Zhang, L.: A Prototype Implementation of the Two-Tier Architecture for Differentiated Services. RTAS99 Vancouver, Canada, 1999.
15. Adaptive Resource Control for QoS Using an IP-based Layered Architecture, AQUILA. http://www-st.inf.tu-dresden.de/aquila/
16. Braden, R.; Zhang, L.; Berson, S.; Herzog, S.; Jamin, S.: Resource ReSerVation Protocol (RSVP) - Version 1 Functional Specification. RFC 2205, 1997.
17. Boyle, J.; Cohen, R.; Durham, D.; Herzog, S.; Rajan R.; Sastry, A.: The COPS (Common Open Policy Service) Protocol. Internet Draft, IETF, 1999.
18. CORBA/IIOP 2.3.1. Specification. http://www.omg.org/corba/cichpter.html
19. OMG Unified Modelling Language Specification, version 1.3. 1999.
20. Gamma, E.; Helm, R.; Johnson, R.; Vlissidis, J.: Design Patterns – Elements of Reusable Object-Oriented Software. Addison-Wesley, 1995.
21. Rosen, E.; Viswanathan, A.; Callon, R.: Multiprotocol Label Switching Architecture. Internet Draft, IETF, 1998.
22. Awduche, D.: MPLS and Traffic Engineering in IP Networks. IEEE Commun. Mag., Vol. 37, No. 12, Dec. 1999.

Seamless and Uninterrupted Interworking of Wireless and Wireline Network Technologies

Konstantinos Vaxevanakis, Sotirios Maniatis, Nikolaos Nikolaou, and Iakovos Venieris

National Technical University of Athens,
Department of Electrical and Computer Engineering,
9 Heroon Polytechniou str, 15773,
Athens, Greece

{vaxevana, sotos, nikolaou}@telecom.ntua.gr,
ivenieri@cc.ece.ntua.gr

Abstract. The Internet Protocol (IP) dominates in today's data communication infrastructures. The network evolution also favors the explosion of the wireless network technologies. The latter have to be combined and interoperate properly with current wireline installations. Although IP assures interoperability, there are open interworking issues at the session, network, and data link layers related to the nature of the wireless environment. The architecture presented in this paper, proposes specific enhancements to an IP infrastructure, that alleviate the deficiencies of wireless networks and propel the efficient interworking between wireline and wireless network technologies.

1. Introduction

The trend in network evolution is not towards a global and uniform environment. It rather favors the appearance of a combination of wireline and wireless networks that complement each other in terms of coverage area, underlying infrastructure and throughput. In this complicated environment, the Internet Protocol (IP) seems to be the common denominator for the information world. IP services are supported by almost all physical interfaces both in the wireless and wireline worlds, thus realizing interoperability of services.

Designed and operated for many years over wireline infrastructure, IP is not optimized for supporting communications over wireless interfaces. Consequently, IP is not always adequate to offer seamless and uninterrupted operation of a mobile host that moves within the diverse environment that is designated by the mixture of various wireless and wireline technologies.

To start with, the mobile host's IP address reveals location information, thus limiting the operating range of the host within the boundaries of the IP domain this address belongs to. Migration to other IP domains is possible only if the host offers the appropriate mechanisms. Besides, migration to a different IP domain might imply that a modification would occur at the physical layer, too. The seamless and uninterrupted roaming between IP domains and physical interfaces is not currently offered by the IP protocol.

Additionally, one major drawback of the wireless networks is that they are susceptible to interruptions of any kind, like short disconnections in case of a temporary signal loss. On the other hand, applications are usually implemented according to the properties of the wireline world, where such phenomena do not exist. Software developers require network transparency, so they cannot be obliged to have the properties of the wireless environment in mind when they design their applications. In order to facilitate end-to-end interconnection and operation – independently of the underlying wireless and wireline infrastructure mixture – specific enhancements to the operation of the IP service must be presented to the wireless side.

This paper introduces an architecture that takes into consideration the aforementioned requirements of the foreseen complicated environment as well as the specific deficiencies of the wireless one, and provides IP-based enhancements in various layers of the OSI model. In this way, applications that exist in wireline environments can be supported in wireless environments too, without any modification, while still efficiently utilizing the basic IP services.

The rest of the paper is organized as follows. Section 2 provides a thorough insight into the proposed architecture. It discusses the functionality of added components, along with the rationale behind them. Section 3 presents a reference environment, which had been setup in order to act as a platform for trials and experiments. Section 4 discusses the results obtained by the trials' execution. Last, Section 5 gives the conclusions and possible future extensions.

2. Proposed Architecture

There are two basic wireless enhancements supported by the architecture, namely the location transparency and the application resiliency to link disconnections. The former refers to the Network and Data Link layers of the OSI model, while the latter belongs to the Session layer. These enhancements are logically placed in the convergence point between the wireless and wireline infrastructures. Physically, the proposed enhancements are introduced in the mobile host as well as in a Gateway/Proxy at the home Intranet.

The address of a node that resides within an IPv4 network, apart from identifying the node itself, contains topological information, thus binding the terminal with a specific location. Consequently, if a mobile terminal attempts to move without changing its IP address, then it will be unable to route packets correctly, while if it dynamically modifies its address, then all active connections will be terminated.

To overcome this problem and allow a mobile terminal to roam freely around the network, while still communicating and maintaining the same IP address, the concept of Mobile IP protocol has been followed. Mobile IP [1] defines two new entities, the mobility agent, which include the Home and Foreign Agent (HA and FA), and the Mobile Host (MH). Mobility agents are located inside the network and are responsible for specific sub-networks, while MH is located inside the mobile terminals. Whenever the mobile terminal is connected to a foreign sub-network, the Mobility Agent (HA) controlling the home sub-network, forwards all the packets destined for the mobile host to a specific FA that resides within the foreign sub-network. The FA in turn delivers the packets to the mobile terminal. In this manner, although the mobile terminal moves to a different network domain, active TCP/UDP sessions are maintained alive and data packets continue to reach the mobile terminal.

Taking into consideration that location transparency must be provided to the terminal, independent from the underlying wireline or wireless medium, we have concluded to the overall architecture, depicted by Fig. 1. It covers both the terminal and the network side, having part of the functionality of MH and FA supported within the mobile terminal, while its counterpart HA functionality is implemented within the Gateway/Proxy. The mobile terminal is equipped with more than one Network Interface Cards (NICs). At least one of them is wireless, while the others could be either wireless or wireline. Depending on the network environment – indoors (e.g. home, corporate premises/office) or outdoors (e.g. when away from home/office) communications – one of the interfaces is active, while when it is necessary, a switching can be performed from one to the other.

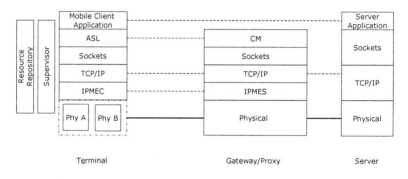

Fig. 1. Proposed Architecture

The IP Mobility Enhancement Client and Server (IPMEC/IPMES) team up with the IP protocol to achieve location transparency and make mobility transparent to higher level protocols (like TCP) and, eventually, to applications.

More specifically, the IPMEC module resides within the mobile terminal and when it is activated it registers with the Gateway/Proxy. IPMEC is mainly responsible for redirecting IP packets over the correct underlying interface. IPMEC has the privilege to modify the routing table of the terminal. More precisely, during switching between physical interfaces, all the routing entries directed over the current interface are deleted and re-directed over the new selected interface.

Whenever the mobile terminal switches from one interface to another, no modification is made with the terminal's IP address. This is one of the prerequisites that enables the terminal to switch between interfaces, without having all the active transport connections terminated. To fulfill this requirement, the IP address of the terminal is either associated with the MAC address of the current interface, or with the MAC address of the Gateway/Proxy. The latter serves the purpose of having the packets, which are destined to the mobile terminal, received by the Gateway/Proxy, and, subsequently, forwarded to the terminal.

In order to achieve such a behavior, the Gateway/Proxy sends to the network, on behalf of the mobile terminal, ARP (Address Resolution Protocol [2]) packets and updates, accordingly, its routing table. This mechanism of sending ARP packets (Request or Reply [2]) to spontaneously cause other nodes to update their ARP cache, is also known as Gratuitous ARP (GARP [3]).

On the network side, the IPMES component resides in the Gateway/Proxy and implements the complementary functionality of IPMEC. More specifically, following the idea of the Mobile IP mobility agent, it is responsible for the processing of the registration/de-registration messages received from IPMEC. In addition, IPMES caters for the appropriate interception of IP packets from the network and the forwarding of these packets to the proper mobile terminal. Finally, IPMES implements the aforementioned GARP mechanism.

The Abstract Socket Layer (ASL) and Communication Manager (CM) pair of modules provides for the control of the link disconnection for data services. Fig. 2 depicts the operation of these two modules.

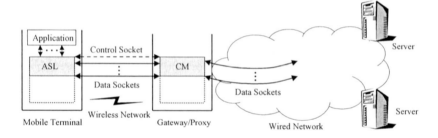

Fig. 2. ASL-CM Operation

ASL resides at the mobile terminal side and provides a socket-like Application Programming Interface (API). In this manner, every socket call initiated by a client application passes through the ASL library. ASL co-operates with the CM module, which is running at the Gateway/Proxy side, using a lightweight proprietary protocol, which operates over UDP (Control Socket). The purpose of this protocol is to have ASL and CM exchange control information that will enable them to manipulate (*OPEN*, *ReOPEN* and *CLOSE*) TCP connections, started by the terminal side. The CM module, upon receiving control messages, interprets them and tries to establish the required connections (Data Sockets) in order to realize the connection that the client application of the mobile terminal requested from ASL.

Figure 3 depicts the exchange of control messages, between ASL and CM, concerning the two most interesting situations. In the first case the application initiates a connection, through a TCP socket, with a server. The socket call is handled by ASL that establishes a connection with the application, registers the essential information of the connection, so as to be in position to duplicate it, and sends an *OPEN* message to CM, through the wireless medium. The CM module consecutively stores the required information that mirrors the connection information kept by ASL and tries to connect both to ASL and the requested server. In case CM successfully establishes both connections, it returns an *Answer* message to ASL, indicating that the requested connection is active. In any other case the Answer message inform the ASL for the connection failure. Following the aforementioned procedure, the ASL and CM modules assemble an end-to-end connection between the application and the appropriate server.

In the second case, Figure 3 depicts the steps that ASL and CM follow to reopen a broken, due to the wireless medium properties, physical connection. ASL sends a *ReOPEN* message to CM, identifying the connection that must be reopened. CM compiles the *ReOPEN* message, retrieves the connection properties from its database, and re-establishes the required connection with ASL. Afterwards, it sends and *Answer* message to ASL, acknowledging the successful reconnection.

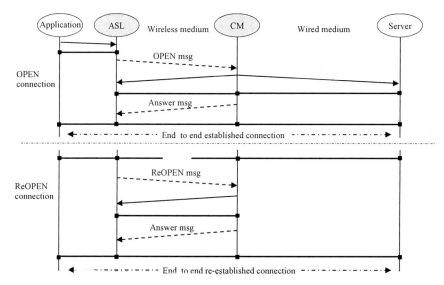

Fig. 3. Exchange of ASL – CM control messages

Splitting the original end-to-end connection between the client and the server application, into three parts (Application-ASL, ASL-CM and CM-Server) isolates the end points from the temporal wireless link disconnections. Whenever an interruption occurs, ASL re-establishes transport connections to CM, and, subsequently, re-associates sessions with physical connections. The terminal's networking applications throughout this period are "frozen", but not terminated.

In order for the aforementioned modules to communicate and co-operate properly, an additional mechanism is introduced in the terminal, called the Supervisor module [4], [5]. It is responsible for monitoring the connectivity status of the underlying wireless/wireline physical interfaces and selecting the most appropriate among them. The selection process takes place according to predefined signal strength thresholds and user preferences.

Moreover, Supervisor maintains a repository of necessary information in the terminal (Resource Repository). There it stores all values related to the operation of the various modules in the terminal architecture that are needed to be communicated between them. For example, connectivity status, signal strength values, IP and MAC addresses are vital parameters for the terminal operation. These parameters are supervised for changes and, in case of that, appropriate notifications to the interested modules are issued.

3. Reference Network Environment

The presented architecture is independent of the wireless interfaces. However, in order to validate its correctness, we have implemented a reference environment based on the two most widespread wireless interfaces in production: an 802.11-compatible wireless LAN (WLAN [6]) and GSM [7]. Alternatively to GSM, we have also tested access through a standard PSTN modem.

The mobile terminal is configured with the IPMEC, ASL, Supervisor and Resource Repository modules, and therefore it offers both location transparency and resilience to short disconnections. Its operation heavily depends on the selection of the underlying wireless medium.

When at home or office, the WLAN is always favored, as it provides more bandwidth than GSM, with low operational cost. On the other hand, WLAN coverage is quite limited. When the terminal moves towards the boundaries of the coverage area and the wireless LAN SNR (Signal to Noise Ratio) falls under a predefined threshold, the system switches to GSM. More particularly, it sets-up a GSM connection and registers the terminal to the Gateway/Proxy, in order to tunnel all terminal packets via the GSM. During the switching period, all running sessions are kept alive so that they are able to continue their operation after the new physical connection is established. A reverse switching is performed, when the WLAN signal rises to operational levels.

The Gateway/Proxy acts as bridge between the wireless and wireline parts of the corporate network. Additionally, the Gateway/Proxy is used to interconnect different sub-segments of the corporate network. It is configured with the IPMES and CM modules of the proposed architecture and manages a pool of modems that can be used to offer access through the GSM/PSTN network.

For the realization of the reference network environment, we have concluded to the utilization of a notebook PC as a mobile terminal, supplied with the appropriate WLAN and GSM/PSTN adapters. A notebook appears to be the preferred solution for a nomadic user, as it gives the capability of preserving a working environment, while moving between different locations. Knowing that the big majority of end-users are familiar with the Microsoft Windows environment, we have selected Windows NT 4.0 as the Operating System (OS) of the terminal. Additionally, in order to prove the generality and implementation feasibility of the proposed architecture, a separate implementation on a UNIX environment (Linux OS) has been also followed. The selection of the OS has, to a large degree, influenced the materialization of the overall architecture, introducing limitations and problems that have not been foreseen.

The adopted Reference Network Environment (Fig. 4) can support four different connectivity scenarios for accessing the corporate network, including indoor (WLAN) and outdoor (GSM/PSTN) access. More particularly:

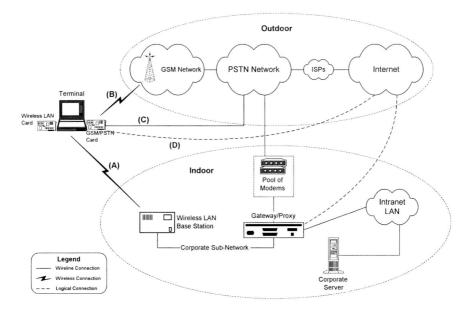

Fig. 4. Reference Network Environment

(A) The mobile terminal is located in the network segment in which its IP address belongs to and access is achieved by means of the WLAN interface.

(B) The terminal is connected to the corporate LAN utilizing a GSM physical link to the GSM service provider. At the GSM service provider's local exchange, traffic is routed, via a direct PSTN connection, to a modem pool, which is managed by the corporate Gateway/Proxy.

(C) The physical connection is achieved through the PSTN network only. In this case mobility is restricted, as the terminal uses a telephone line to connect to the corporate network. However, it may have lower cost, compared with scenario "B".

(D) A logical connection to the Internet is established utilizing the services of an Internet Service Provider (ISP). The physical link of the connection is achieved through the GSM or the PSTN interface. This scenario is quite close to cases "B" and "C". The main difference is that in scenario "D", an ISP router comes in between the terminal and the corporate Gateway/Proxy.

333

4. Trials and Results

In order to be able to test the implementation of the proposed architecture, a testbed has been established in the lab, but also a few trials have been performed in actual corporate environments. The testbed, following the Reference Network Environment, consists of the components listed below:

- *Mobile Terminal*: it is a notebook that uses Windows NT 4.0 OS and is configured with all the modules that our architecture proposes. Alternatively, we also used a notebook with Linux OS installed.
- *Gateway/Proxy*: it is a PC with Linux OS that is connected to corporate LAN. It is configured with the IPMES and the CM modules that are required from the architecture. Additionally, it serves as a dial-up server for the incoming calls from the mobile terminals.
- *WLAN Base Station*: it is the base station for the wireless LAN. It is connected to the same sub-network with the Gateway/Proxy.

The testing scenarios were based on standard networking applications, including Telnet and Web browsing (Internet Explorer for Windows NT and Netscape for Linux) and videoconference application (Microsoft NetMeeting for Windows NT). Additionally, a custom-built FTP-like application that utilizes directly the ASL library was used in the Windows NT environment. The use of this application separated the testing of ASL-CM pair of modules from the OS inherent restrictions. The testing scenarios that were used are described beneath (Fig. 5):

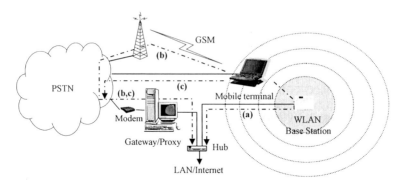

Fig. 5. Testbed Infrastructure

Scenario A: The mobile terminal is connected to the network through path (a). The selected networking applications are executed and operate normally. At this point, an automatic transition from WLAN to GSM link is performed, due to a deliberate degradation of the WLAN SNR. After the switching, network traffic is routed through the path (b). Alternatively, the PSTN dial-up connection, path (c), can be utilized. The desired result for the running applications is a seamless handover.

Scenario B: It is the same with scenario A, only this time the reverse transition from the GSM link to the WLAN is performed (from path (b) or (c) to (a)), due to an increase of the WLAN SNR. The uninterrupted operation of the selected networking applications is the desired result.

Both scenarios have been tested for the cases where the mobile terminal was configured with the Windows NT OS and the Linux OS. In the former case, regarding scenario A, all the applications continue their normal operation unhindered. However, during the handover of scenario B, the telnet and videoconference applications fail to retain their connections. On the contrary, the custom application continues its normal operation and completes the file download. Additionally, the browsing application also continues its operation seamlessly.

The results with the Windows NT OS have strengthened the necessity to develop the ASL-CM pair due to the following reason. It has been observed that during switching from GSM to WLAN, TCP connections are lost, which did not occur in any of the experiments in the Linux OS. More particularly, while the terminal is operating over GSM (or PSTN) the Point-to-Point protocol (PPP [8]) is utilized. As soon as the Windows NT OS senses that the GSM link is down, it releases all resources related to the PPP protocol, with the imminent result of terminating any previously active TCP connection. Incorporating the ASL-CM pair solves this drawback.

In the case that the mobile terminal operates under the Linux OS, two representative networking applications were used: a common telnet application and the Netscape browser. The results, after applying scenarios A and B, are that both applications continue their operation unhindered, even if the switching of media occurred, while downloading an image (browsing).

According to the presented results, we see that switching between media, while maintaining the sessions at both the TCP/UDP and IP levels, occurs successfully in almost all examples, apart from the cases that are restricted by the operating system's limitations. Moreover, some performance measurements that have been performed during the execution of the testing scenarios, reveal that the time to setup a GSM connection can be very large, as it is about 20 seconds on average. This big duration can also result in the termination of existing TCP connections (due to time-out), in which case the existence of the ASL-CM pair provides a good solution for the applications that can make use of the socket-like API it offers.

Regarding the ASL-CM operation, some performance measurements have been executed using of the proprietary file transfer application. The average measured time to initially setup the end-to-end communication over the ASL and CM modules is 0.463 seconds, instead of the measured 0.152 seconds without the use of ASL-CM.

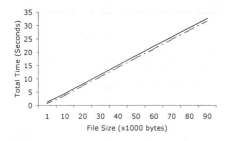

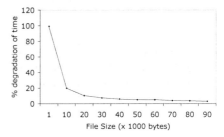

Fig. 6. Total Time needed to transfer files of various sizes with and without (dotted line) the use of the ASL-CM pair

Fig. 7. The percentage of the degradation of the total time needed to transfer a file with ASL-CM compared to the case that the file is transferred without the ASL-CM pair

It has also been detected a degradation on the communication speed even when no interruption occurs (Fig. 6), especially for files that are small in size. Nevertheless, the degradation is decreasing very quickly to accepted values (Fig. 7). For example, for files larger than 20 Kbytes the degradation percentage is less than 10%.

In case of a disconnection, the application is frozen for as long as the physical connection is re-established and the ASL-CM pair re-associates sessions with TCP connections. The time for the former operation can be very large (20 seconds as mentioned above), while the average measured time for the latter operation is about 0.3 seconds. In this sense, the degradation is mostly owned to the establishment of the physical connection.

5. Conclusions

The architecture presented in this paper provides enhancements, both in the terminal and the Gateway/Proxy side, in order to support seamless and uninterrupted interworking between wireless and wireline interfaces. As a consequent, the end-user is not aware of the adjustments performed for the correct interworking of the underlying mediums.

The proposed architecture has been implemented under two different Operating Systems (Windows NT and Linux). Extensive experiments have been performed with standard networking applications and proved the correctness of the solution. In order to alleviate limitations posed by the OS, further enhancements (ASL/CM) are required for the case of Windows NT.

In addition to the functionality described in the paper, the architecture could be extended to various directions. To start with, although the implementation involves two specific wireless technologies, namely WLAN and GSM, it can be extended to include others, like the General Packet Radio System (GPRS [9]), which is the packet-based equivalent of GSM. Since the overall architecture is built in a modular manner, it can easily accommodate other wireless interfaces that fit better with the specific user needs.

Another extension, stemming from the ASL-CM pair, would be the encryption of the packet's contents before transmitting it to the CM and vice versa. This would apply a supplementary level of security over the security inherent in the WLAN or GSM radio transmission path. The charge for this is the probable degradation of transmission speed and delay, according to the selected encryption and decryption algorithm.

Acknowledgement

This work was performed in the framework of NOTE (NOmadic Teleworking business Environment - EP27002 [10]) project, co-funded by the European Community under the ESPRIT Programme. The authors wish to express their gratitude to the other members of the NOTE Consortium (Thomson CSF, Lucent Technologies NL, Archetypon, Solinet GmbH, Intrasoft International and Nationale Nederlanden) for valuable discussions.

References

1. C. Perkins, "IP Mobility Support", RFC 2002, October 1996.
2. D. Plummer, "An Ethernet Address Resolution Protocol: Or Converting Network Protocol Addresses to 48.bit Ethernet Addresses for Transmission on Ethernet Hardware", RFC 826, November 1982.
3. W. Richard Stevens, "TCP/IP Illustrated, Volume 1: The Protocols", Addison-Wesley, Reading, Massachusetts, 1994.
4. S. Maniatis, I.S. Venieris, R. Foka, "Nomadic Teleworking Business Environment", EMMSEC99, Stockholm, Sweden, Jun. 1999
5. K. Vaxevanakis, S. Maniatis, T. Zahariadis, N. Nikolaou, I.S. Venieris, N.A. Zervos, "A Software Toolkit For Providing Uninterrupted Connectivity Over a Variety of Wireless Physical Interfaces", SoftCom99, Croatia & Italy, Oct. 1999.
6. IEEE 802.11 - Working Group for Wireless Local Area Networks, URL: http://grouper.ieee.org/groups/802/11/index.html.
7. ETSI, "General Packet Radio Service, GSM specification 03.60, version 6.2.0", Oct. 1998.
8. W. Simpson, "The Point-to-Point Protocol (PPP) for the Transmission of Multi-protocol Datagrams over Point-to-Point Links", RFC 1331 May 1992.
9. ETSI, "General Packet Radio Service, GSM specification 03.60, version 6.2.0", October 1998.
10. ESPRIT NOTE – "NOmadic Teleworking business Environment, EP27002", URL: http://www.telecom.ntua.gr/note.

Modelling and Performance Evaluation of a National Scale Switchless Based Network

Josep Solé-Pareta, Davide Careglio, Salvatore Spadaro, Jaume Masip,
Juanjo Noguera and Gabriel Junyent

Universitat Politècnica de Catalunya - UPC,
Advanced Broadband Communications Lab. - CCABA[1]
Jordi Girona, 1-3, Mòdul D6 (Campus Nord),
08034 Barcelona, Catalunya, Spain
{pareta, careglio, sspadaro, jmasip, jnoguera}@ac.upc.es and
junyent@tsc.upc.es

Abstract. In a packet switching based scenario, optical technologies can not completely overcome the problem of insufficient network capacity, due to limitations produced by the presence of electronic switches and regenerators in the fibre path. As a new alternative approach, the SONATA project proposed a switch-less based network with the triple objectives of avoiding the need for large and fast switching nodes and drastically simplifying both the network structure and the layer architecture within the network. This paper describes the simulation model and presents essential results from the performance evaluation of a Network Controller for a large-scale SONATA network. The simulations were carried out on a Silicon Graphics Origin-2000 using TeD (Telecommunication Description language) and the main objective was to assess the feasibility of switchless based networks to be used as large IP backbones.

1   Introduction

The SONANA network is based on the switchless network concept, which was defined and demonstrated in the SONATA (Switchless Optical Network for Advanced Transport Architecture) AC351 ACTS project. The "switchless" network concept is based on a mixture of WDMA (Wavelength Division Multiple Access based on wavelength tuneable transmitters and receivers) and TDMA (Time Division Multiple Access) methods to allow point-to-point and multi-point-to-multi-point interconnections between passive optical networks without the need for electrical switch nodes within the network. The general structure of this network is shown in Figure 1, and is described in detail in [1].

The key element of SONATA is the Network Controller. Figure 1 shows that a Network Controller is attached to the PWRN (Passive Wavelength Routing Node). This device is responsible for the resolution of access conflicts, for the allocation of

---

[1] Centre de Comunicacions Avançades de Banda Ampla

transmission resources to user (SONATA terminals) requests, and for the configuration of the network connectivity using wavelength converter arrays.

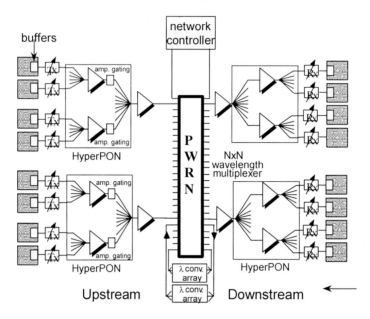

Fig. 1. SONATA network structure.

SONATA has an optical fibre dedicated to communicate with the Network Controller. To access the Network Controller through this fibre, each PON has at least one wavelength assigned. Thus, all the terminals attached to the same PON have to share the wavelength assigned to that PON. For the resolution of access conflicts, a Medium Access Control (MAC) protocol is needed. Such a MAC protocol has to manage the signalling channels that allow the communications from the terminals to the Network Controller and from the Network Controller to the terminals. The MAC protocol adopted within the SONATA project was based on the Time Division Multiple Access (TDMA) method. Note that with this strategy, signalling channels can become a bottleneck when increasing the number of terminals. To overcome this problem, either the number of fibres to communicate with the network controller or the number of signalling wavelength per PON has to be increased.

Although globally optimal algorithms to allocate transmission resources to the user requests in the switchless network viewed as a five-stage T-λ-λ-λ-T switch are possible, within the SONATA project a much simpler approach to resource allocation and contention resolution at the network controller was proposed [1], [3]. The simple approach is based on the decoupling of the time dimension from the wavelength dimension. This means that the resource allocation problem was split into two sub-problems: *scheduling of user requests* in the time domain given a PON-to-PON channel assignment, and *design of the network connectivity* by properly assigning wavelengths to PONs via the arrays of wavelength converters (logical topology design). While scheduling is a simple task, which must be done slot-by-slot, the logical topology design is a much harder task, which can be performed at a lower rate.

This paper focuses on the performance evaluation of the algorithm proposed within SONATA for *scheduling the user requests*. The evaluation was done in terms of the *Request Loss Rate* and the achieved *Resource Occupancy*. Since the objective of the project was to evaluate the performance and global behaviour of the Network Controller for a large-scale (national and metropolitan scale) SONATA network, a theoretical study was not sufficient, but a simulation study was needed. The main advantage of the simulation process is that once the simulation elements are incorporated in the simulation layer, then easier comparison and optimization processes can be accomplished. Due to the adoption of a TDMA-AC protocol, simple traffic assumptions were sufficient.

The simulation environment used for this work was based on parallel computing simulation techniques. We used TeD [2], which stands for Telecommunications Description language. TeD is a network modelling language that is transparently mapped on top of a high-performance parallel computer. As a high-performance parallel computer we used a Silicon Graphics 2000.

This paper is organised as follows. In Section 2 the scheduling algorithm for the Network Controller proposed within SONATA is described. Section 3 is devoted to the simulation scenario we modelled to carry out our performance evaluation study. In Section 4 some samples of the simulation results are discussed. Finally Section 5 concludes the paper.

2   Scheduling Algorithm

This section is devoted to explain the basic concepts of the concrete scheduling algorithm we used within the SONATA Network Controller. The details of this scheduling algorithm can be found in [3].

The scheduling algorithm deals with *requests* and *frames of requests*. Each SONATA terminal generates requests for connection, and using a TDMA based approach all requests from SONATA terminals connected to the same control wavelength enter to the Network Controller as a frame. The Network Controller, using the scheduling algorithm, will decide which of the incoming requests are served or not.

The implementation of this scheduling algorithm is based on a set of matrices, which represents the network available resources. These matrices are (see Figure 2):

- **Matrix W:** Each entry of this matrix points out to a list of records describing the wavelength channels that provide connectivity between the corresponding PON-to-PON pair. The number of rows and columns is equal to the number of PONs.
- **Matrix S:** This matrix indicates, for each wavelength channel (λi), which slots of the frame are already assigned.
- **Matrix Utx:** This matrix indicates for each transmitter (Tx) of , which slots of the frame are already assigned.
- **Matrix Urx:** This matrix indicates for each receiver (Rx), which slots of the frame are already assigned.

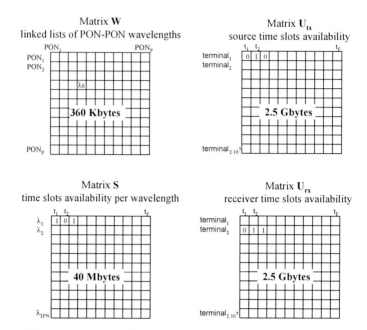

Fig. 2. The several matrices (data structures) used by the scheduling algorithm.

Based on the contents of these matrices, the scheduling algorithm is specified in four steps, which will be executed for each incoming frame. These four steps are the following:

1. Select the wavelength channel on which request is attempted; scan the linked list of wavelength channels associated with the pair (source PON, destination PON) in matrix W.
2. Select the time slot for use within the frame; logically AND the row of matrix S corresponding to the selected wavelength channel with the complement of the rows of source and destination matrices Utx and Urx.
3. Assign the number of requested slots to the requesting terminal, using a round-robin scan on the vector resulting from the previous step of the algorithm.
4. Update matrices S, Utx and Urx and descriptor in matrix W.

3 Simulation Scenario

Here we completely decouple the two problems of scheduling and logical topology design, assuming that the wavelength converters are reconfigured of-line and do not vary during the simulation time, and that the scheduling algorithm operates on a fixed logical topology. According to this assumption, a simulation model for the network control unit has been defined. This model consists of (see Figure 3) N_P request generators representing the PONs, which access the control unit through N_C TDMA signalling channels each, and the scheduling algorithm module. The requests generators were implemented as random sources based on the Bernoulli random process, the number of signalling channels was set to $N_C > 1$ in order to avoid the signalling bandwidth bottleneck, and the scheduling algorithm module was implemented based on the above described slot-by-slot scheduling algorithm. Since what we need to simulate is the case where each slot carries a request with probability **p** and no request with probability **q** = 1 − **p**, the Bernoulli processes perfectly model the TDMA signalling channels.

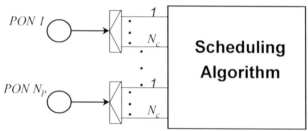

Fig. 3. Simulation model for the SONATA network controller.

Two types of traffic modes were implemented, namely the balanced and the unbalanced types. For the balanced type, a uniform distribution of the destination PONs was used (all PONs has the same probability to be destination of the traffic), while for the unbalanced type a non-uniform distribution of the destination PONs was used.

3.1 Network Configuration

In order to reproduce a real scale configuration of the SONATA network that could fit in the bounds of the simulation platform, we could not consider the FTTH solution where end-user and SONATA terminals are equivalent terms. We assumed either the FTTCab or the FTTK/B solutions with 200 end-users per SONATA terminal. According to this assumption, in the simulator we set up a network configuration with 100 PONs (N_P = 100) and 1000 SONATA terminals per PON (N_{ST} = 1000). These values lead to 20 millions customers (100 x 1000 x 200 = 2 10^6), which within the project was considered as a national range, but reduces the amount of memory required by matrices W, S, Utx and Urx. The rest of the configuration parameters was the following: 1000 signalling mini-slots per frame (K_S = 1000, one per terminal in each PON), N_C = 3 or 10 (depending on the simulations) signalling channels per PON. 100 data slots per frame (frame duration, 1 ms.), and a transmission bit rate per channel of 622 Mbps Finally, the number of dummy ports was fixed either to 0, 100, 200, 300 or 400 (0, 1, 2 ,3 or 4 additional wavelengths per PON-to-PON pair).

3.2 Traffic Characteristics

We established the following traffic conditions: Only Internet traffic (no connection oriented traffic load). Internet traffic requests modelled by the Bernoulli generators with an offered load per PON **p** = 0.6, 0.75, and 0.9. Concerning the distribution of the data traffic destination PONs. 1) For the balanced traffic type, a uniform distribution among all the PONs with mean 1/100 was used. 2) For the unbalanced traffic type, a Gaussian distribution, with mean = 50 (this means that numbering the PONs between 1 and 100, the most probable destination of the traffic would be the PON number 50) and standard deviation = 20.

4 Experiments and results

Under the configuration and traffic conditions described in the previous section, we carried out a set of simulations assuming two different approaches. A per packet request approach (requests were issued by the IP layer, one per IP packet to be transmitted), and a per packet-flow request approach (requests were issued from the TCP layer, one per TCP session). In each case we simulated 10 seconds of the network operation time, which required an execution time of around 36 hours.

4.1 Per Packet Request Approach

In the per packet request approach, we set the number of slots required per request, based on the IP packet length statistics provided in [4]. Thus, we considered that 70% of the requests would ask for 1 slot and the rest (30%) would ask for 2 slots.

Within this approach, the Network Controller serves the requests in a frame by frame basis, i. e., a request can only be served if the 1 or 2 slots which that request needs can be allocated in the data frame which is being scheduled at that moment. If not the request is lost.

Figure 4 shows a sample of the simulation results obtained for the case of unbalanced traffic type, $N_C = 10$, and up to a total number of 200 dummy ports.

Note that **p** is the requests offered load (to the signalling channels), while ρ is the throughput requested from the network (i. e. the offered load in terms of resources required from the network over the maximum resources available). For the case of 200 dummy ports, the maximum value of ρ is 0.43. This is because the total capacity of this network configuration is:

[(100 direct ports + 200 dummy ports) 100 wavelength per port] 622 Mbps = = 18.66 Tbps

while the maximum requested capacity (when **p** = 1) is:

10^9 requests per second x 1.3 required slots per request x 6220 bit per slot = = 8.086 Tbps (1.3 is the average of required slots per request = 0.7 x 1 + 0.3 x 2)

thus,

ρ (max) = 8.086 Tbps/18.66 Tbps = 0.43.

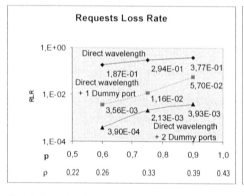

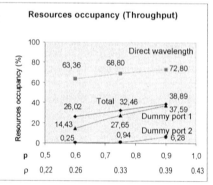

Fig. 4. Sample of Request Loss Rate and Resource Occupancy when using a per packet request approach (using two wavelength converters plus the direct wavelength per PON).

Figure 4 shows that the SONATA network exhibits low efficiency of its transmission capacity. This is mainly due to the fact that the simulation model is based on the per packet request approach. Using this approach, the signalling bandwidth easily becomes a bottleneck. In this particular case (using two wavelength converters plus the direct wavelength), we can only ask for a 43 % (ρ = 0.43) of the total capacity of the network. This is a little bit better than when only a single wavelength converter is added to the direct wavelength (see Figure 5). In this case, a higher resource occupancy can be reached (up to 65 %), but then the Request Loss Rate performance degrades drastically.

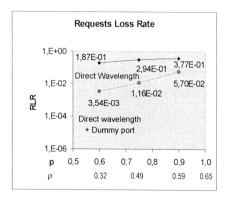

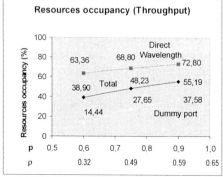

Fig. 5. Sample of Request Loss Rate and Resource Occupancy when using a per packet request approach (using a single wavelength converter plus the direct wavelength per PON).

To solve this problem without trying to enhance the MAC protocol established for signalling, either we increase the number of signalling wavelengths (N_C) or we reduce the amount of signalling by issuing per packet flow requests instead of per individual packet requests. The former is easy to implement, but then in practice we can come to the problem that, in such a big network, the computing requirements to execute the scheduling algorithm exceed the current available limits. Thus, as mentioned before, in our simulator we decided to implement the per packet-flow request approach option. This option is also not realistic since (at least as of today) signalling can not be implement at TCP/UDP level –only *ftp* announces the amount of data to be transferred when the TCP session is set up. We any way assumed such a visionary scenario because it completely avoids the bottleneck-signalling problem, and facilitate to cope with our main objective, namely to test the SONATA potential capabilities.

4.2 Per Packet-Flow Request Approach

In order to carry out the simulations to assess the feasibility and performance of the network control unit using a per packet-flow request approach, we assumed that the traffic in the Internet has the following composition: 42% of the traffic is *http* (35%) plus *ftp* (7%). 45% of the traffic is *irc* (23%) plus *telnet* (7%) plus *smtp* (5%) plus games (5%) plus other applications (5%) which typically implies low data transfer. The rest (around 13%) is IP over IP traffic (tunneling). We also assumed that within the *http* plus *ftp* traffic, 20% of flows in average require 2 slots, 21% of flows in average require 13 slots, 6% of flows in average require 130 slots, and finally 0.5% of flows in average require 1300 slots. For the *irc* plus *telnet* plus *smtp* plus games etc. traffic we assumed that 100% of flows require a single slot. And finally for the IP over IP traffic we considered as it has the same composition than the ordinary IP traffic (i. e., 70% of packets require 1 slot while 30% require 2 slots). According to these assumptions, the number of required slots per request distribution we adopted in our simulations was 46% of requests 1 slot, 23% of requests 2 slots, 23% of requests 13 slots, and 8% of requests 130 slots.

In contrast to the per packet request approach, in the per packet-flow request approach, the requests are served by allocating the requested slots in consecutive data frames. For example, consider the case of a request needing for 2 slots. If this request can be served, the first slot is allocated in the data frame, which is being scheduled at that moment, and for the second slot a reservation is made to allocate it in the next data frame.

Figure 6 shows a sample of the results obtained in the per packet-flow request approach simulations. These results correspond to the case of unbalanced traffic type, number of signalling channels $N_C = 3$, and up to a total number of 400 dummy ports. The total capacity of the network configuration, in this case, is:

[(100 direct ports + 400 dummy ports) 100 wavelength per port] 622 Mbps =
= 31.10 Tbps

while the maximum requested capacity (when $p = 1$) is:

3×10^8 requests per sec. x 14.31 required slots per request x 6220 bit per slot =
= 26.702 Tbps (14.31 is the average of required slots per request, i.e.,
0.46 x 1 + 0.23 x 2 +0.23 x 13 + 0.08 x 130)

thus,

ρ (max) = 26.702 Tbps/31.10 Tbps = 0.86.

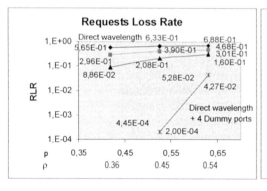

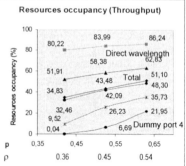

Fig. 6. Sample of Request Loss Rate and Resource Occupancy. Per packet-flow request approach and unbalanced traffic type.

Figure 6 shows that using the per packet-flow request approach, the network capacity can cope with the offered load, i. e. in this case ρ is closer to **p** than in the case of per packet request approach. Nevertheless, the performance is again very bad. As an example in Figure 6 it can be seen that for the case of 51.1% of the resource occupancy (a very low occupancy), the Request Loss Rate is 4.27 10^{-2}, which is a very bad figure. This is due to the fact that the wavelengths of the dummy ports were uniformly assigned to the destination PON, while the destination traffic distribution was Gaussian with a traffic concentration between PONs from 30 to 70. Figure 7, which includes the per PON resource occupancy distribution, shows the effect of this traffic concentration.

Both diagrams of Figure 7 show that wavelengths of the dummy ports assigned to PONs from 30 to 70 are overloaded, while wavelengths of the dummy ports assigned to PONs from 1 to 29 and from 71 to 100 are almost unused.

Obviously, a better performance can be obtained by an unbalanced assignment of the wavelengths to PONs, i. e., assigning more wavelengths to reach PONs from 30 to 70 than to reach PONs from 1 to 29 and from 71 to 100. This is shown in the next

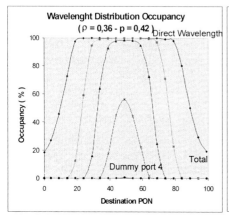

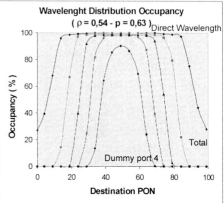

Fig. 7. Per PON wavelength occupancy distribution.

experiment, which results are summarised in Figure 8. Figure 8 includes a sample of the simulation results obtained with the per packet-flow request approach for the case of balanced traffic type, $N_C = 3$, and up to a total number of 200 dummy ports.

Figure 8 shows that, when the wavelengths of the dummy ports to the destination PONs assignment fits with the distribution of the traffic destination PONs, SONATA exhibits a very good performance. Note that even when reaching a high resource occupancy, as for example 80%, the Request Loss Rate stays below 10^{-2} ($7.5 \cdot 10^{-3}$ in the case of a total resource occupancy of 80%). This is, in fact, the effect that would result from a joint collaboration of the scheduling algorithm and a dynamic logical topology design algorithm, which would be able to adapt the wavelength converters to the instantaneous traffic requirements.

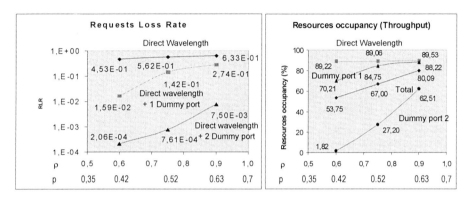

Fig. 8. Sample of Request Loss Rate and Resource Occupancy. Per packet-flow request approach and balanced traffic type.

Figure 8 depicts, in our opinion, the most representative performance figures of the potential capabilities that can be obtained from, in general, a large scale switchless based network solution, and in particular from SONATA using a heuristic scheduling algorithm.

5 Conclusions

SONATA Network is a switchless single layer optical transport network designed to cover national and metropolitan areas, which exemplifies a possible vision of future optical Internet backbones. In this paper we have presented some performance evaluation results obtained by simulation, which provide a picture of the potential capabilities of a national scale configuration of SONATA network based on a simple heuristic algorithm of scheduling the user transmission requests that was adopted within the SONATA project.

This picture shows that the SONATA approach is feasible and can provide good Resource Occupancy and Request Loos Rate figures. Nevertheless, In order to reach these figures, apart from overcoming the signalling-bottleneck problem, a dynamic assignment of the dummy ports wavelength to the PONs (dynamic logic topology design) algorithm has to be used, jointly with the scheduling algorithm, to implement the Network Controller. Such an algorithm was already proposed last year of the

SONATA project and is currently incorporated to the SONATA simulation platform. A future version of this paper will include the simulation results that will be obtained from the new platform.

Acknowledgements

This work is part of the SONATA project (AC351), partially funded by the European Commission under the Advanced Communications Technology and Services (ACTS) program. This work was also partially funded by CICYT (Spanish Ministry of Education) under contract TIC99-0572-C02-02. The authors thank CEPBA (the European Center for Parallelism of Barcelona) for allowing them to use the Silicon Graphics 2000. The authors also thank Dr. Alan Hill of British Telecom for his comments and suggestions.

References

1. Nuncio P. Caponio, Alan M. Hill, Fabio Neri and Roberto Sabella, "Single Layer Optical Platform Based on WDM/TDM Multiple Access for Large Scale "Switchless" Networks", European Transactions on Telecommunications, Vol. 11, No. 1, January-February 2000.
2. S. Bhatt, R. Fujimoto, A. Ogielski, K. Perumalla, "Parallel Simulation Techniques for large-scale Networks", IEEE Communications Magazine, pp. 42-47, August 1998.
3. A. Bianco, E. Leonardi, M. Mellia, M. Motisi and F. Neri. "Initial Considerations on the Implementation of Control Functions in SONATA", Working Document, ACTS project AC351 SONATA, Politecnico di Torino, 1998.
4. Kevin Thompson, Gregory J. Miller and Rick Wilder, "Wide-Area Internet Traffic Patterns and Characteristics", IEEE Network, November-December 1997.

Broadband over Inverse Multiplexed xDSL Modems

Einar Edvardsen

Telenor R & D
PO Box 83, 2027 Kjeller, Norway
Einar-paul.edvardsen@telenor.com

Abstract. The paper describes a new scenario for how to transform the existing telephone access network into a real broadband network. The objective is to introduce a low-cost network evolution scenario, which makes maximal use of the huge investments connected to this kind of networks. The existing network is sub-divided into a small cell network, each covering an area with an approximate radius of 1 kilometre. Inverse multiplexing of xDSL systems is used to aggregate bandwidths in the range of 100 to 1000 Mb/s between nodes in the network. VDSL modems able to carry about 25-50 Mb/s, are used on the last section from the node to the customer premises. These modems provide all households within each geographical cell with real broadband access to a large common bandwidth.

Keywords: Inverse multiplexing, ATM, xDSL, twisted pairs, FSAN, ITUNET

1   Background

At the present time there is a common understanding that the vision of 'Fibre-To-The-Home' (FTTH) still belongs to an undefined future. This is both due to the fact that the demand for broadband services has not matured yet, that the related investments is supposed to be enormous and that such a huge civil work will take long time to accomplish. Other intermediate solutions are therefore looked upon to be more realistic.

FTTC (Fibre-To-The-Cabinet) is one of the approaches that have been adopted by the FSAN (Full Service Access Network) group. The idea behind FTTC is to utilise the existing telephone access network from the cabinet to the customer, but use optical fibre cables on sections between the cabinets and the broadband switch. By doing it this way, optical fibre cables will only be installed on sections were the costs can be shared by large numbers of customers, while the existing infrastructure will be used on the rest. In practice this means that ADSL or VDSL technology is supposed to play a major role in the access network in order to make available the necessary bandwidth on the last mile to the customers.

Though the FTTC is flexible and less expensive than the FTTH scenario, it still involves huge investments. In areas were cable ducts do not exist, optical cables will have to be buried in the ground. Even in cases were such ducts do exist, the cost of fibre cables and related installation work will reach considerable amounts and it will take time to accomplish – time that telecom operator companies possibly do not have.

New competitors entering the market are offering broadband access over power-lines, radio, satellites and CATV networks. To maintain their market shares, the owners of the telephone cable network will have to meet this challenge by upgrading their networks. As we shall see, their potential performance is remarkable, and their possibility to still keep ahead in the race for broadband provisioning is rather good.

1.1 The capacity of telephone cables

Though each single twisted pair in a telephone cable has a limited bandwidth, the total capacity of a telephone cable can nevertheless be very large. This is due to the fact that telephone cables contain large numbers of copper pairs, each of them can by utilising xDSL technology, provide a relative large bandwidth over a certain distance. If the capacity of each single pair could be summed, the aggregated capacity would be huge – see figure 1 below. Inverse multiplexing is the technology that enables such aggregation of capacity from a number of lower bandwidth digital channels.

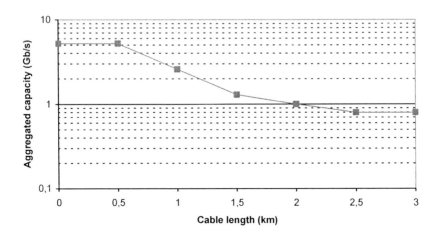

Fig. 1.   Capacity of one hundred twisted pairs as a function of the distance

The figure shows the aggregated bandwidth of a telephone cable with 100 twisted pairs as a function of line length by using standard xDSL modems. A telephone cable with one hundred twisted pairs are in this context a small cable. The graphs indicate also that within reasonable cable lengths, the capacity of one hundred twisted pairs is in the range of several gigabits, i.e. a capacity that today is far beyond what is needed in residential access networks.

These brief calculations are of course not exact. Noise and cross talk influence on the obtainable bandwidth, resulting in less performance. Nevertheless they do indicate that the telephone cables are much more powerful than we are used to believe. The question is whether the bandwidth of this network can be exploited in an efficient way, thus opening a new possibility of how to provide real broadband access to the general public.

2 Technology

2.1 Inverse multiplexing

Inverse multiplexing is used to aggregate bandwidth from a number of 'lines with smaller bandwidths' as shown in figure 2 below:

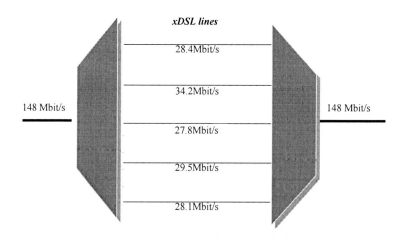

Fig. 2. The principle of inverse multiplexing

The incoming traffic stream from the left is distributed sequentially over a number of lower bitrate lines, and reassembled at the other end.

Inverse multiplexing is standardised by a number of standardisation organisations, such as ITU, ETSI and ATM Forum. ATM Forum's AF-PHY-0086.001 covers inverse multiplexing E1/DS1 links to aggregate bandwidths up to E3/DS3 rates. Though it from a technical point of view is possible to multiplex channels running various bitrates, this has so far not been covered by any official standard yet. Since xDSL modems provide bandwidths deviating from the standardised hierarchy, the standards cannot be applied unless it is used on circuit emulated E1/DS1 channels.

It exists several methods of how to perform inverse multiplexing of packet based traffic (ATM). One of them is shown in figure 3, where the principle of sequence numbering is used.

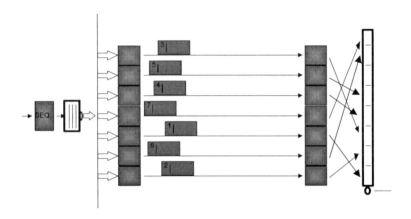

Fig. 3. Inverse multiplexing by use of sequence numbering over xDSL lines

Each cell arriving from the left is given a sequence number before transmitted over one of the available lines. Though not indicated on this figure, the bandwidth of each line may differ from each other. One line can perform 13 Mb/s, while an other may perform 19,3 Mb/s. The cells will therefore arrive out of order at the receiver end, and has to be intermediately stored in the receiver buffer on right side in order to re-establish the sequential order. Since it is mandatory that inverse multiplexers should be transparent for traffic streams, and that the ATM cell itself does not contain any field useful for carrying the sequence number, it must be transferred as a tag to the cell. The cell length will therefore deviate from the ATM standard (53 bytes). However, the interface between the two terminals of the inverse multiplexer can be looked upon as an internal interface, and a deviation from the standards is therefore acceptable.

Mixing of lines with low speed and different bandwidths may create delay and cell delay variation (CDV) that are unacceptable for certain types of traffic. To meet requirements from the various traffic types, each of them may have to use dedicated line groups with properties that match their demands.

2.2 The network concept

The proposed network concept, from here of named the IMUX concept, is based upon using the existing telephone access network infrastructure. In practice it means that the existing telephone cables are used to provide broadband access to the public. In the general structure of the new network, new access nodes have to be installed near the existing street cabinets. These network nodes must perform both inverse multiplexing to establish the transmission links to other nodes, and statistical multiplexing to concentrate the traffic from the users. Figure 4 below shows a functional description of such a network node.

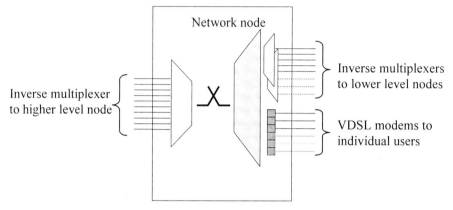

Fig. 4. The network Node

Figure 5 below shows the principal structure of the new network. The nodes in the network are installed near the distribution cabinets in the old telephone network in order to have easy access to the necessary copper pairs. The nominal distance between the nodes is about 2 kilometres. Each node covers an area with an approximate radius of 1 kilometre, which makes it possible to provide 25-50 Mb/s to each of the customers.

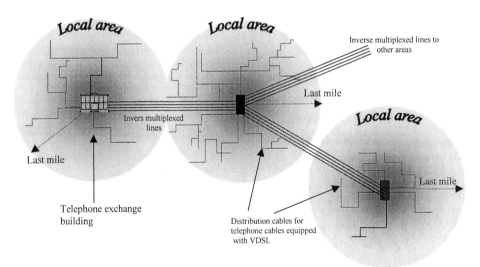

Fig. 5. The architecture of the IMUX network

Inverse multiplexing over a number of copper pairs running VDSL modems is used to aggregate the requested bandwidth between the main ATM switch located at the telephone exchange building, and nodes in the network. The link capacity needed between the nodes is estimated to be in the range of 100 to 1000 Mb/s, which can be obtained by inverse multiplexing of a few tens of lines.

Each network node performs both inverse multiplexing and statistical multiplexing of traffic flows from the individual users. The users are connected to the nearest node by VDSL modems, giving them access to a large, common bandwidth shared among them. The nodes will be equipped with standard ATM signalling enabling use of SVCs (Switched Virtual Connections) with QoS (Quality of Service) as defined by relevant standards.

The structure of this network is similar to the recommendation from the FSAN (Full Service Access Network) consortium. The difference is that FSAN recommends optical fibres between the nodes, while this approach is based upon copper. However, due to the similarities between the two concepts, it is easy to adopt the technology that fits the need best in each case. On sections were optical cables can be installed at low cost, fibre cables are the natural choice. But on sections where optical cables cannot be installed at a reasonable cost, inverse multiplexing is the choice. The two approaches go hand in hand – they will complement each other in a perfect way that both the operators and the public may profit on.

3   The ACTS project AC309 ITUNET

To promote the network concept and to pave the way for commercial products needed to implement this kind of network, the EU financed ACTS project AC309 ITUNET was initiated. The project started in the middle of March 1998 and was terminated at the end of March 2000.

The main objectives of the project was to study how the existing access network infrastructures can be upgraded using xDSL technologies to form a cost-efficient integrated service network providing the necessary capacity and functionality for broadband services.

In order to identify eventual limitations of the technology and the concept, an experimental access node was developed. The node performed inverse multiplexing of up to 16 VDSL lines and statistical multiplexing of up to 28 user modems.

Secondly, since the existing infrastructure imposes strong requirements to the location of network nodes, it was of special interest to carry through implementability studies from real networks in order to see if the network could be transformed according to the concept. Were there copper pairs enough on all sections? Were the line lengths within acceptable ranges? How about power supply?

And thirdly, it was necessary to see if it was possible to realise the necessary bandwidth. Video services like TV broadcasting (multi-casting) and Video-on-Demand were chosen to be the killer applications that every household wanted to subscribe to.

The project also carried through in-depth calculations and evaluation of the cost related to establishing access networks based on inverse multiplexing.

The project counted telecom operators and industrial partners from France, Switzerland and Norway.

4   Evaluation

In order to have a rough evaluation of the implementability of such a network and also some imagination about the cost to establish it, several case studies were performed. The case studies covered various areas in Norway and Switzerland. Rural areas as well as urban areas were investigated.

In the following sub-sections one of these case studies is described in details, while only summary information about the others is given.

4.1   The upgraded network

NorVillageA is a small village in a rural area with population consisting mainly of farmers and private households. The village had 728 telephone subscribers in total. 384 of them lived closer to the telephone exchange building than 1000 metres, thus one could provide 25-50 Mb/s to each of them directly from the exchange building with single user VDSL modems. The remaining subscribers, 344, were connected over remote access nodes. See figure 6 below.

To connect all customers, 18 access nodes had to be installed. The average number of customers connected to each access node was 19. Due to the topology and the existing cable infrastructure access nodes had to be installed even at locations with very few customers (<5). As a result of this, the number of network nodes became higher than expected.

The case study also uncovered that due to the above non-optimal location of the network nodes, up to 4 cascaded nodes were needed to reach the most distant users.

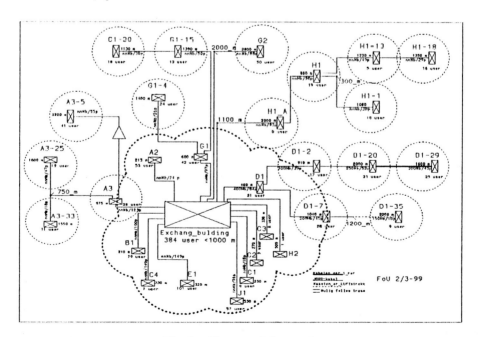

Fig. 6.   The upgraded network

4.2   Cost comparison

Since the FTTC and the IMUX architectures only differ as regards one item, optical fibres vs. inverse multiplexers on the feeder link, the cost comparison could be made very simple. The number of VDSL modems needed in the two concepts is almost equal since both concepts recommend use of modems between the access node and the customers. The number of end-user modems is much higher than the number of modems used for establishing link capacities. It is assumed that IMUX concept needs 10-15 % more modems than the FTTC concept. Consequently, it was not

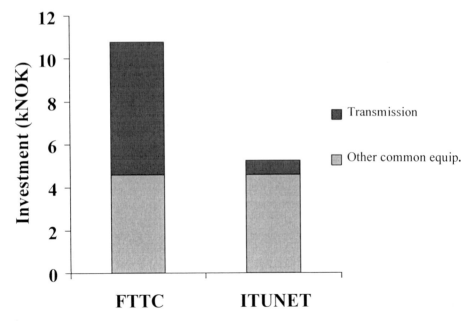

Fig. 7.   Comparison of investment cost

The result from the cost comparison (Figure 7) shows that the investment to establish link capacities between the nodes in the network can be reduced by approximately 90 % by utilising IMUX technology. This percentage has been verified by a number of case studies in different countries as well as different topologies

5   Recommendations

The IMUX concept is introduced as an independent strategy for how to build a broadband access network able to provide both todays and tomorrow's services to the general public. This does not mean that other solutions should be excluded. The concept is well adapted to other approaches, such as the above-mentioned FTTC approach. These two strategies are complementary in the sense that one easily can combine sub-nets based upon the two technologies without large influence on the overall network structure. Both of the two approaches rely on utilisation of the existing telephone network. Optical fibres can supplement a network based upon the IMUX idea, as well as inverse multiplexers can supplement a network based upon in the FTTC concept.

The IMUX concept is an inexpensive and fast method to establish a broadband access network. The investment cost to establish link capacities can be reduced by 90 % compared to the FTTC concept and the network can be realised very fast since civil work is reduced to a minimum. The will be no need to acquire ground and getting permissions to dig cable trenches.

The IMUX concept only requests a moderate increase of the total number of modems in the network. The estimated increase is approximately 10-15 %, which is judged to be insignificant as regards maintenance cost.

An optical cable has more capacity than a telephone cable, but the bandwidth of the latter one will in most cases be sufficient. Bandwidths in the range of Gb/s can relatively easily be achieved.

None of the concepts need to influence on the telephony service. Both analogue and digital telephony can be transmitted in a separate frequency band on the twisted pairs.

The complexity of the remote network nodes in the two cases will only be slightly increased since inverse multiplexers will have to be integrated into the IMUX equipment.

Global Service Control for Multi-media Networks

Tanja de Groot [1], Robert Mathonet [2], David Stevenson [3]

[1] Alcatel CIT, 10 rue Latécoère,
78141 Vélizy, France
tanja.de-groot@alcatel.fr
[2] Alcatel TITN Answare, 1 Rue Ampère,
91747 Massy Cedex, France,
robert.mathonet@alcatel.fr
[3] Alcatel, 280 Botany Road,
Alexandria 2015, Australia
david.stevenson@alcatel.com.au

Abstract. The evolution of the market towards broadband and IP-based services is a major driving factor for network operators towards the integration of their so far separate data and telephony networks. This paper presents an approach to realize a mode of Global Service Control for telecommunication network operators that provides simplified automated management of services over a heterogeneous, multi-vendor network infrastructure. Combining existing management paradigms with new integration techniques, the Global Service Control approach makes it possible to manage carrier-class quality services over the global multi-service multi-media network.

1   Introduction

The trend of many incumbent operators to migrate to global multi-service networks and to provide ATM or IP-based services to business customers reflects a major change in strategy with respect to the up-to-now separate voice, data and IP networks. At the same time, operators are investing more and more in future-safe, flexible technology based operations systems (OS) that promise to provide them with

- maximum automation and control of service management solutions.
- flexibility in service definition
- speed in service activation
- security in service assurance

Defining, designing, implementing, testing and operating services on the multi-media, multi-vendor network infrastructure are as many complex tasks that impact various aspects of the operator's business, including strategy, business processes, people and technology.

For incumbent operators, the success of an evolution of his existing network to one that can support the new types of multi-media services depends on one hand on the ability to move their current business environment to the target convergent network porting narrowband, broadband and IP-based business environment. Investments in network evolutions will only create the expected benefits if services are successfully implemented from the perspective of the operator's customers.

This paper presents a network and service management analysis and integration approach that can serve both as the basis for the evolution of an incumbent operator's business, or for defining and realizing the business of a new operator. It is referred to as the Global Service Control solution.

The realization of a Global Service Control solution depends on the availability of an effective Integrated Network Management (INM) system. This INM system needs to provide a coherent view and control of the network resources available for services, and to provide automated processes for service activation and surveillance across the entire network.

Several Global Service Control implementations have been realized for major operators worldwide. This has made it possible to refine the approach described here to encompass multi-technology, multi-media and multi-vendor network and service management.

2   The End-to-End Service Management Challenge

The challenge for operators is to provide services that offer more value to their customers than those of their competitors, that may run over both public and private networks, and that provide an optimal cost/revenue ration despite the fact that these networks are in general multi-vendor, multi-technology and multi-operator in nature. This requires an operator to build a Global Service Control solution that allows him to compete efficiently in terms of:

- rapid service introduction and flexible service provisioning;
- presentation, pricing and packaging of services, which will change as market demand evolves;
- added-value services, thus reducing customer churn.

A Global Service Control solution will be the differentiating factor for the operator, making it possible to integrate a wide range of different services, and to provide a one-stop, high-speed shopping point to his customers.

The three key operational processes for achieving Global Service Control are:

- *Service provisioning:* Allows the operator to economically and rapidly provide a wide variety of services. This includes sophisticated pre-order processes to reliably predict what can be delivered to customers and when.
- *Service assurance:* Includes flexible quality of service management, allowing the operator to offer a predictable service at a reasonable cost.

- *Billing:* collection and analysis of the usage data of the services, in order to charge customers for services in accordance with their actual needs and to create personalized bills.

Two conditions must be satisfied to meet these challenges: the integrated management system must provide a network-wide and customer-oriented view of the services, and the processes mentioned above must be automated as much as possible.

A phased process is required to define an operator's required business model and to achieve an actual Global Service Control solution supporting and enforcing this business model. This process must concentrate on high-speed product time-to-market, product flexibility, return on investment for the operator, and end-customer satisfaction.

3   Divide and Conquer the Multi-technology Network

The way to achieve a Global Service Control solution is based on a three step methodology:

Step 1: Target Business Analysis
In this step, the operator's high value products and target services (e.g. voice over packet) are clearly identified. For incumbent operators, this includes the existing products and services that first have to migrate to Global Service Control environment. Achieving this maximizes the operator's return on investment and clearly positions the operator in the customers' minds as the carrier of choice.

Step 2: Definition of the Operational Process Model
The goal of this step is to define the operator's target operational process model. In case of incumbent operators, it might include the analysis of existing operational professes. The target operational process model shall:

- provide flexible, high-speed support for the introduction of customer-oriented services;
- be easy to use (pushbutton) by the operator's employees and, optionally, by the operator's customers;
- be a smooth evolution of the existing model (incumbent operators only).

Step 3: Service Evolution Realization through INM
Step 3 requires the controlled and coordinated execution of all actions needed to realize a Global Service Control system for the selected services, as defined in step 1, and according to the target process model defined in step 2.

The goal of the Global Service Control solution is to provide a service management support according to the operators business model, that hides the complexity and diversity of the underlying network and that proposes automated processes that comply with the service operator's process model.

Implementing the Global Service Control solution means bridging the gap between customer-oriented service management and network-oriented technology management. This requires a detailed knowledge of the actual implementation of services using the different underlying network technologies involved. A Global Service Control solution can only be realized through Integrated Network Management (INM).

4 From Technology (TMN) to Business Management (TMF)

Until recently, the main worry of many operators was principally the well functioning of the (mono- technology) network. Today's management systems for telecommunication networks fully support the operator's in this function. A main set of requirements for network management have been defined by a set of ITU-T recommendations.

In particular, standard M.3010 [1] defines a Telecommunication Management Network (TMN) architecture with different management layers. The Element Management Layer (EML) and Network Management Layer (NML) deal with the technology aspects of the network. They are dedicated to management of the detailed equipment aspects and network connectivity, respectively.

Furthermore, the TMN model defines the Service Management Layer (SML) and Business Management Layer (BML) to handle service activation and assurance processes and customer-related aspects and business support systems (e.g. workforce management) respectively.

The TMN model works applies well when used with mono- technology networks, such as Synchronous Digital Hierarchy (SDH) or Asynchronous Transfer Mode (ATM) networks. Technology-dedicated standardized information models specify in detail the way in which such networks should be managed at the EML level.

Only a few standards exist today that define the management of a particular technical domain at network level (NML). An example is the ATM Forum M4 Network View model that defines ATM service management on a network basis rather than on an element basis.

Furthermore, virtually no TMN standards are available today for the service (SML) and business (BML) management layers.

Finally, a major drawback of today's TMN standards is that they address technology management rather than operational activities. In particular, they do not provide service- and customer-oriented management views that are required to perform effective service provisioning, activation and surveillance.

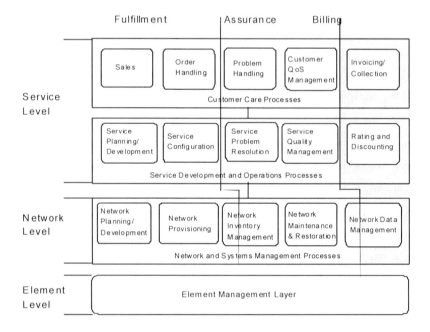

Fig. 1. TMF Operational processes on TMN architecture (TMF Operations Map)

For these reasons, the TMN model has been complemented by an operation model description, the Telecommunication Management Forum (TMF) Operations Map (TOM) ([2], see Figure 1). The TOM addresses the management problem starting from service management and operational requirements point of view.

More and more, the TOM model is used as the main reference by operators. This does not mean that the TMN functional specifications and network object-oriented models have become invalid. It mainly means that these functions now are seen as a basic given of a Network Management System, and that they serve as the basis for supporting the TOM operational processes.

5   Service Management: from Dedicated to Cross-technology

Apart from the fact that the accent for operator's is moving from TMN defined technology management to TMF operational process support, another shift in focus is taking place due to the arrival of convergent voice and data network architectures. The related management issue, that is not covered by today's TMN standards, is the case of managing services that may cross one or more technology boundaries.

Relations for managing the dependencies between different technologies supporting the same end-service are not defined by any standard. As an example, the impact of a fiber-optic cable break on the customers of an IP data transfer service carried over an ATM transport layer using an SDH network infrastructure is not defined anywhere.

A combination of TMN, TMF and new models is needed to achieve a process-oriented service management approach that integrates the different technology domains of the multi-vendor network into end-to-end network views and operations. Only with such a combined approach can a network management system fully support the requirements that allow an operator to achieve Global Service Control.

6 Global Service Control Implementation

According to the requirements listed in the previous sections on TMN technology management, TMF operational process support and cross-technology service management, Global Service Control relies on the implementation of an Integrated Network Management (INM) system, which allows to map between the objects present in the complex network infrastructure (e.g. switches, cards, ports) and the services provided on top of this infrastructure (e.g. end-to-end connections, virtual private networks (VPNs)), as shown in Figure 2.

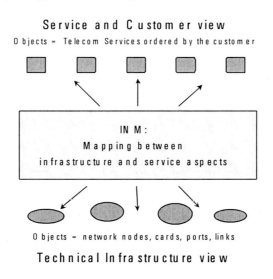

Fig. 2. The role of INM domain managers

This feature is mandatory to perform end-to-end service provisioning and activation, and to measure quality of the offered services. For example, it allows operators to determine the impact of technical faults on affected services and customers. Management processes are adapted to Service Level Agreements (SLA), helping to optimize/maximize service revenues by minimizing service outages relative to the contracted SLAs.

The key concept of such INM solutions is the Domain Manager: the tool that bridges the complexity gap between the service-oriented view and the network-oriented views.

A Domain Manager represents an abstraction function that allows disparate technologies within the network to be managed to be handled according to business processes, rather than by technological processes.

In most cases, several levels of abstraction (2 or 3) are needed to master the complexity of the full multi-technology multi-vendor network. This is done by dividing the full network into service oriented entities, each of which is handled by a dedicated Domain Manager that makes the network to service abstraction for that part of the network.

Each Domain Manager provides the hooks down and upwards that are needed to reach the required level of service management and supporting the main operational processes for its part of the network.

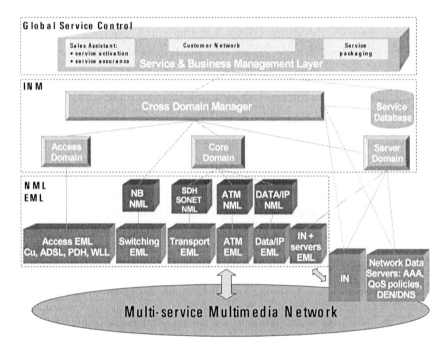

Fig. 3. Example of a Global Service Control architecture

By creating the appropriate number of Domain Managers, according to the size and complexity of the network, a divide and conquer strategy is used to master the service provisioning, assurance and billing processes across the complex network.

Thus, using the concept of Domain Manager, the shift in focus from technology management to service control is achieved. The top-level Domain Manager, also referred to as the Cross Domain Manager, At the top-level, the Cross Domain Manager interfaces with service dedicated or business management products, or with the operator's existing Operations Support System (OSS).

Figure 3 illustrates the architecture of such an INM system. This example uses three Domain Managers and a Cross Domain Manager to handle the complex network. Such solutions have been implemented by Alcatel for major telecommunication companies. It focuses on end-to-end service management while allowing for the use of either parallel or convergent voice and data networks.

Alcatel's experience in Domain Manager based approaches (i.e. ALMA Vision) has proven the need for the presence of carrier class technology managers (i.e. element and network technology managers, or EML/NML) as the basis on which to build the INM support (Domain and Cross-Domain Managers) needed to realize a Global Service Control solution.

Initiatives, such as the TMF Catalyst project on generic connection management, use the concept of Domain Manager as their basis.

The tendency is to use CORBA-based software platforms that enable the different modules of the solutions (Technology, Domain and Service/Business Managers) to be connected together in a flexible way. TMF CASMIM [3] is defining a service-oriented generic network model in Interface Definition Language (IDL), which complies with TMF and TMN requirements. These interfaces are also meant to be the used to communicate with OSS applications.

The main set of functions supported by a Domain Manager are related to circuit or path management, topology management, alarm management, with a specific focus on alarms affecting circuits or paths; path performance monitoring, traffic statistics recording and SLA management.

With this kind of architecture, an operator will be able to control services over yesterday's, today's and tomorrow's network infrastructure. The approach allows an operator

- to capitalize on his existing management systems, by integrating them under a Domain Manager
- to flexibly extend or add new networks and their management systems under appropriate Domain Managers
- to make service and business management independent of the networking technology
- optionally, to migrate existing services smoothly to the convergent network architecture

7 Conclusions

This paper has presented the Global Service Control concept, which is crucial for operators to develop their business in a competitive environment characterized by very rapid evolution of telecom technology towards convergent networks, and by the diversity and sophistication of services, especially those proposed to corporate customers.

A methodology has been presented to achieve Global Service Control in a progressive way, starting from the operator's current situation. It has been shown that the implementation of Global Service Control requires the deployment of a carrier class INM solution to perform the mapping between technology-oriented and customer service-oriented management objects and features.

As illustrated in the paper, the Global Service Control approach, which has been realized by Alcatel for large network operators, covers the handling

- from the service definition and customer interaction,
- through the realization of an INM system using intermediate points of control (Domain Managers) for managing service provisioning and activation, service assurance (including identification and authentication, service level agreement management, directory services, etc), and billing,
- all the way down to the technical management of the network resources.

Building an actual Global Service Control support system, requires a wide range of experience covering, amongst others, different networking techniques, service and business management and operational processes.

The methodology to reach Global Service Control applies both to today's existing services on convergent networks, as well as to the services that will be available on tomorrows multi-media networks.

The current tendency of operators to move existing services to packet based networks, and to define new services on those networks, does not change the requirements for the management of these service and the corresponding required control of the network resources and of the Quality of Service delivered by this network.

Thus, the requirements for Global Service Control remain valid in the currently foreseen network evolutions.

References

1. ITU-T: M.3010 Principles for a Telecommunications Management Network
2. Telecommunication Management Forum (TMF): TMF Operations Map
3. TMF web: http://www.tmforum.com/ for catalyst project information

Applicability of ABR Service to Internet Applications

Mika Ishizuka

NTT Information Sharing Platform Laboratories
3-9-11, Midori-cho, Musashino-shi, Tokyo, 180-8585 Japan
ishizuka.mika@lab.ntt.co.jp

Abstract. Many studies on the performance of the ABR service category have been done. Most of these were simulation studies, and most evaluated performance only in the ATM layer. However, simulations do not provide a complete picture of how ABR service is used by actual applications running on existing networks. Few studies of how applications operate in a real environment have been done. We have used a testbed ABR network to examine the applicability of Internet applications. In this network, We tested four types of application: an FTP application (bulk data transfer), and a WWW application (interactive flow), a Video on Demand (VOD) application that used Motion JPEG, and an IP telephony application. As a result, We found that ABR service can be applied to these applications and determined how to set the best ABR parameters to ensure the best performance of each application.

1. Introduction

Asynchronous Transfer Mode (ATM) is widely used to achieve high-speed communication. ATM has a number of service categories. Best effort service is suitable for Internet applications because these applications send unpredictable and burst traffic. A typical example of a best effort service category is Unspecified Bit Rate (UBR) service. However, UBR service uses no form of traffic control, so unless each application controls its traffic flow properly, application performance will be degraded by cell loss when a network is congested. To solve this problem, Available Bit Rate (ABR) service category was designated in the ATM Forum [1]. ABR service can avoid ATM layer congestion and minimize cell loss, because it uses a closed loop feedback rate control mechanism. In ABR service, a source end system calculates the Allowed Cell Rate (ACR) based on feedback information from a network. ABR service can also provide a Minimum Cell Rate (MCR) guarantee.

Many studies on the performance of the ABR service category have been done [2, 3]. Most of these were simulation studies, and most evaluated performance only in the ATM layer, (for example, ATM link utilization or the maximum queue length in ATM switches). However, simulations do not provide a complete picture of how ABR service is used by actual applications running on existing networks. Few studies of how applications operate in a real environment have been done.

We have used a testbed ABR network to examine the applicability of Internet applications. In this network, I tested four types of application. Two were TCP applications: an FTP application (bulk data transfer), and a WWW application (interactive flow). The other two were real-time applications: a Video on Demand (VOD) application that used Motion JPEG, and an IP telephony application.

The paper is organized as follows. In section 2, we explain ABR feedback control mechanism. In section 3, we depict configuration of our experiment. Then in section 4, we show experiment results and discuss theses results. Finally, in section 5, we conclude the paper.

2. ABR behavior

In this section, we briefly describe ABR feedback control mechanism. In ABR service, a Resource Management cell (RM cell) is used to form closed loop feedback control system. A source end system (SES) is given information about congestion through RM cells. According to these information, an SES calculates allowed cell rate (ACR), and adjusts sending cell rate to ACR. An ATM switch indicates information about congestion in one of three ways: EFCI, Relative Rate, and Explicit Rate (ER) marking. In this experiment, ATM switches supporting only EFCI marking are used. Therefore we explain EFCI marking here. In [1, 4], the others are described in detail.

- An SES generates an RM cell every after Nrm-1 cells.
- An ATM switch sets the EFCI bit in a data cell when its queue length exceeds its threshold.
- A destination end system sends the RM cell back to the SES, when an RM cell reaches it. At this time, congestion indication bit in the RM cell is set when the EFCI bit in the last received data cell was set.
- When an SES receives an RM cell, the SES calculates ACR as follows:
- if the CI bit in the RM cell is set
  ACR := max(ACR- RDF*ACR, MCR),
- else
  ACR := min(ACR+RIF*PCR, PCR).

3. Experiment configuration

3.1. Network Configuration

The network configuration we used is shown in Fig. 1.

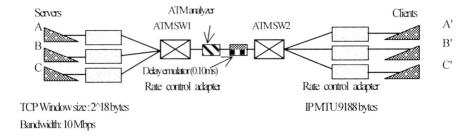

Fig. 1. Network configuration

- In each ATM switch, cells transmitted to the same output port share buffer memory, which size is configurable. In this configuration, it was set to 4k cells. Threshold for EFCI marking was set to 127 cells.
- As for end systems, SS20+Solaris2.4 (except C), and Ultra2+Solaris 2.5.1(C) were used. Solaris 2.4 supports window scaling option [5] by applying consulting-option.
- Rate Control Adapters (RCA) work as ABR end systems. ABR parameters were set based on ATM Forum TM 4.0.
- PVCs were used to set up ATM connections.
- Delay was set by emulator. Default setting was 10ms.

3.2. Applications setting

Traffic from A was sent to A', and that from B was to B'. These two connections offered background traffic using an FTP application. As for C to C', three applications were used, WWW, VoD, and IP Telephony.

In a WWW application, thirty identical WWW pages, which include a html (600bytes) and a gif file (300kbytes), were downloaded. Cache setting at a WWW client was disabled.

Table 1 shows other settings concerning to each application.

Table 1. applications used in this experiment.

AP	protocol	parameters CODEC, etc.
FTP	TCP	TCP maximum window size 2^{18} bytes
WWW	TCP	TCP maximum window size 9148 bytes
VoD	UDP	Motion JPEG frame rate = 24fps
IP Telephony	RTP(UDP)	Freephon[6] (with recovery techniques [7]).

3.3. Performance metrics

- Common among three applications : cell rate between ATM SW1 and 2, and ACR of each connection.

- FTP: throughput.
- WWW : response time. In this paper, interval between a request passing the ATM analyzer and the last cell of response data passing it is defined as response time.
- VOD : loss ratio that is indicated by the application.
- IP telephony : packet interval.

4. Results and analysis

4.1. WWW

Three scenarios were tested to examine applicability of ABR to a WWW application: (1) A UBR connection for a WWW application without background traffic, (2) Three UBR connections sharing bandwidth, one of these connections for a WWW application and the others for an FTP application, (3) Three ABR connections sharing bandwidth, which use the same application as (2).

For each scenario, cell rate between ATM switches is shown in Figs. 2,3, and 4 (measurement interval is 100ms).

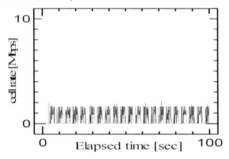

Fig. 2. Cell rate between switches for a UBR connection without background traffic

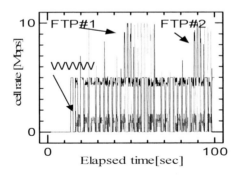

Fig. 3. Cell rate between switches for UBR connections with background traffic

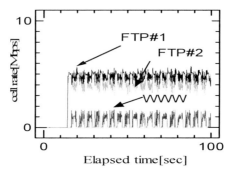

Fig. 4. Cell rate between switches for ABR connections with background traffic

From these figures, We found ABR is better than UBR, because the available bandwidth can be fairly assigned to each connection. Note that, when we use the ABR, a WWW application can be assigned a fair allocation of bandwidth even though its traffic is less than others. This is because the larger a VC shares bandwidth, the larger rate reduction the VC suffers.

Then, a response time and ftp throughput are shown in Tables 2, and 3. In Table 3, ABR parameters, RIF and RDF, are changed to determine how to set the best parameters for a WWW application. In both tables, response time is calculated about a html and a gif file respectively.

Table 2. WWW responce time and FTP throughput for a UBR connection.

	No background traffic	UBR
Html response time[ms]	2.9±0.1	3.7±1.5
Gif response time[s]	3.22±0.15	3.55±0.94
ftp throughput[Mbps]	8.9	7.7

Table 3. WWW response time and FTP throughput for a ABR connection.

RDF	RIF	1/16	1/64	1/256
1/16	Html[ms]	4.1±0.8	4.7±2.0	4.4±1.5
	Gif[s]	2.66±0.08	2.68±0.10	2.67±0.08
	ftp[Mbps]	8.2	7.6	7.3
1/64	Html[ms]	4.2±0.9	4.2±0.8	4.1±0.8
	Gif[s]	2.65±0.07	2.64±0.08	2.63±0.06
	ftp[Mbps]	9.0	8.8	8.7
1/256	Html[ms]	4.3±0.9	4.2±0.8	4.2±0.8
	Gif[s]	3.59±0.83	2.67±0.09	2.66±0.06
	ftp[Mbps]	5.0	8.1	8.8

: cell loss is observed.

As is shown in Table 3, for bulk data transfer such as an FTP application, ABR is the best service category because the available bandwidth can be fairly assigned to each connection and cell loss can be minimize. As a result, a TCP connection can fully utilize the available bandwidth by minimizing the overhead caused by TCP retransmission of lost data. The ABR parameters must be set to avoid both buffer starvation and buffer overflow; this means aggressive parameter setting is desirable for this kind of application.

As for a WWW application, when we use no flow control, that is, when we use a UBR, a WWW application cannot be assigned a fair allocation of bandwidth, so cell loss occurs for a WWW connection even though the WWW application uses less bandwidth than the background traffic. As a result, use of a UBR degraded the response time in presence of the background traffic. Furthermore, response time becomes unstable, due to data loss, when many segments are retransmitted. Actually, about 4% of data segment are retransmitted in case of UBR. These two problems can be solved by use of ABR service, that is, short and stable response time is enabled. To avoid TCP retransmission by using congestion control mechanism in ABR keeps the quality of interactive TCP applications high. ABR parameters must be set to keep the queue length short to reduce delay and to avoid cell loss. Therefore, contrary to the case of an FTP application, parameters setting for a WWW application should be conservative.

Considering above results, to ensure good performance for both applications, FTP and WWW, (RIF,RDF)=(1/64,1/64) is the best in this environment (number of connections is 3, and ICR=PCR).

Finally, a response time and ftp throughput for three ICR settings are shown in Table 4. In addition, snapshot of ACR is shown in Fig. 5 (RDF=1/64 ,RIF=1/64, ICR=PCR/10) As is shown in Fig. 5, when ACR>ICR, ACR is reset to ICR before next WWW page is transmitted Therefore, a response time becomes longer by setting smaller ICR, though queue length can be kept low. From these results, small ICR setting is better, as long as a response time, when background traffic is not heavy, is not so worse.

Table 4. WWW response time and FTP throughput for an ABR connection when ICR is changed.

ICR	RIF	1/16	1/64	1/256
PCR	Html[ms]	4.2±0.89	4.2±0.8	4.1±0.8
	Gif[s]	2.67±0.07	2.65±0.08	2.63±0.06
	ftp[Mbps]	9.0	8.8	8.7
PCR/3	Html[ms]	4.8±0.8	4.6±0.2	4.7±0.2
	Gif[s]	3.21±0.13	3.21±0.17	3.21±0.23
	ftp[Mbps]	9.0	8.8	8.7
PCR/10	Html[ms]	11±0.4	11±0.3	11±0.5
	Gif[s]	3.23±0.19	3.23±0.13	3.20±0.16
	ftp[Mbps]	9.0	8.8	8.7

The following is the summary of this subsection.
[Applicability of ABR service for TCP applications]
- For an FTP application, ABR is the best service.
- For an interactive application such as a WWW application, ABR service also works well, because short and stable response times are achieved. ABR parameters must be set to keep the queue length short to reduce delay and to avoid cell loss

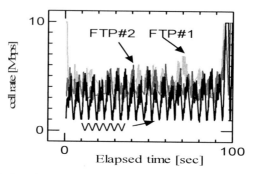

Fig. 5. Snapshot of ACR when RIF=RDF=1/64, and ICR=PCR/10

4.2. VoD

We made our measurements with a VoD application that uses UDP for data transmission. The CODEC of our VoD application is Motion JPEG. We consider three conditions for underlying networks: (1) A UBR connection for the VoD application without any other background load, (2) A UBR connections for the VoD application and two other UBR connections for background TCP traffic during 10 seconds, and (3) An ABR connections for the VoD application and two other ABR connections for background TCP traffic during 10 second. By observing video movies at the receiver side, we found the following results.

- In case of UBR connections with background traffic, we observe suspension of movie due to losses of movie frames. After the congestion period, the movie restart playing but lost part of the movie is skipped and is never displayed. As a result, we observe continuous loss of movie more than the half of the congestion duration.
- In case of ABR connections with background traffic, we observe delay in displaying movies during congestion periods. This is because ABR rate control adapter limits the transmission rate of video stream. After finishing the congestion period, the rate control adapter transmits the cells stored in its buffer at PCR and this results in high frame arrival rate at the receiver host. This causes overflow of the frame buffer at the receiver host and we observe approximately 10% loss of the movie at that period.

We show the transient cell rate characteristics in 100 ms time unit between the switches in Fig. 6 (UBR without background traffic), Fig. 7 (UBR with background traffic), and Fig. 8 (ABR with RDF=1/64 and RIF=1/64).

In Fig. 7, we can see that the share of VoD, which is UDP, is dominant and TCP traffic can obtain only a small share of the whole bandwidth. The total throughout for the two TCP connections was 0.7 Mbps. In case of Fig. 8, we used ABR connections and the total bandwidth is fairly shared by the three ABR connections. As a result, the total throughout for the two TCP connections was 6 Mbps.

As we mentioned above, when we use ABR, we observe loss in frames of VoD movie at the time when the background traffic terminated. To solve this traffic phenomenon, we propose two approaches. The first one is the reduction of video frame rate. The other one is the increase of MCR.

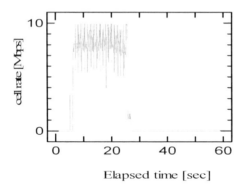

Fig. 6. Cell rate between switches for a UBR connection without background traffic.

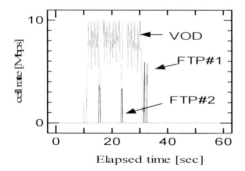

Fig. 7. Cell rate between switches for UBR connections with background

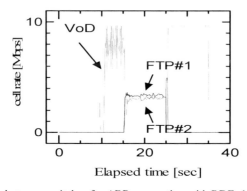

Fig. 8. Cell rate between switches for ABR connection with RDF=1/64 and RIF=1/64

(1) Reduction of video frame rate
We reduce frame transmission rate from 24 fps by 1 fps steps and search the maximum frame rate at which the VoD application does not indicate any frame loss. We found 9 fps is the maximum one. We show the results of the transient cell rate characteristics between the switches in Fig. 9 (UBR without background traffic), Fig. 10 (UBR with background traffic), and Fig. 11 (ABR).

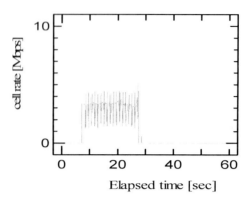

Fig. 9. Cell rate between switches for 9 fps VoD and a UBR connection without background traffic

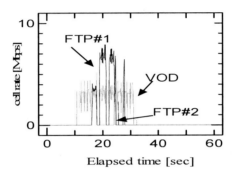

Fig. 10. Cell rate between switches for 9 fps VoD and UBR connections with background traffic

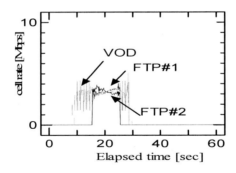

Fig. 11. Cell rate between switches for 9 fps VoD and ABR connection with RDF=1/64 and RIF=1/64

In case of UBR (Fig. 10), we could not solve unfairness among UDP and TCP connections and the total throughput for two TCP connections is 2.0 Mbps. We also note that in this case, we observe suspension of movie at the receiver host and the

VoD application indicates 10% of frame losses. In case of ABR with 9 fps for the video stream, although the video quality degrades due to the reduction of frame rate, we did not perceive any degradation in actual displayed movie comparing to the case for UBR without background traffic. We also note that we can maintain zero frame loss at application level for ABR with 9 fps. As we can see from the Fig. 11, by reducing the frame rate to 9 fps that is 3/8 of the original 24 fps, the corresponding cell rate is also approximately 3/8 of the original cell

rate. This suggests a possibility to the dynamic rate adaptation for VoD. By investigating the maximum bit rate that does not cause frame loss for the original movies and dynamically controlling the frame rate proportional to (ACR/bit rate of movie), we shall maintain the quality of VoD even when the available bandwidth changes time to time.

(2) Increasing MCR

We investigate the effect of MCR by changing MCR from 0Mbps by 1 Mbps steps. The minimum MCR that does not cause loss in VOD frame is 7 Mbps. We show the cell rate between switches for this case in Fig. 12.

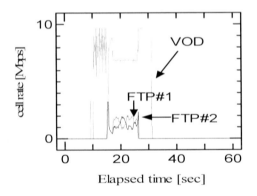

Fig. 12. Cell rate between switches for ABR connections with MCR=7Mbps for VoD

As we explained in section 4.1 with regard to WWW, the higher the transmission rate of a VC is, the larger rate reduction the VC suffers. Thus, the residual bandwidth is 3Mbps in our condition but the bandwidth share for VoD is less than that of the other VCs with MCR=0 Mbps. By observing the actual movies, we found 7 Mbps is the smallest allowable MCR for VoD traffic to maintain smoothness of the video. This means that by allocating large MCR for VoD traffic, we can increase tolerance for the VoD traffic against background load. Here, the PCRs and the durations of the background connections specify the background load. We can find the maximum value of the parameters for background ABR connections to avoid a frame loss in VoD application as follows.

(R - cell rate for the VoD connection during congestion) * T/(PCR − R) < receiver buffer size.

Here, R [Mbps] denotes the cell rate for the VoD connection when there is no background load, and T [s] denotes the duration of congestion. We can approximate cell rate for the VoD connection during congestion as MCR.

In the above relationship, the receiver buffer size is unknown but we could estimate it from measurements.

The followings are the summary of the above results.

[Applicability of ABR service for VoD]

(1) No unfairness occurs when we use ABR even if there coexists FTP traffic using TCP and VoD traffic using.

(2) We observed that the bandwidth used by the VoD is approximately 8.5 Mbps so that it is larger than the fair share of the bandwidth. In that case, we can calculate allowed background load for it. By allocating MCR enough, we can enlarge the allowed background load.

(3) By using the currently available bandwidth information, such as ACR, we can maintain movie quality by dynamically updating transmission frame rate (namely, transmission cell rate). As we mentioned above, frame rate is approximately proportional to the cell rate. Thus we suggest the effectiveness of periodical frame rate updating method by using mean bit rate of a video and mean available bandwidth in a certain interval.

4.3. IP Telephony

When RIF and RDF are set to 1/64, cell rate between switches is shown in Fig. 13. For comparison, Fig. 14 shows cell rate between switches when no background traffic is offered. From these figures, we can confirm share of each application is fair.

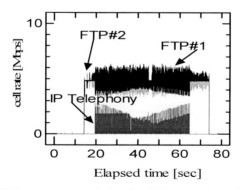

Fig. 13. Cell rate between switches for ABR connections when RIF=RDF=1/64

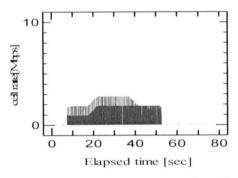

Fig. 14. Cell rate between switches for a UBR connection without background traffic

When we use a UBR, many silent periods occured. This is because the error recovery techniques in such an application do not work well when burst data is lost.

On the other hand, the ABR worked well in that we could hear undegraded sounds. This is because IP telephony uses relatively little bandwidth and has recovery techniques to be used in an Internet environment that does not guarantee any quality of service. The ABR parameters setting, as in the case of a WWW application, must keep the queue length short to reduce delay.

Delay variation, however, is caused by use of ABR, because cell rate is adjusted dynamically. Actually, packet interval for an IP Telephony without no background traffic differs from that using an ABR connection with background traffic at the order of milliseconds. Thus, only when a real-time application has some robustness for delay variation, ABR service is useful for the application. Furthermore, considering queueing delay in rate control adapter, the smaller bandwidth the application uses the better performance is achieved.

5.   Conclusion

Overall, our findings show that the best ABR parameter setting should depend on applications. However, when users cannot set appropriate ABR parameters for each application, conservative setting to keep queue length low and to avoid buffer starvation is effective in most cases.

Even though ABR service is not primarily intended for real-time applications, the ABR can be effective for these applications when some conditions are satisfied. In addition, we have found that harmonization of ABR flow control and application level flow control increases the variety for which the ABR can be effectively used.

References

1. ATM forum traffic management specification 4.0, The ATM Forum, (1996)
2. Ishizuka, M. et al.: Performance Analysis of TCP over ABR in High-Speed WAN Environment, IEICE Tarns. COMMUN, vol. E80-B, no.10, (1997)
3. Ohsaki, H. , Murata, M., Miyahara, H.: Parameter Tuning of Rate-Based Congestion Control Algorithm for ABR Service Class in ATM Networks, Int. Journal on Communication Systems, vol.16, no.7, (1998) 103-128
4. Saito, H., Kawashima, K., Kitazume, H., Koike, A., Ishizuka, M., Abe, A.: Performance Issues in Public ABR Service, IEEE Communication Magazien, vo.34, no.5, (1996) 40-48
5. Jacabson, V., Braden, R., Borman, D.: TCP Extensions for High Performance, RFC1323, (1992)
6. http://www-sop.inria.fr/rodeo/fphone/
7. Bolot, J. C., Gare, A. V.: The case for FEC-based error +control for packet audio in the Internet, to appear in ACM Multimedia Systems

Deploying Advanced IP Services on a Community Network

João Paulo Firmeza, Francisco Fontes

Portugal Telecom Inovação, SA
Rua Eng. José Ferreira Pinto Basto, 3810 Aveiro, Portugal
{pfirmeza, fontes}@ptinovacao.pt
Tel: +351 234 403 311, Fax: +351 234 420 722

Abstract. Community Networks have been implemented for the last few years on several cities, most of them supported by local governments or by experimental projects. Portugal was not an exception and since the early 1990s some small-scale projects were deployed. By the end of 1998, the Portuguese government sponsored a new wave of community networks under the scope of the project 'Digital Cities', promoting the use of new technologies and the development of enhanced services for the citizen. The city of Aveiro was the first of a new generation of Digital Cities in Portugal. This paper describes, in detail, network implementation options for this community network, concerning both its core and access enabling technologies, as well as the services that are being provided, including description and architectures for each of the relevant ones.

1. Introduction

Aveiro Digital City is the name of a large-scale community network sponsored by Portugal Telecom and by the Portuguese government, located in the region of Aveiro, 60 km in the south of Oporto. To support this community, an IP multi-services platform was set-up, requiring services ranging from typical IP narrowband to broadband and real-time services to coexist peacefully with the best possible quality.

All those services are supported by a common platform providing users with a set of basic services such as a portal service, e-mail, Web hosting (both institutional and personal pages), personalization services, discussion forums, IRC, multi-point H.323 conferences, and games. For users and projects with special needs a set of advanced services is also available like an open e-commerce platform, broadband IP streaming (MPEG1/2 VoD streaming and IP based Interactive TV) and high bitrate H.323 conferencing services.

This community network in place for almost one year and will grow to support in the near future more than 20.000 users, mostly accessing via Plain Old Telephony System (POTS), Integrated Services Digital Network (ISDN) and Asymmetric Digital Subscriber Line (ADSL). Very high speed Digital Subscriber Line (VDSL) and Fibre-To-The-Home (FTTH) technologies are also available from a local trial and are being used to connect a number of special users. Core network technology is based on Asynchronous Transfer Mode (ATM) switches and IP routers.

Based on the experience obtained with Aveiro Digital City, and supported by its success, new similar experiences will grow in Portugal during the next few years. Besides improving existing Digital Cities, new ones will require the integration of new services and network technologies, based on particular user profiles and environment characteristics.

2. The Network and its Basic Services

2.1 The Core Network

Core network was defined in order that bandwidth limitations should not become a problem, at least in the medium term. For that purpose, Ethernet and ATM network technologies were considered and evaluated in terms of advantages and disadvantages. Ethernet (100 Mbit/s and 1 Gbit/s) is a cheap and simple technology that offers very good performance, but it is a pure local area network (LAN) technology, not suitable for a wide area network (WAN) application and for interconnection scenarios (very likely to exist in different flavours on our network). Additionally, quality-of-service (QoS) management is very limited.

On the other hand, a pure ATM solution is more complex, more expensive and less performing when used in a LAN environment, but it offers very good QoS mechanisms and is very well suited for the required interconnection scenarios. Moreover, ATM is required as transport technology for some of the available access technologies that were intend to be used, such as ADSL and VDSL.

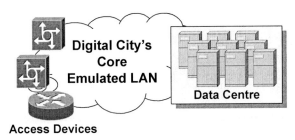

Fig. 1. The Core ELAN and connected devices: Data Centre Services and Access Devices

The final adopted solution was a hybrid one: switched and dedicated Fast Ethernet (100 Mbit/s) on the server farms and ATM for the core and access device interconnection. Since all the services are IP based, IP is the only protocol supported on this network, over Ethernet and ATM technologies.

On this hybrid core network, an ATM emulated LAN (LANE) was established, thus allowing the seamless integration of Fast Ethernet hosts and native ATM

381

equipment over a single infrastructure. For this, a Cisco Catalyst 5500 switch equipped with one redundant ATM STM-4 card and a router switch module was used in the core network. This LANE is the main 'cloud' of Aveiro's Digital City network; the services data centre and some of the access/edge devices are directly connected to this cloud over ATM or Fast-Ethernet. ATM connected access devices reach the Emulated LAN through a Cisco LS1010 ATM switch (Figure 1). Besides the core LANE-based virtual LAN (VLAN), other VLANs are configured on this network: one for service management, one to support a special project and several to support RFC1483 bridged ADSL clients, as described later on. The Catalyst route switch module assures IP routing between those VLANs. A schematic diagram of the core network is shown in Figure 2.

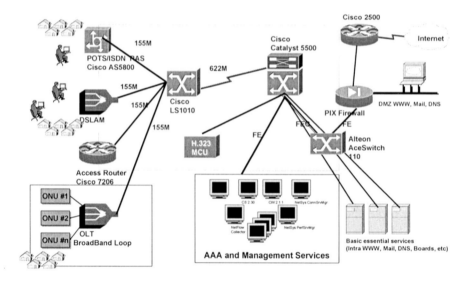

Fig. 2. Aveiro Digital City's Network Diagram

2.2 Access Infrastructure

This network supports several access options, narrowband and broadband. Narrowband access comprises POTS and ISDN dial-up at rates up to 56 kbit/s on analog modems or up to 128 kbit/s on ISDN. A Cisco 5800 access server, featured with 12 primary rate lines and 10 MICA modem cards, is being used to terminate up to 720 simultaneous PPP sessions. User authentication and accounting uses a Lightweight Directory Access Protocol (LDAP) enabled Remote Authentication Dial In User Service (RADIUS) server, which is part of the service platform described later on. Broadband access is provided through three different technologies: ADSL, VDSL and FTTH.

ADSL service is being provided by two Digital Subscriber Line Access Multiplexer (DSLAM), one from Alcatel and one from Siemens. They are located in the Aveiro's central telephone office.

The service is provided in 2 different configurations, according to particular requirements. Some users/projects use ADSL in an always-on basis; in this case IP services use RFC1483 bridged over ATM PVC connections terminating on the core switch, which comprises a VLAN for each of these user groups/projects (Figure 3a). This is the case of a kiosks project described later on.

Other users/projects are using PPP/PPPoA/PPPoE[1] sessions terminated on a BroadBand Remote Access Server (BBRAS). In this case PVCs carrying PPP sessions are established from user premises ATU-R devices and terminated on the BBRAS. Those sessions are authenticated using the same AAA server as for the narrowband users (Figure 3b).

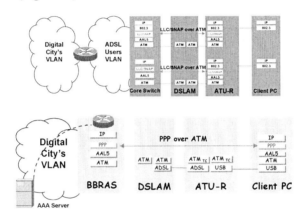

Fig. 3. Diagram of ADSL access scenarios. a) Access using RFC1483 bridged b) Access using PPP/PPPoA

For users connecting through ADSL, special broadband services are available on the network, such as Video-on-Demand or local IP-based television programmes, as shown in Figure 4.

VDSL and FTTH technologies are supported by the BroadBand Loop project's field trial in Aveiro. BroadBand Loop (BBL) was a project supported by the EU ACTS programme (AC038). It addresses access network solutions based on ATM-passive optical network (ATM-PON)/VDSL technology, paving the way for the widespread usage of broadband services in the local loop. As an evolution step from narrowband topologies towards broadband service features, BBL followed a hybrid fibre-copper approach.

---

[1] PPPoA: PPP over ATM. Defined in RFC 2364
PPPoE: PPP over Ethernet: Defined in RFC 2516

Fig. 4. Video-On-Demand and High-bitrate video streaming integrated on Digital City's service Portal using an ADSL access.

The Portuguese field trial system installed in Aveiro is based on a hybrid fibre/copper transport system providing end-to-end ATM transport. The optical distribution network (ODN) is based on a PON network comprising a single head-end unit, the optical line termination (A-OLT) with 16 remote optical network units (ONUs) located in street cabinets or directly on the customer's premises. A reference architecture is shown in Figure 5 and Interface definitions are given in the Table 1.

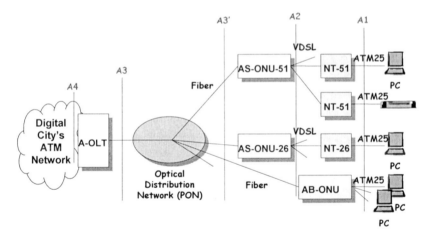

Fig. 5. A PON reference architecture

Whenever the ONU is located in a street cabinet, the last drop to the customer is based on two forms of VDSL technology supporting short-range (200 m) or medium range (800 m) data transfer at fixed bandwidths. The VDSL lines are terminated using an active network termination (NT) function supporting up to two ATM-Forum compliant ATM-25 interfaces. ONUs and NTs are named on the basis of their VDSL delivery capabilities. An AS-ONU-26 supports four VDSL 26/3 (Mbit/s) lines, each terminated by an NT-26, whereas an AS-ONU-51 supports eight VDSL 51/1.6 (Mbit/s) lines each terminated by an NT-51. In all other situations, where the ONU is located in the customer's premises, a business ONU (AB-ONU) supporting a direct fibre termination and two ATM-25 interfaces is deployed. AS-ONUs are used in fibre-to-the-cabinet (FTTC), fibre-to-the-kerb (FTTK), and fibre-to-the-exchange (FTTE) topologies and AB-ONUs are used in a fibre-to-the-home (FTTH) topology.

Table 1: PON architecture Interface definitions

A1	ATM-25
A2	VDSL-51, VDSL-26
A3'	Split fibres
PON downstream	VC-4 (STM-1, G.707)
PON upstream	TUG-A (proprietary)
A3	Single fibre
A4	STM-1 or STM-4

Figure 6 shows a BBL AS-ONU mounted on a street cabinet and a VDSL modem (NT-51). The entire BBL network consists of 16 ONUs, 10 of them are AB-ONUs, two are AS-ONU-51 and four are AS-ONU-26s, giving a total capacity of connecting 42 clients (NT-26s and AB-ONUs provide to each one 2 ATM-25 ports).

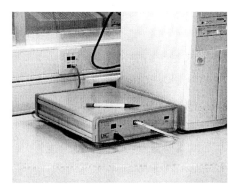

Fig. 6. BBL AS-ONU mounted on a street cabinet and a VDSL modem (NT-51).

IP services are being delivered over ATM PVCs via optical line terminations (OLTs) on the ATM switch to Aveiro Digital City's ATM core network. Since bandwidth limitation was not a problem on the PON network, all the connections were configured with constant bit rate (CBR) class of service, allowing the most sensitive services, such as MPEG2 VoD and H.323 videoconferencing to behave properly. Each of the connections has a peak cell rate of 20 Mbit/s on each of the client ATM-25 adapters and traffic shaping was applied to limit upstream bandwidth (3 Mbit/s for the NT-26, 200 kbit/s for the NT-51 and 9 Mbit/s for the AB-ONU).

IP transport uses IP-over-ATM (RFC1483) routing. For each ONU drop point, an IP subnet was assigned; each one of these subnets has a virtual interface defined on the access router allowing each user's traffic to be routed to the core network, from where services are being provided.

2.3 Data Centre Service Architecture and Generic Services

A generic IP services platform provides the building blocks for all the services developed for this community. This platform was designed with the objective to require a minimum effort to develop any specific high-level service for any specific sub-community or project. This main goal led to the following sub-objectives:

- Simplified process of user registration and automatic provisioning of customised services.
- Reduced end-user administration, operations, and support costs.
- Easy creation and deployment of new services.
- Easy and decentralised server management.

The generic service platform and its supporting information system were implemented using both Microsoft Commercial Internet Services (MCIS) and LINUX based open-source solutions. A particularly important service is the personalisation and membership directory and its back-office database. The membership directory provides a LDAP based hierarchical directory optimised for storing the whole of service's and user's information.

The membership directory can be used to store user properties that are leveraged by the personalisation features of MCIS. In addition, Digital City services use the membership directory to store user authentication credentials and permissions as well as other information required for some community services. With the membership the following features are available:

- extensible LDAP schema, allowing creation of new attributes for existing users or to define entirely new classes of directory objects;
- LDAP interface with a well-known service interface (Active Directory Services Interface (ADSI)) that enables easy access to directory objects and attribute manipulation;
- a directory security model.

Authenticated services, such as remote dial-up accounts, e-mail accounts, protected web-content, or user accounts for content uploading, are all supported by the membership directory authentication service. Standard services such as WWW, e-mail, RADIUS and file transfer protocol (FTP) can use the membership as an authentication provider through membership's LDAP interface. Figure 7 illustrates the membership's architecture.

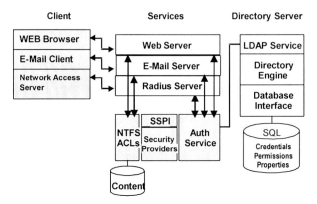

Fig. 7. Membership Architecture

386

2.3.1 Information system integrated services

The basic services provided by the IP platform are: web hosting, Internet relay chat, web-based discussion boards, software archive, H.323 conferencing, video streaming and AAA. Additionally RADIUS, e-mail (SMTP, POP3 and IMAP4), personal web pages, and some protected web based services (like web based front end for e-mail service) are available, using authentication based on the membership authentication provider using its LDAP interface.

Taking as an example the case of the e-mail service, the interaction with the membership and storage area where user mailboxes are located is done like this: when a client tries to access its POP3 or IMAP4 server, the server authenticates the user against the membership directory to verify if the user can be logged on. If the authentication is successful, another LDAP query returns the location of the user's mailbox directory. The mailbox can either be located in a local server directory or on a remote file store. A SMTP server will accept mail for local delivery (inbound) from any host from which it accepts an SMTP connection. For outbound mail, the SMTP server can be set up to accept only connections from authenticated users, from specific IP-address ranges or both, in order to cut down spamming.

Figure 8 shows the basic architecture of the Digital City's Data Centre.

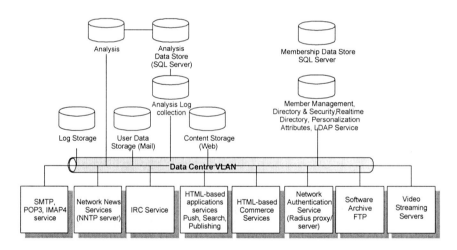

Fig. 8. Data Centre Service Architecture

2.3.2 Internet Access Service

From the point-of-view of the community network, Internet access is just another service. Internet is accessible via a firewall placed between Digital City's VLAN and a small unprotected LAN. Both network address and port address translation (NAT and PAT) are used to translate private IP addresses from the community network to public IP addresses assigned to Aveiro Digital City. In addition, a demilitarised zone is used to place some Internet services, like the public Domain Name System (DNS) or public Web servers. To increase performance and conserve bandwidth on the Internet connection (a E1 connection), Internet Web access is made through a transparent proxy/caching service.

387

A traffic redirector was configured on the router that handles traffic from Digital City's intranet to the Internet firewall; all hypertext transfer protocol (HTTP) requests from inside to external addresses are sent to the transparent proxy/cache servers. Caching service is implemented using Squid cache software running on two Linux boxes, which are placed beyond a L4 switch from Alteon Web Systems. This switch can load balance Web requests to the existing cache servers.

2.3.3 Multi-point conferencing service

To support multi-point audio/video conferences, an H.323 MCU is available on the data centre. This device is configured to be running, permanently, a free access conference. Like other services, the conferencing service uses the same LDAP based directory platform, allowing users to be registered when they are available for conversation. Conference names are also registered on the directory.

To manage closed conferences, a web-based conference management interface is being developed allowing special users with the right privileges to create, edit and view those conferences on the conferencing system.

Typical clients for this service are using Microsoft NetMeeting 3.0 for low quality conferencing and Intel ProShare conferencing kits for high quality conferences.

3. Projects and Enhanced Services

Apart from supporting generic IP services, this network is also the place where platforms and specific content belonging to some of the projects running on the Digital City initiative are located. In some cases, special services or extensions to existing platforms, were developed to support these projects. Projects on the Digital City are segmented in terms of intervention area, such as public information services, school and education communities, municipality support services, health care services, services for commerce and industry and social exclusion aid services.

Targeting those intervention areas, Aveiro Digital City initiative has currently 40 projects running. Four of these projects were selected for a briefly description in the next sections, based on the particular services that were developed to support them or on the special features implemented over the common network.

3.1 Project EIRÓ

EIRÓ is a project that consists of 12 outdoor kiosks, placed around the city, to give citizens and tourists public city information such as vectorial browseable/clickable city maps (based on a geographic information system), weather information, cultural events, local news and tourism information, with high quality videos (MPEG2) and audio.

Those kiosks are connected together and with the core, over the ADSL network, using RFC1483 bridged connections into a special VLAN supported on the core infrastructure. A special server farm located at the data centre building gives exclusive services to this VLAN. MPEG-1 and MPEG-2 video streaming up to 6 Mbit/s is being supported on Microsoft NetShow Theater platform, while lower bandwidth video and audio streaming is being provided by Microsoft Windows Media Services platform. User interface and dynamic data is based on an HTTP server and a Structured Query Language (SQL) server. Kiosk Graphical User Interface (GUI) was developed using ASP technologies, eXtensible Markup Language (XML) and Dynamic HyperText Mark-up Language (DHTML). A special user interface was designed for back-office operators allowing them to update the system SQL based information (with news, events, etc) or to update the multimedia library (audio/video).

In addition to typical kiosk functionality, each of these street-mounted public boxes also acts as a public telephone post. This service uses Portugal Telecom's Intelligent Network in combination with voice-over-IP on the Digital City's network. A user owning a virtual calling card (identified by a card number and a PIN) can dial from a public kiosk to any public network telephone number, by using a GUI-based dialer pad. These calls are routed in IP over the project's VLAN, then routed to Digital City's VLAN and finally handled by a H.323 gateway placed on the core network. EIRÓ's network architecture is shown on Figure 9.

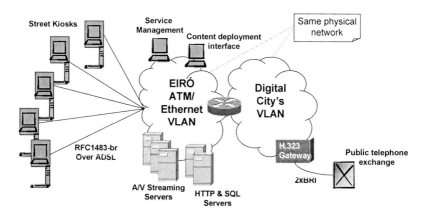

Fig. 9. Project 'EIRÓ' network architecture

3.2 Project AVEIRO MEGASTORE

In this e-commerce project a shop for the community network was developed. 18 existing traditional commerce shops participate on this project and are connected to Digital City's core network mostly by ADSL and ISDN dial up. These shops are able to use all the generic services and in addition are able to access the commerce platform to exchange transactions (orderings, payments etc), and also to update their own product catalogues or place promotions. Microsoft Site Server was used to build the e-commerce platform, being used exclusively by the Aveiro Megastore project. However this platform has capacity to support as many e-commerce sites as needed. Secure transactions use Secure Socket Layer (SSL) and are based on certificates issued by an internal Certification Authority (CA).

3.3   Project EABL.NET

This project has built a Virtual Private Network (VPN) for farmers and cow producers in the region of Aveiro. This VPN runs over Digital City's network and uses secure IP-tunnelling technologies (L2TP and IPSec). This network uses a topology where all the farmers connect via ISDN dial-up to a central office where information systems for agriculture and cattle-breeders are located (Figure 10).

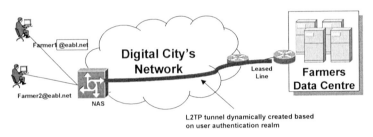

Fig. 10. Project 'EABL.NET' network architecture

4.   Conclusions

Based on the experience acquired with Aveiro Digital City, and supported by its success, new similar experiences will grow in Portugal during the next few years. A new official programme called Digital Portugal was already approved extending the digital city's concept to the entire country. Besides improving existing Digital Cities, new ones will require the integration of new services and network technologies, based on user profiles and surrounding environment needs. Connection to actual and future broadband ATM and IP back-bones, with a national coverage, will allow the deployment of existing and emerging services based on core network functionality.

Support for user communities services, based on the VPN concept, will allow a coherent distribution of components between individual community network and operator networks. Figure 11 shows the integration of other city communities on Aveiro's common service platform.

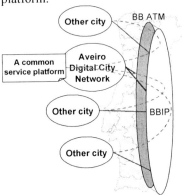

Fig. 11. Integration of other city communities on Aveiro's common service platform

Reusing Aveiro's platform to host new communities will be our target for the future. In that scenario QoS provision mechanisms and more powerful management procedures must be further developed. This may require the integration of new network solutions, QoS aware, like integrated services and differentiated services, together with monitoring and management tools. The ability to differentiate between different user application areas will allow better usage of network resources. Adopting QoS policies, will distinguish between user services and applications requiring different network levels of commitment.

References

1. Baldonado, M., Chang, C.-C.K., Gravano, L., Paepcke, A.: The Stanford Digital Library Metadata Architecture. Int. J. Digit. Libr. 1 (1997) 108–121
2. Bruce, K.B., Cardelli, L., Pierce, B.C.: Comparing Object Encodings. In: Abadi, M., Ito, T. (eds.): Theoretical Aspects of Computer Software. Lecture Notes in Computer Science, Vol. 1281. Springer-Verlag, Berlin Heidelberg New York (1997) 415–438
3. van Leeuwen, J. (ed.): Computer Science Today. Recent Trends and Developments. Lecture Notes in Computer Science, Vol. 1000. Springer-Verlag, Berlin Heidelberg New York (1995)
4. Michalewicz, Z.: Genetic Algorithms + Data Structures = Evolution Programs. 3rd edn. Springer-Verlag, Berlin Heidelberg New York (1996)

Author Index

Lecture Notes in Computer Science

For information about Vols. 1–1865
please contact your bookseller or Springer-Verlag

Vol. 1902: P. Sojka, I. Kopeček, K. Pala (Eds.), Text, Speech and Dialogue. Proceedings, 2000. XIII, 463 pages. 2000. (Subseries LNAI).

Vol. 1903: S. Reich, K.M. Anderson (Eds.), Open Hypermedia Systems and Structural Computing. Proceedings, 2000. VIII, 187 pages. 2000.

Vol. 1904: S.A. Cerri, D. Dochev (Eds.), Artificial Intelligence: Methodology, Systems, and Applications. Proceedings, 2000. XII, 366 pages. 2000. (Subseries LNAI).

Vol. 1905: H. Scholten, M.J. van Sinderen (Eds.), Interactive Distributed Multimedia Systems and Telecommunication Services. Proceedings, 2000. XI, 306 pages. 2000.

Vol. 1906: A. Porto, G.-C. Roman (Eds.), Coordination Languages and Models. Proceedings, 2000. IX, 353 pages. 2000.

Vol. 1907: H. Debar, L. Mé, S.F. Wu (Eds.), Recent Advances in Intrusion Detection. Proceedings, 2000. X, 227 pages. 2000.

Vol. 1908: J. Dongarra, P. Kacsuk, N. Podhorszki (Eds.), Recent Advances in Parallel Virtual Machine and Message Passing Interface. Proceedings, 2000. XV, 364 pages. 2000.

Vol. 1910: D.A. Zighed, J. Komorowski, J. Żytkow (Eds.), Principles of Data Mining and Knowledge Discovery. Proceedings, 2000. XV, 701 pages. 2000. (Subseries LNAI).

Vol. 1912: Y. Gurevich, P.W. Kutter, M. Odersky, L. Thiele (Eds.), Abstract State Machines. Proceedings, 2000. X, 381 pages. 2000.

Vol. 1913: K. Jansen, S. Khuller (Eds.), Approximation Algorithms for Combinatorial Optimization. Proceedings, 2000. IX, 275 pages. 2000.

Vol. 1914: M. Herlihy (Ed.), Distributed Computing. Proceedings, 2000. VIII, 389 pages. 2000.

Vol. 1916: F. Dignum, M. Greaves (Eds.), Issues in Agent Communication. X, 351 pages. 2000. (Subseries LNAI).

Vol. 1917: M. Schoenauer, K. Deb, G. Rudolph, X. Yao, E. Lutton, J.J. Merelo, H.-P. Schwefel (Eds.), Parallel Problem Solving from Nature – PPSN VI. Proceedings, 2000. XXI, 914 pages. 2000.

Vol. 1918: D. Soudris, P. Pirsch, E. Barke (Eds.), Integrated Circuit Design. Proceedings, 2000. XII, 338 pages. 2000.

Vol. 1919: M. Ojeda-Aciego, I.P. de Guzman, G. Brewka, L. Moniz Pereira (Eds.), Logics in Artificial Intelligence. Proceedings, 2000. XI, 407 pages. 2000. (Subseries LNAI).

Vol. 1920: A.H.F. Laender, S.W. Liddle, V.C. Storey (Eds.), Conceptual Modeling – ER 2000. Proceedings, 2000. XV, 588 pages. 2000.

Vol. 1921: S.W. Liddle, H.C. Mayr, B. Thalheim (Eds.), Conceptual Modeling for E-Business and the Web. Proceedings, 2000. X, 179 pages. 2000.

Vol. 1922: J. Crowcroft, J. Roberts, M.I. Smirnov (Eds.), Quality of Future Internet Services. Proceedings, 2000. XI, 368 pages. 2000.

Vol. 1923: J. Borbinha, T. Baker (Eds.), Research and Advanced Technology for Digital Libraries. Proceedings, 2000. XVII, 513 pages. 2000.

Vol. 1924: W. Taha (Ed.), Semantics, Applications, and Implementation of Program Generation. Proceedings, 2000. VIII, 231 pages. 2000.

Vol. 1925: J. Cussens, S. Džeroski (Eds.), Learning Language in Logic. X, 301 pages 2000. (Subseries LNAI).

Vol. 1926: M. Joseph (Ed.), Formal Techniques in Real-Time and Fault-Tolerant Systems. Proceedings, 2000. X, 305 pages. 2000.

Vol. 1927: P. Thomas, H.W. Gellersen, (Eds.), Handheld and Ubiquitous Computing. Proceedings, 2000. X, 249 pages. 2000.

Vol. 1929: R. Laurini (Ed.), Advances in Visual Information Systems. Proceedings, 2000. XII, 542 pages. 2000.

Vol. 1931: E. Horlait (Ed.), Mobile Agents for Telecommunication Applications. Proceedings, 2000. IX, 271 pages. 2000.

Vol. 1658: J. Baumann, Mobile Agents: Control Algorithms. XIX, 161 pages. 2000.

Vol. 1766: M. Jazayeri, R.G.K. Loos, D.R. Musser (Eds.), Generic Programming. Proceedings, 1998. X, 269 pages. 2000.

Vol. 1791: D. Fensel, Problem-Solving Methods. XII, 153 pages. 2000. (Subseries LNAI).

Vol. 1799: K. Czarnecki, U.W. Eisenecker, Generative and Component-Based Software Engineering. Proceedings, 1999. VIII, 225 pages. 2000.

Vol. 1932: Z.W. Raś, S. Ohsuga (Eds.), Foundations of Intelligent Systems. Proceedings, 2000. XII, 646 pages. (Subseries LNAI).

Vol. 1933: R.W. Brause, E. Hanisch (Eds.), Medical Data Analysis. Proceedings, 2000. XI, 316 pages. 2000.

Vol. 1934: J.S. White (Ed.), Envisioning Machine Translation in the Information Future. Proceedings, 2000. XV, 254 pages. 2000. (Subseries LNAI).

Vol. 1935: S.L. Delp, A.M. DiGioia, B. Jaramaz (Eds.), Medical Image Computing and Computer-Assisted Intervention – MICCAI 2000. Proceedings, 2000. XXV, 1250 pages. 2000.

Vol. 1937: R. Dieng, O. Corby (Eds.), Knowledge Engineering and Knowledge Management. Proceedings, 2000. XIII, 457 pages. 2000. (Subseries LNAI).

Vol. 1938: S.Rao, K.I. Sletta (Eds.), Next Generation Networks. Proceedings, 2000. XI, 392 pages. 2000.

Vol. 1939: A. Evans, S. Kent, B. Selic (Eds.), «UML» – The Unified Modeling Language. Proceedings, 2000. XIV, 572 pages. 2000.

Vol. 1940: M. Valero, K. Joe, M. Kitsuregawa, H. Tanaka (Eds.), High Performance Computing. Proceedings, 2000. XV, 595 pages. 2000.

Vol. 1942: K. Masanori, R. Popescu-Zeletin (Eds.), Active Networks. Proceedings, 2000. XI, 424 pages. 2000.

Vol. 1945: W. Grieskamp, T. Santen, B. Stoddart (Eds.), Integrated Formal Methods. Proceedings, 2000. X, 441 pages. 2000.

Vol. 1948: T. Tan, Y. Shi, W. Gao (Eds.), Advances in Multimodal Interfaces – ICMI 2000. Proceedings, 2000. XVI, 678 pages. 2000.